OFFSHORE STRUCTURES ENGINEERING

OFFSHORE STRUCTURES ENGINEERING

Proceedings of the International Conference on Offshore Structures Engineering held at COPPE, Federal University of Rio de Janeiro, Brazil, September 1977

Edited by:
F. L. L. B. Carneiro and A. J. Ferrante
Federal University of Rio de Janeiro
C. A. Brebbia
University of Southampton

GULF PUBLISHING CO
Book Division
Houston, London, Paris, Tokyo

This edition published 1979
by Gulf Publishing Company,
Houston, Texas.

Published simultaneously
by Pentech Press, Limited.
Estover Road, Plymouth, England.

ISBN 0-87201-608-0

Library of Congress Catalog Card No. 78-74102

Printed in Great Britain by
Billing & Sons Ltd.,
Guildford and London

Contents

PREFACE

In the last few years oil has become a very valuable natural resource making economical the development of offshore oil fields. Consequently it is now necessary to design, build and operate different kinds of devices and equipment to look for and to extract oil from the sea. New engineering problems have been encountered, which in turn have stimulated many new research projects.

Brazil is one of the countries most interested in taking advantage of the oil resources in its continental shelf. In view of this the Federal University of Rio de Janeiro, through its graduate centre of engineering, COPPE, organized the International Conference on Offshore Structures Engineering, which took place in Rio de Janeiro, Brazil, on the 12th–15th September, 1977. This publication is a result of that Conference, and includes the main papers presented, conveniently edited.

The Conference was supported by the Brazilian National Council for Scientific and Technological Development (CNPq), the British Council and the Ministry of Trade (London).

The Conference was the first of a series of bi-annual meetings on this subject, the second will be an International Symposium on Offshore Structures to be held jointly with the 33rd Meeting of the Permanent Committee of RILEM in Rio de Janeiro, in the first week of October, 1979.

Organizing Committee:

Prof. Fernando Luiz Lobo B. Carneiro, Head Civil Engineering Department at COPPE, Federal University of Rio de Janeiro
Dr. Carlos A. Brebbia, University of Southampton
Prof. Agustin J. Ferrante, COPPE, Federal University of Rio de Janeiro
Prof. Paulo S. M. Cotta, Head, Naval and Oceanographic Dept. at COPPE, Federal University of Rio de Janeiro.

OPENING ADDRESS

I wish, initially, to express my gratitude to the organizers of this Conference and to the sponsoring institutions, the National Council for Scientific and Technological Development (CNPq), the British Council and the Ministry of Trade, for the honour conferred upon me, to preside over the opening session of this important event.

The Organizing Committee is to be congratulated on the timely and suitable choice of the subjects to be treated. Technological developments related to the exploration of marine oil fields have, from the beginning, been making large strides, and as President of Petrobrás S.A., I can say that the introduction and realisation of new techniques is of the greatest importance to us. In that sense it is a duty at this moment, to point out that the Federal University of Rio de Janeiro, through its Coordination for the Postgraduate Programs in Engineering (COPPE), has shown commendable foresight in starting a research and development program in the engineering of offshore structures.

I also wish to thank Prof. T. Patten, Director of the Institute of Offshore Engineering, Heriot-Watt University, Edinburgh, Scotland, Dr. G. L. England, King's College, London, invited by the British Council; Dr. C. A. Brebbia, from Southampton University and Dr. R. Eatock-Taylor, from University College, London, invited by COPPE, Mr. P. Edge, from Wimpey Laboratories, Dr. J. Lamb, from Atkins Research and Development, and Dr. W. Schum, from Lloyds Register, invited by the Ministry of Trade, and Dr. B. Steinvisk, from Aker Engineering, Norway for participating.

According to the Technical staff of the Exploration and Production Department of Petrobrás, the best prospects of finding productive oil fields is in the continental shelf. This alone justifies the previous statement regarding the opportuneness and importance of this Conference.

The final production system for the Campos basin, which will be ready at the end of 1980 or beginning of 1981, will produce, according to proven reserves, 220,000 barrels per day. However, due to the urgency of the Government in initiating production in the shortest possible time, a provisional system was projected, with know-how from Lockeed, to reach a production of 45,000 barrels per day using the nine most promising wells. Since the availability of the system will be delayed by 8 to 10 months, an emergency system was conceived, which consisted simply of immobilizing a semi-submersible platform, with production equipment. This platform was placed on top of the well which, according to information received, had the largest production capacity, and is presently reaching a production of 10,000 barrels per day. Obviously the production more than covers the cost of immobilization of the equipment and was, therefore a commercially sound measure.

Finally, on behalf of the Organizing Committee, I welcome all present today, who have come to participate in these important discussions, and to initiate this Conference I invite Director Marques Neto, Chief of the Exploration Department of Petrobrás, to start his presentation.

General Araken de Oliveira
President of Petrobrás S.A.

PART I OFFSHORE DEVELOPMENTS

PRESENT AND FUTURE OF OFFSHORE DEVELOPMENTS IN BRAZIL

J. Marques Neto

Petrobrás

Prof. Sidney Santos, Vice-Dean of the Federal University of Rio de Janeiro, representing the Dean of this University; General Araken de Oliveira, President of Petrobrás and Chairman of this session; Professor Hélio Fraga, former Dean of the University; Professor Paulo Rodrigues Lima, Deputy Dean of the Technological Center; Professor Sérgio Neves Monteiro, Director of COPPE; Professor Júlio Coutinho, Director of the School of Engineering; Mr. Jeremy Hills, of the British Council; members of the Organizing Committee; members and representatives of international petroleum organizations; faculty members and students of this University; my colleagues at Petrobras here present; ladies and gentlemen:

It is common knowledge that petroleum technology has developed substantially in the world of our times since the first pioneer oil-well was drilled in 1859 near Titusville, Pennsylvania. It is also common knowledge that the oil industry gained a new and formidable impetus in respect of operations and profitability when the world's financial power centers shifted in 1973.

There are some 20 oil and natural gas exporting countries in the world today, and a further 80 countries which may yet become exporters or will have to continue to import oil for their primary power requirements.

An analysis of the world situation in terms of population, according to a recent United Nations publication issued in 1977 which summarizes the papers submitted in Geneva from 10th to 20th November 1975, discloses that out of a total world popula tion of 3.74 billion inhabitants, about 916 million live in 10 countries that import oil or produce some oil, while about 531 million inhabitants live in 75 petroleum importing developing countries that produce no oil. Among the first group of countries there are some that will very likely become oil producers and even oil exporters. Looked at in another way, these figures also show that 25% of the world's population lives in the above 10 importing countries, while in the 75 remaining countries 15% of the population may continue to be dependent on oil. If we simply refer to the 85 importing countries, we find that 10 have no oil prospects at all, 21 have limited prospects, while another 10, including Brazil, India, Mexico and Peru, with a total population of 739 million inhabitants, have a reasonable chance of supplying their own needs and becoming exporters.

Experts in the petroleum industry have been chiefly concerned with the successful finds and developments that have taken place since 1947 when the first offshore well was drilled in the Gulf of Mexico, off the coast of Lousiana. They were also concerned with the 40 countries that must find oil and develop new sources of energy supplies and of which 31% are dependent today on imports of oil and natural gas. Brazil, with the world's eighth largest population and the fourth largest continental land area, is among these countries. However, on the basis of strong technical evidence, we Brazilians alone are not the only ones who believe that there are additional oil reserves in our country's continental shelf and land basins.

The publication mentioned earlier is called "Petroleum - Cooperation among Developing Countries" and was issued in 1977. On page 25, under the heading "Future Activities Exploration", we read: "Eighty-five petroleum importing countries are included in this survey. Ten of them, with a population of 36 million have no petroleum prospects, 21 others with a population of 108 million, have only extremely limited prospects. Approximately 10 countries, including Brazil, India, Mexico and Peru, with a combined population of 739 million, have a reasonable chance of supplying their own needs or even becoming exporters in the next ten years.

The problem of exploration, continues the publication, is thus raised for the remaining 40 countries which have neither the financial means nor the human capabilities to undertake prospection on their own and whose petroleum potential is not good enough to attract financing from outside.

I will now pass on to a description of the status of oil-well drilling in Brazil by comparing offshore prospects and accomplishments.

The basins on the continental shelf along Brazil's coastline have not yet been satisfactorily sampled. Most of them have not been subjected to an adequate degree of exploration. Their investigation has been limited to semi-detailed seismic surveying and some require additional exploratory drilling.

It is important to be aware of the risks involved in exploration, and to avoid both excessive optimism, leading to exaggeration, and pessimism, leading to underdevelopment, undervaluation and stagnation.

To achieve a more adequate treatment of the current stage of exploration, a reassessment of the potential prospects of Brazil's continental shelf was undertaken, including the establishing of relative priorities based on a comparative study of probable recoverable volumes of oil, from analyzing geological potentials. An exploration strategy was planned in the form of a short-term program which, while assigning priorities, would also investigate the most promising prospects. This program also aims at proving new reserves in the short run, thanks to a massive, rational technically and financially planned, well-managed effort. It may be termed aggressive and, thus, highly stimulating, both technically and professionally. For the implementation of this program, funds, human resources, and equipment are being mustered as basic elements for the support of the exploration prospects that are to be developed using the best technique and most advanced knowledge available.

EVALUATION VERSUS POTENTIALITY CRITERIA

The potential of the sedimentary basins on Brazil's continental shelf was evaluated by using two distinct methods: comparison with West African basins, excluding Nigeria, and a systematic study of the parameters responsible for generation and accumulation of hydrocarbons on the Brazilian continental shelf.

Using the two methods mentioned above, figures of potential of the same order were obtained.

These figures are not predictive, but serve as a basis for assigning investment priorities, since the figures mentioned are not used as predictive determining factors, they merely support economic evaluation of investments for basis assignment of priorities.

The common genesis of Brazil's coastal water basins and of West Africa's has been evidenced, and according to H.D. Klemme they fall into the category of Type V stable coastal basins.

This type of basin is relatively widely distributed throughout the world, but only scanty data have been published and, though the characterization of these basins is easy, it calls for progressive evaluation studies to determine their actual potential. There are a number of areas in the world similar to the basins along the Brazilian coast, that are oil-producing. The oil-producing capacity of such basins has been acknowledged and accepted worldwide, yet there is no notable presence of giant fields with reserves in excess of 500 million barrels or their gas equivalent, since merely 6 of the 280 to 300 giant fields in the world are located in this class of basin.

The West African basins which have the same geological origin as the basins of Brazil's continental shelf have a recoverable oil volume of 4.1*billion barrels.

Nigeria's recoverable volume, amounting to approximately 19.5* billion barrels, was intentionally excluded from this analysis because it occurs in Tertiary delta sediments, a type of formation that has not yet been thoroughly investigated in Brazil's continental shelf. Of course, West Africa's estimated overall oil-bearing potential is still dependent on ascertaining the potentials of countries that are still developing their oil prospecting activities, such as the Ivory Coast, Liberia, Ghana, the Cameroons, Togo, Equatorial Guinea, etc. Gabon,* Congo,* Cabinda-Angola* and Zaire* are already producing oil commercially and their recoverable oil volumes are respectively 2.1 billion barrels, 0.28 billion barrels, 1.2 billion barrels and 0.5 million barrels.

Thus, it is estimated that West Africa's total recoverable oil volume lies between 10 and 15 billion barrels, the first figure being a conservative estimate and the second a realistic or a somewhat optimistic figure. Thus, the geological parameters of the continental coastlines are comparable, except for the existence of large deltas, which in Brazil's case remains to be confirmed, in addition to the contingency of more difficult prospecting because most of the area to be explored is located in deep water. The development of estimates of the potential of the offshore basins is perfectly feasible using data for the African coast. On this basis, our oil-bearing potential, which is to be proved, should range from 10 to 15 billion barrels, the two figures representing respectively the conservative and the more optimistic limits.

These comments are a general description of the geological status or the geological investigation of the coastal sedimentary basins of Brazil's continental shelf.

We describe next the activities that involve this administration and for which PETROBRÁS represents the direct action taken by the Country, for the discovery and proving of additional oil reserves in Brazil.

As stated earlier, Brazil is one of the four countries that have a reasonable chance of becoming additional oil producers and perhaps exporters. It is also known that only 5 countries in the world today are not dependent on oil and natural gas as

* Oil and Gas Journal (Dec. 1976)

major sources for their primary energy requirements: India, Korea, Turkey, Zambia and Mozambigue.

It should be noted that coal, which played an important role in the past as the world's primary source of energy, may shortly regain a prominent position. The noble products derived from oil and natural gas will continue to be of more importance than they were up to year 1973. The need to increase exploration for oil and output in developing and industrialized countries is, therefore, crucial and vital.

It is also interesting to note that the increase in oil prices since 1973 has made other basins attractive, such as, for instance, those located in deep offshore areas where exploration was previously restricted for economic reasons.

In Brazil, sedimentary areas are distributed over the continent and along the coastline both in submerged and continental regions.

I shall refer only to the submerged continental sites, in keeping with the agenda.

I will first present the figures already achieved by Brazil and the first short-term exploration program.

Figure No. 1 shows a schematic distribution of sedimentary basins in Brazil. On the right we can see the activities scheduled on dry land. These are part of an exploration strategy planned for 4 years, the time considered necessary to prove new reserves.

Note that, based on the considerations made earlier, a 25% recovery factor has been assumed for these sedimentary basins.

With regard to Brazil's offshore areas, the north coast includes the basins of Cassiporé and the mouth of the Amazon, with an area of 150,000 square kilometers and composed of Tertiary sediments with a recoverable oil potential of 77 million cubic meters. Northwast Coast I includes the States of Ceará, Rio Grande do Norte and Maranhão, an area of 190,000 sq. kilometers, with Aptian geological formations and an oil-bearing potential of 105 million cubic meters. Northeast Coast II includes the States of Paraíba, Pernambuco, Alagoas, Sergipe, North Bahia, with an area of 40,000 sq. kilometers, Aptian and pre-Aptian formations and a recoverable oil potential of 84 million cubic meters. The East Coast, including South Bahia and the State of Espirito Santo, with pre-Aptian formations, an area of 70,000 sq. kilometers and recoverable reserves of 56 million cubic meters. The Campos basin, which today is the major additional reserve-proving short term requirement, has an area of 30,000 kilometers, composed of Albio-Cenomanian sediments with a

EXPLORATORY DRILLING

BASIN LOCATION (EXTENSION IN km²)	NUMBER OF WELLS	RESOURCES-MILLION Cr$ EXPL. DRIL.	%	REF
HIGH AMAZON 600 000 km²	1	30,6	2	1
BARREIRINHAS 13 000 km²	12	159,8	10	2
MARANHÃO 700 000 km²	9	150,9	10	3
SERGIPE/ALAGOAS 7 200 km²	52	406,8	26	4
TUCANO 14 000 km²	9	69,5	4	5
RECÔNCAVO 11 500 km²	61	446,7	28	6
SOUTH BAHIA ———	10	81,0	5	7
ESPIRITO SANTO 7 200 km²	24	189,5	12	8
OTHERS 22 500 km²	4	55,6	3	—
TOTAL	182	1590,5	100	—

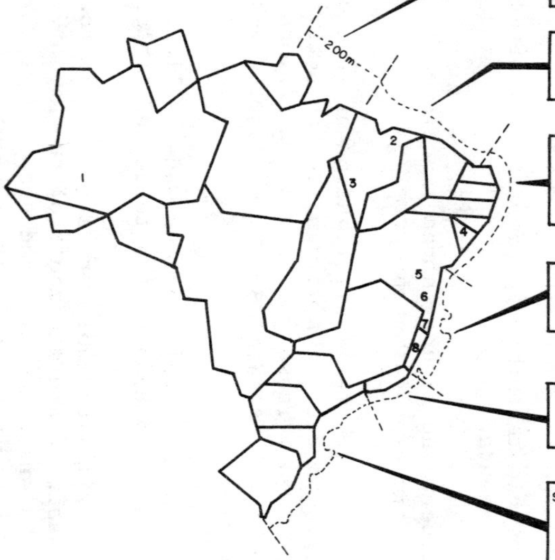

BASIN LOCATION (EXTENSION IN km²)	NUMBER OF WELLS	OIL POTENTIAL TARGET	REC. OIL (10^6 m³)	%	RESOURCES-MILLION Cr$ EXPL. DRIL.	%
NORTHERN COAST CASSIPORÉ AMAZON MOUTH (150 000 km²)	34	TERTIARY	77	11	2976	16
NORTHEASTERN COAST I MA/CE/RÑ (190 000 km²)	54	APTIAN	105	15	2623	15
NORTHEASTERN COAST II PB/PE/AL/SE NORTH BA (40 000 km²)	30	APTIAN PRE-APTIAN	84	12	1351	7
EASTERN COAST SOUTH BA / ES (70 000 km²)	49	PRÉ-APTIAN	56	8	1898	11
CAMPOS (30 000 km²)	110	ALBO-CENOMANIAN	280	40	6552	36
SOUTHERN COAST SANTOS FLORIANOPOLIS PELOTAS (320 000 km²)	48	ALBO-CENOMANIAN	98	14	2755	15
TOTAL (800 000 km²)	325	——	700*	100	18 125	100

* CONSIDERING 25% AS RECOVERING FACTOR: 700 X $10^6 m^3$ = 4,4 BILLION bbl.

— (CR$ 14,85 / US$ 1,00)

FIG. 1 EXPLORATION STRATEGY
OIL POTENTIAL & BUDGET DISTRIBUTION

recoverable potential of 280 million cubic meters. The South Coast, which includes the cities of Santos, Florianópolis and Pelotas, likewise has Albio-Cenomanian sediments and a recoverable potential of 98 million cubic meters. The above total covers a submerged sedimentary area of 800,000 sq. kilometers.

The drilling of 325 exploratory wells is scheduled to obtain a better sampling of these sedimentary basins located on Brazil's continental shelf.

Figure 1 shows the approach taken for the distribution of the wells and also the percentage distribution of oil-bearing potential estimated on the basis of geological ages, in addition to the funds allocated for exploration in each area.

Observe and take notice of the importance of Campos, which accounts for 40% of our expected geological potential. The Northeastern Coast I comes second with 15%, followed by Santos, Florianópolis and Pelotas with 14%, the Northeastern Coast with 11%, the Northeastern Coast II with 12%, and the Eastern Coast with 8%. This is, of course, the distribution of our oil-bearing potential according to the geological strata on which offshore exploration in Brazil is based.

This aggressive oil prospecting activity scheduled over the four-year period from 1978 through 1981 will require an investment of 18,125* billion cruzeiros and it will cover the sampling of some prospects and complete investigation of other priority high-yield prospects.

We now pass on to a timely analysis of what has been done and the rewards for Brazil's offshore achievements.

Figure 2(b) shows successes on the left vertical axis, and the years from 1968 through 1980 are shown on the horizontal axis with projections after 1977. The right vertical axis represents the number of discoveries made during the period from 1968 through 1977 with statistical projections thereafter. The dashed line indicates the start of offshore drilling in Brazil in 1968. Note the remarkably high success factor of about 50% for the first year. Compare the upper curve (Figure 2(a)) showing the distribution of exploration wells drilled and note the growth of drilling. Note that offshore exploratory drilling which began in 1968 has grown to a remarkable total of about 50 offshore wells on the continental shelf in 1977.

Exploratory successes in 1971 and 1972 were under a peculiar situation because PETROBRÁS was engaged more in oil production than in drilling at that time. Exploratory successes increased after 1972, and the 1975/1976 period was influenced by the need

* Cr$ 14.85/US$ 1.00

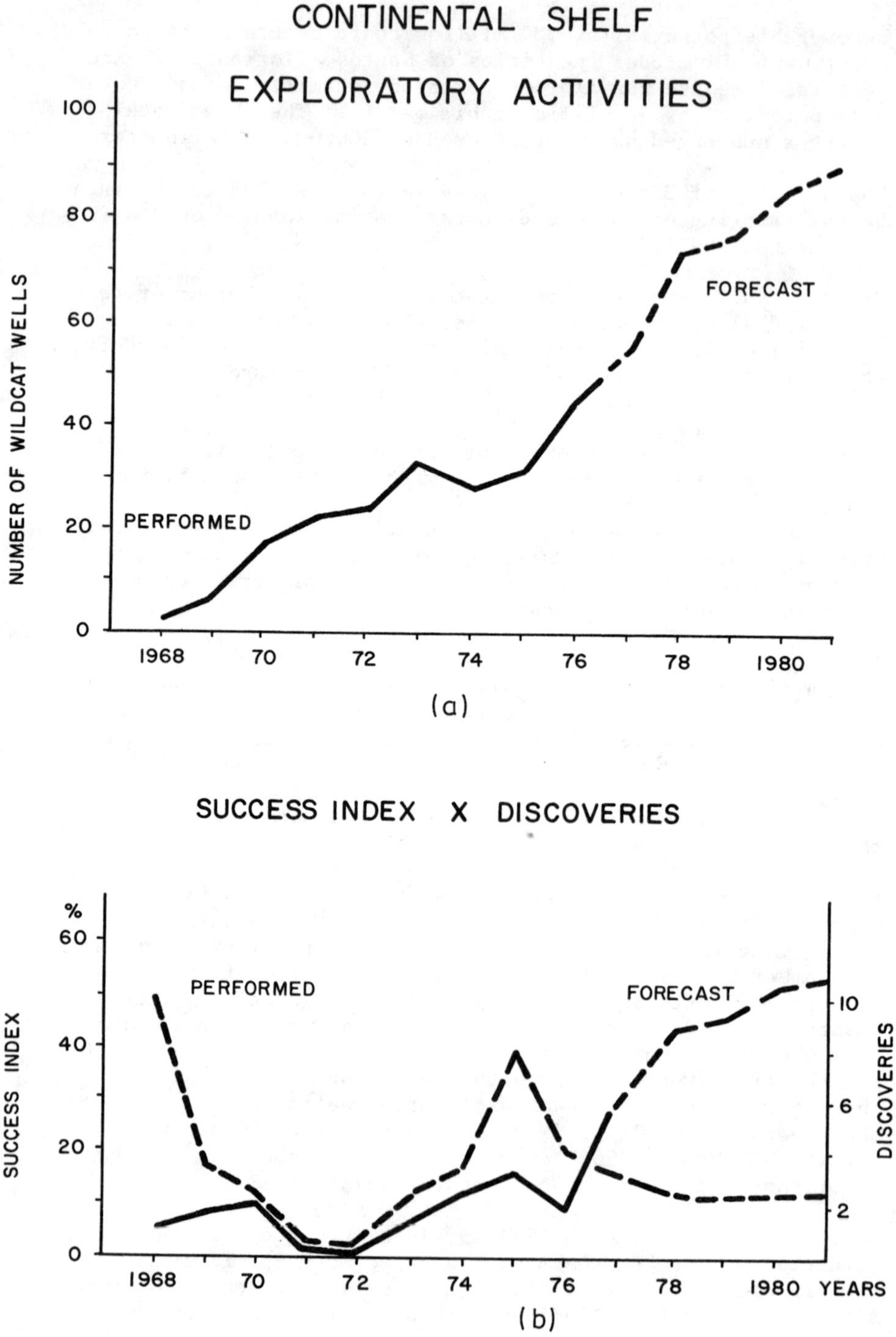
CONTINENTAL SHELF
EXPLORATORY ACTIVITIES
NUMBER OF WILDCAT WELLS
100
80
60
40
20
0
PERFORMED
FORECAST
1968
70
72
74
76
78
1980
(a)
SUCCESS INDEX X DISCOVERIES
%
60
40
20
0
SUCCESS INDEX
PERFORMED
FORECAST
DISCOVERIES
10
6
2
1968
70
72
74
76
78
1980 YEARS
(b)

FIG. 2

to prove reserves. It is expected that the success ratio will remain only at 12%, despite a world success rate of 10%. Although our success ratio, despite its fluctuations, could be as high as 18%, as it was during the period between 1974 and 1976, a conservative ratio of 12% has been adopted for forecasting purposes. The forecasting of offshore investments, as in the case of any investment, must be a determining factor so that the prospects may be investigated in a concrete and, above all, objective manner. It is expected that, with a conservative ratio of 12%, discoveries will be made from year 1977 to 1978, which are assumed to be 3, 6, 9, and 10 discoveries. It is not possible to predict whether the discoveries will be large or small, since the areas to be explored have been only semi-detailed geological surveying.

It is hoped that important medium to large fields may be discovered, since there is still no answer to the question as to whether large fields exist. The upper curve shows the growth rate of the exploratory drilling program. This program is to be very strongly implemented and will grow substantially from 1978 when it is intended to drill 80 exploratory wells, only exploratory ones, to 1980, when 90 exploratory wells are scheduled. This last figure is not impressive when compared with offshore drilling carried out throughout the world. However, if we regard well-drilling as a science based on feedback information, this program, despite its aggressive approach, will benefit from the feedback of data that will be obtained during the next 4 years and the system receiving the feedback data will benefit from an increase in offshore investments.

More specifically, Brazil's exploratory offshore areas are defined in Figure 3 in a similar way as Figure 1, but divided into Northern, Northeastern I, Northeastern II, Eastern, Campos, and Southern.

The total area of Brazil's continental shelf to be explored is 800,000 sq. kilometers. From 1969 to this date, Brazil has drilled 236 exploration wells on its continental shelf. This number is not sufficiently expressive to make a comparison on what could be termed as a high prospection basis, but only a relatively weak one when compared with oil-prospection carried on throughout the world.

Figure 3 shows the scheduled pioneer exploration program for 325 wells with which it is intended to bring the total number of exploration wells to 561. For comparison purposes it has been intended to show drilling density per square kilometers, which system is widely used to indicate the exploration density or exploratory phase of each area. For July 1977, we have only 5,500 sq. kilometers per exploratory well in the Northern area: 5,000 sq. kilometers per exploratory well in the Northeastern I area; 660 sq. kilometers per exploratory well in the

CONTINENTAL SHELF

WILDCAT WELLS

1968 — JUL / 77

COASTAL AREAS	STATES/ TERRITORIES	AREA (km^2)	WILDCAT WELLS			DENSITY (km^2/ well)	
			PERFORMED	FORECAST	TOTAL	PERFORMED	TOTAL
NORTHERN	AP / PA	150.000	27	34	61	5.550	2.460
NORTHEASTERN I	MA-CE-RN-PI	190.000	38	54	92	5000	2070
NORTHEASTERN II	PB- PE-AL- SE- BA/North	40.000	61	30	91	660	440
EASTERN	BA/SOUTH-ES	70.000	63	49	112	1110	630
CAMPOS	RJ	30.000	39	110	149	770	200
SOUTHERN	SP-PR-SC-RS	320.000	8	48	56	40.000	5.710
TOTAL.	BRAZIL	800.000	236	325	561	3.390	1.430

FIG. 3 SOURCE : DEXPRO/ DIPLAN

Campos area; and 40,000 sq. kilometers per exploratory well in the Southern area. Upon completion of the program for drilling 325 exploratory wells, these densities will have been reduced to 2,460 sq. kilometers per well in the Northern area; 2,070 sq. kilometers per well in the Northeastern II area; and so on. In the case of Campos, a very important area today, we shall have 200 sq. kilometers per well.

Figure 4 provides a comparison in round numbers of drilling density in the United States, as published in the AAPG. Note that exploration on an industrial scale in the United States began in 1920, that is 57 years ago, and that 41,000 wells were drilled in that country last year alone.

Figure 5 shows planned investment for exploration, drilling and production, representing the effort put into the development of the offshore areas.

This figure shows the growth of investment for the 1975-1981 period. In 1975, 56.2 billion cruzeiros were invested and slightly over 6.2 billion cruzeiros in 1976. After 1976 and until 1980, investment shows increases of 52%, 31%, 8% and 21%, respectively. An explanation of the forecast and the reasons for increases of 52% from 1976 to 1977, and 31% from 1977 to 1978, the rather conservative growth in 1978 and 1979 as well as the 0.2% decrease in 1981, will be given later.

The need to drill 325 wells in Brazil's continental shelf without discontinuing in terms of success or lack of success was defined as the first alternative under the overall investment program for 1978-1981. This means that, if all the 75 wells scheduled for the first year yield negative results, the program will still be continued in order to obtain the minimum sampling required for an adequate geological survey of the basins.

Figure 5 also shows the distribution of investments over the 1978-1981 period on an 87% offshore and 13% onshore basis. Note that this ratio was 72% offshore and 28% onshore in 1975. Offshore investments continue to increase because the continental shelf provides greater expectations of higher yields.

Figure 6 reflects the investment program in terms of basic items and representative amounts. It provides a break-down of the exploration program and the funds to be invested in exploration, in exploratory drilling, and in provisional systems, including the systems currently being implemented at Enchova, also known as the "Early Production System" which the President of PETROBRÁS has already described. The purpose of this system is to obtain data on the reservoir, the productivity of each well, and the mechanical working conditions of the rock for normal production to determine the size of the final projects in the

DRILLING DENSITY

COUNTRY	SEDIMENTARY AREA (km^2)			WELLS	DRILLING
	LAND	SEA	TOTAL	TOTAL	DENSITY
U. S. A. (1)	7.420.559	969.780	8.390.339	2.500.000	3.36 km^2/well
BRAZIL (2)	3.250.000	—	—	4404	730 km^2/well
	—	800.000	—	561	1430 km^2/well
			4.050.000	4779	850 km^2/well

SOURCES: (1) AAPG - 1976

(2) PETROBRÁS / DEXPRO

FIG. 4

CONTINENTAL SHELF AND LAND BASINS

EVOLUTION AND FORECAST OF INVESTMENTS IN THE PERIOD 75/81

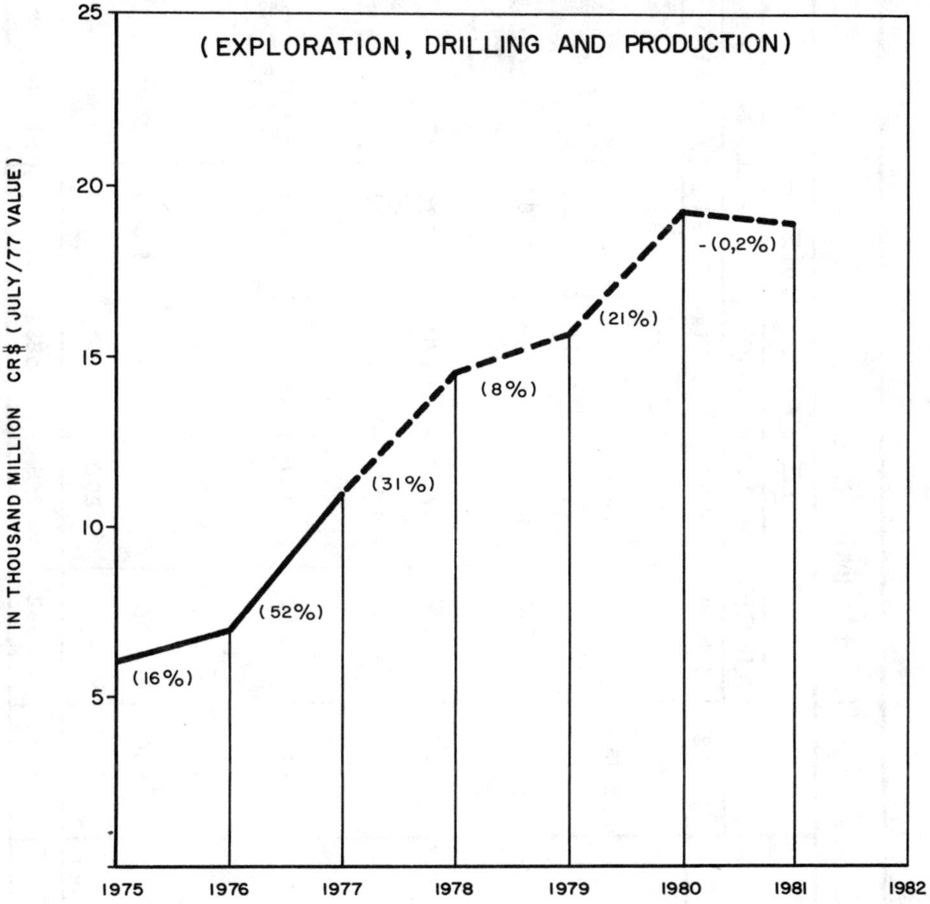

NOTES:
- 1975/76 - ACHIEVEMENTS
- 1977 - PROGRAMMED INVESTMENTS, APPROVED
- 1978/81 - GLOBAL INVESTMENTS PROGRAM — ALTERNATIVE I
- (%) - AVERAGE PERCENT INCREMENT
- THE INVESTMENTS DISTRIBUTION FOR THE 78/81 PERIOD SUPPOSES A RATIO 87% SEA TO 13% LAND (1975=72%/28%, 1976=80%/20%, 1977=82%/18%)
- CR$ 14,85 / US $ 1,00

FIG. 5

SOURCE: DEXPRO / SERPLAN
JUL/77

DEXPRO INVESTMENTS DURING 78_81 — SEA

ACTIVITIES AND PROJECTS	CONTINENTAL SHELF — RESOURCES_ALTERNATIVE I (MILLION Cr$)					
	1978	1979	1980	1981	TOTAL	REMARKS
1- EXPLORATORY INVESTIGATION	290	320	350	370	1 330	- SEISMIC, TECHNICAL (GEOLOGICAL) SUPERVISION.
2_EXPLORATORY DRILLING	5 030	4435	4496	4554	18 515	- GEODESY, DTM, DRILLING, ELECTRICAL LOGGING, LOGISTIC SUPPORT / /ADMINISTRATION, EVALUATION AND COMPLETION I.
3_PROVISIONAL SYSTEMS	371	522	744	1 116	2 753	- SEMI_SUBMERSIBLE, LOGISTICAL SUPPORT, EQUIPMENT ADMINISTRATION, TANKAGE SERVICE.
4_DEVELOPMENT DRILLING IN DISCOVERED AREAS (INCLUDING INJECTION, SPECIAL)	2662	2 085	1 254	1 286	7 287	- GEODESY, DTM, DRILLING, ELECTRICAL LOGGING, ADMINISTRATIVE AND LOGISTICAL SUPPORT, EVALUATION, COMPLETION I, WELL EQUIPMENTS (VALVES, TUBING, ETC) AND SIMULATION
5_DEVELOPMENT DRILLING IN AREAS TO BE DISCOVERED	—	—	728	1 650	2 378	- DITTO
6_ENGINEERING WORKS AND INSTALLA_TIONS IN DISCOVERED AREAS (CONNECTED RESOURCES)	3 292	5 490	5 274	2 277	16 333	- PLATAFORM, PRODUCTION FACILITIES AND FLOW SYSTEM (GP, NA, EN, ETC)
7_ENGINEERING WORKS AND INSTALLA_TIONS IN DISCOVERED AREAS (RESOURCES TO BE APPROVED)	250	310	2 190	1 900	4 650	- PLATFORM, PRODUCTION FACILITIES AND FLOW SYSTEM, ARTIFICIAL ELEVATION SYSTEM AND SECONDARY RECOVERY (GP, NA, UB, RB, ETC)
8_ENGINEERING WORKS AND INSTALLA_TIONS IN AREAS TO BE DISCOVERED	—	—	1 080	2 780	3 860	- DITTO
9_ADDITIONAL EQUIPMENT	400	400	400	400	1 600	- RIGS, COMPRESSORS, TREATERS, MOTO_PUMPS ASSEMBLY, ETC.
10_OTHER ENGINEERING WORKS (CR$ 14,85/US$ 1,00)	—	—	200	250	450	- NATURAL GASOLINE PLANT (CAMPOS)
TOTAL	12 295	13 562	16 716	16 583	59 156	

FIG. 6

JULY_77

area, and to achieve a substantial production of oil.

These figures add up to a total of 59 billion 156 million cruzeiros, or 1 billion 140 million dollars.

Figure 7 shows the overall investment program for both the continental shelf and onshore basins.

The investments scheduled and approved are part of the effort that can be regarded not only as challenging, but above all stimulating, both technically and professionally, as stated earlier.

The management time schedule given in Figure 8 shows 4 alternative schedules for exploration wells. Alternative I has been approved and represents the managerial goals to be reached. Like the other curve, the first curve represents two figures. The top figure is the number of exploration rigs scheduled for drilling a certain number of wells. Under alternative I, 20 rigs will be used to drill 325 exploration wells. The number of rigs is constant mainly because of support logistics, the equipment allocated for support, shall be quantified rationally to meet program requirements in an orderly way over 4 years. The bottom figure is the daily drilling rate per rig. This figure gradually increases from a daily rate of 35 meters per rig in 1978, to 38 meters per rig in 1979, then to 42 meters per rig in 1980, and finally to 45 meters per rig in 1981. Experts familiar with exploratory drilling are well aware that these are highly challenging rates, particularly when compared with international drilling rates, but they can be achieved. Alternatives II and III, which are subdivided into IV, aim at obtaining higher drilling rates using the same number of drilling rigs, beginning with a daily rate of 38 meters per rig in 1978, increasing to 42 meters per rig in 1979, increasing still more to 45 meters per rig in 1980, and keeping to this same rate in 1981. These rates should produce a substantial saving in cost and require less time. Thus, the program would be completed in only 3.7 years and the investment required would amount to 17 billion 370 million cruzeiros as compared to 18 billion 125 million cruzeiros under alternative I. Finally, the most attractive alternative in terms of investment would be the one where 325 wells could be drilled within a period of only 3 years, but which would require an average daily drilling rate per rig starting at 40 meters and rising to 45 meters in the second year. We have the technology, we believe the administrative attraction of the program is challenging and we expect to operate with alternatives so as to complete them within the shortest possible time.

RISK CONTRACTS

Brazil now has an additional tool for handling offshore drilling

PETROBRÁS
PETROLEO BRASILEIRO S.A.

EXPLORATION, DRILLING AND PRODUCTION INVESTMENTS GLOBAL PLAN 78-81

CONTINENTAL SHELF				ACTIVITIES AND PROJECTS	ONSHORE BASINS			
RESOURCES		PARCEL IN NATIONAL CURRENCY	REFERENCE TABLE		REFERENCE TABLE	RESOURCES		PARCEL IN NATIONAL CURRENCY
Cr$ MILLIONS	%					Cr$ MILLIONS	%	
1.330,0	2	70	—	1-EXPLORATORY INVESTIGATION	—	418,5	5	90
18.515,0	31	55	—	2-EXPLORATORY DRILLING	—	1.590,5	18	94
2.753,0	5	50	—	3-PROVISIONAL SYSTEMS	—			
7.287,0	12	65	—	4-DEVELOPMENT DRILLING IN DISCOVERED AREAS (INCLUDING INJECTION AND SPECIAL WELLS)	—	3.995,6	43	94
2.378,0	4	75	—	5-DEVELOPMENT DRILLING IN AREAS TO BE DISCOVERED	—	539,5	6	94
16.333,0	27	79	—	6-ENGINEERING WORKS AND INSTALLATIONS IN DISCOVERED AREAS (CONNECTED RESOURCES)	—	73,0	1	90
4.650,0	8	80	—	7-ENGINEERING WORKS AND INSTALLATIONS IN DISCOVERED AREAS (RESOURCES TO BE APPROVED)	—	1.970,0	21	90
3.860,0	7	80	—	8-ENGINEERING WORKS AND INSTALLATIONS IN AREAS TO BE DISCOVERED	—	215,0	2	90
1.600,0	3	60	—	9-ADDITIONAL EQUIPMENTS	—	400,0	4	60
450,0	1	80	—	10-OTHER ENGINEERING WORKS	—			
59.156,0	100	68	—	TOTAL	—	9.202,1	100	91

NOTES:

- THIS TABLE SPECIFICALLY REFERS TO ALTERNATE I; AS TO THE OTHER ALTERNATIVES THE PRINCIPAL INVESTMENTS PRESENT APPROXIMATE FIGURES.
- ITENS 1 AND 2 REFER TO THE STRATEGIC EXPLORATION PROGRAM: 325 WELLS ON THE CONTINENTAL SHELVES AND 182 WELLS IN LAND SEDIMENTARY BASINS.

ABSTRACT		INDEX
CONTINENTAL SHELF	Cr$ 59.156 MM	87
LAND BASINS	Cr$ 9.202 MM	13
DEXPRO INVESTMENTS IN PERIOD 78/81	Cr$ 68.358 MM	100

REMARKS

- CRUZEIROS (Cr$) VALUES WERE ESTIMATED AT PRICES OF JULY 1977.
- THE EXCHANGE RATIO WAS Cr$ 14,85 PER DOLLAR.

FIG. 7

SOURCE: DEXPRO SERPLAN JULY-77

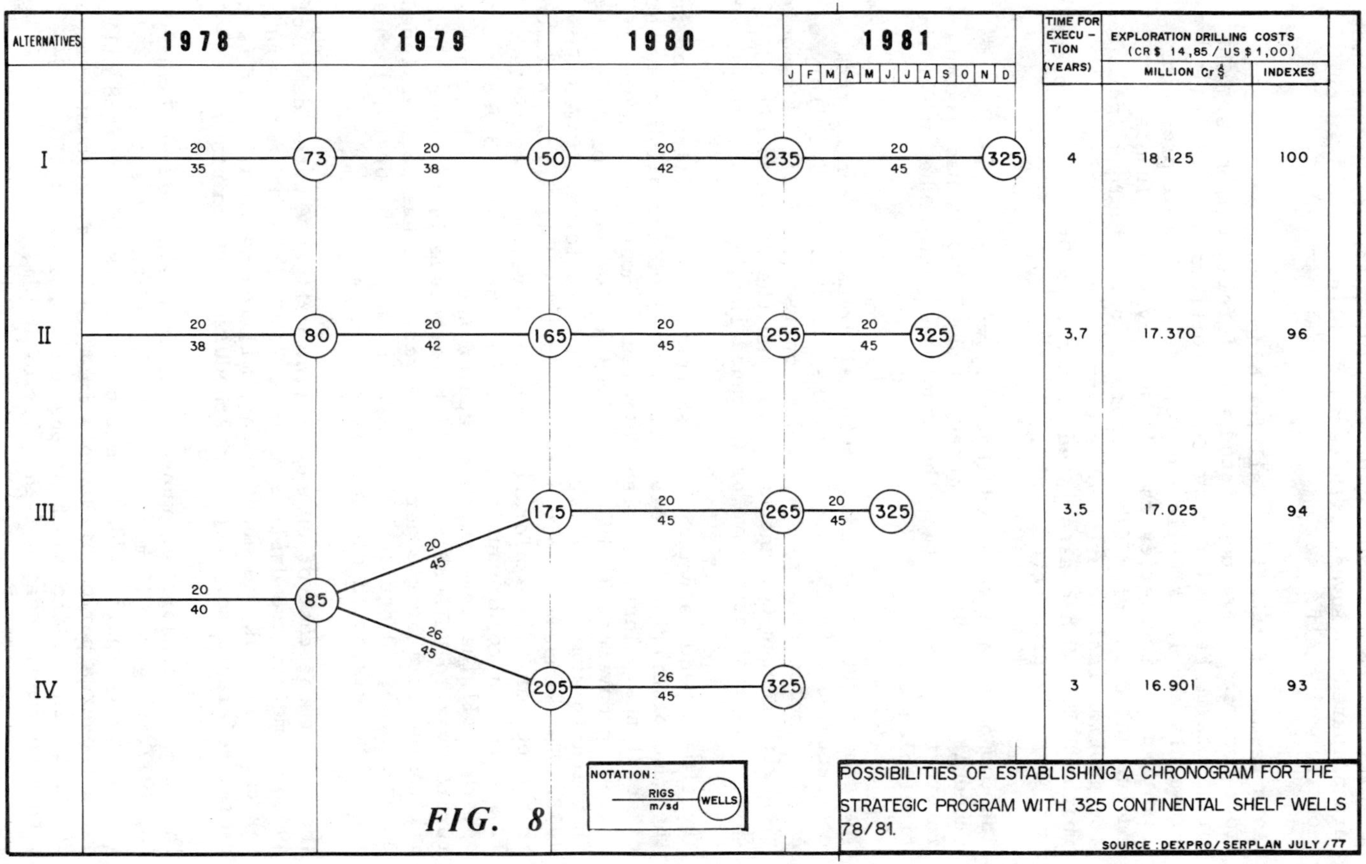

FIG. 8

POSSIBILITIES OF ESTABLISHING A CHRONOGRAM FOR THE STRATEGIC PROGRAM WITH 325 CONTINENTAL SHELF WELLS 78/81.

SOURCE: DEXPRO/SERPLAN JULY/77

for oil: Service Contracts with a Risk Clause. By entering into risk contracts, PETROBRÁS can operate indirectly in priority areas in association with international groups that are prepared to invest with us in drilling for oil in our continental shelf.

As a result of the first call for bids issued by PETROBRAS, four contracts were negotiated with foreign companies. The prospects to be explored by the consortiums are important prospects and it is believed they are capable of producing significant discoveries as well as providing valuable additional data for offshore exploration. Figure 9 indicates the geographical location of the areas in connection with which PETROBRÁS has associated itself with British Petroleum and Exxon to prospect for oil.

The companies are required to comply with operating time schedule under the risk contracts and PETROBRÁS has maximum interest in ensuring that the deadlines established under these timetables are kept.

The same figure shows the additional areas covered by risk contracts: one in the North and the other in the Territory of Amapá. One of the areas is under contract to Shell in association with Pecten Co. and Ensearch, and the other to Elf/AGIP. Thus, four contracts have been signed involving four areas.

The time schedules under the risk contracts, as illustrated in the case of British Petroleum in Figure 10, establish a number of deadlines. Thus, British Petroleum, whose contract was signed on November 9, 1976, is given a time limit of 15 months in which to start its first well scheduled for early February 1978. The company is given three years to define its position in respect of this exploration and the investment committed. If it desires to discontinue operations before the 3 years are up, then all the investment scheduled and not made shall be turned over to PETROBRÁS. British Petroleum is to invest 10.5 million dollars over a period of 3 years and has been assigned an area of 5,500 square kilometers in the Santos basin, located 150 kilometers south of the city of Santos.

The investment cost of an exploration well at a water depth of 40 to 50 meters and drilled down to 2,500 to 3,000 meters in search of a strike is about 5 to 6 million dollars. This is a useful reference for comparing scheduled investments.

The Elf-AGIP consortium, whose contract was signed on January 28, 1977, is given 15 months in which to start the first well, i.e. in early April 1978. The consortium is to invest 8 million dollars over a period of 3 years in a 3,050 sq. kilometer area of the basin off the mouth of the Amazon, located 570 kilometers north-northeast of the city of Belém.

AREAS UNDER RISK CONTRACTS

1 ■ ELF / AGIP
2 ■ SHELL / PECTEN / ENSERCH
3 ■ ESSO
4 ■ BP

FIG. 9

RISK CONTRACTS – TIME SCHEDULES · FIG-10 ·

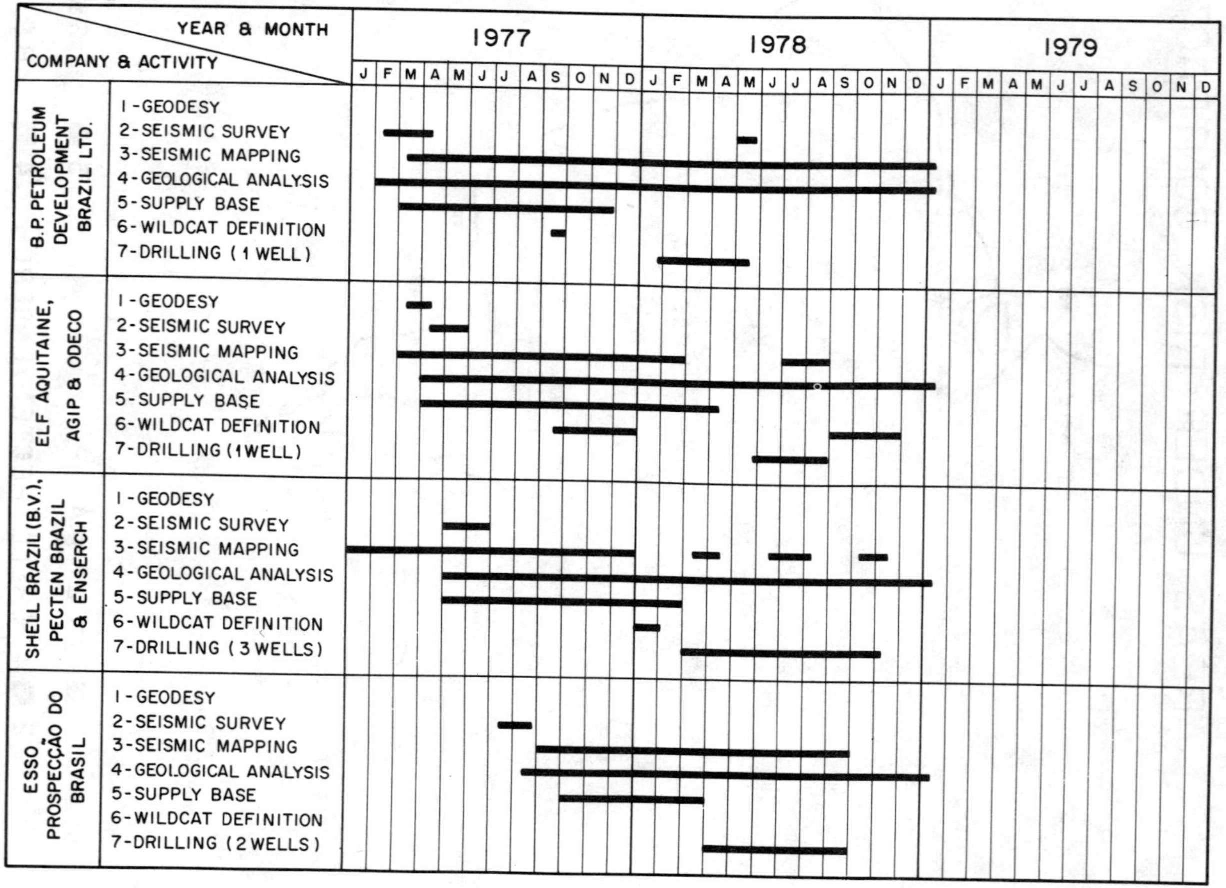

The contract signed with the consortium composed of Shell (50%), Pecten (25%) and Ensearch (25%) on December 22, 1976, also requires spud-in of the first well within 15 months, i.e., by the end of March 1978. The consortium is to invest 20 million dollars and has been assigned an area of 6,150 sq. kilometers in the basin off the mouth of the Amazon, 420 kilometers to the northeast of the city of Belém. Finally, the last contract which was signed recently with Esso on April 26, 1977, also calls for a completion time of 3 years and the investment of 16 million dollars in Area No.3 with 5,650 sq. kilometers, located in the Santos basin, 60 kilometers south of Rio de Janeiro.

The time schedule for these activities given in Figure 10 shows this work developing effectively on the stages for exploratory processing, seismic surveying, experimental seismic processing, complete processing, interpretation, site selection, establishment of supply bases, and the drilling of the first and second well. All activities have been completed on schedule up to August 1977 as shown in the figure.

The above is a general description of what PETROBRÁS has accomplished on Brazil's continental shelf.

In conclusion, I would like to call your special attention to the tremendous effort PETROBRÁS is making in order to implement in a short time its dynamic program for the discovery of oil reserves in Brazil.

A GENERAL REVIEW OF UK NORTH SEA OFFSHORE DEVELOPMENTS

Professor Tom Patten

Institute of Offshore Engineering, Heriot-Watt University

INTRODUCTION

North Sea proven reserves have a potential peak production rate exceeding 200 million tons of oil per annum. In overall energy terms, Britain now receives 98% of its gas requirements and approximately 50% of its oil needs from the UK sector of the North Sea, this oil production rate having been built up in the short span of two years. A recent official statement of reserves was published by the UK Government in April 1977 (Reference 1).

The main purpose of this conference is to provide a forum for expert discussion on the analysis of offshore structures. The illustrative material supporting this paper, as verbally presented, was selected to stress that structure, in this context, can include subsea pipeline, stinger, flare tower single point moorings of various kinds, semi-submersible and floating production systems, as well as the more conventional fixed steel or concrete structures.

An important concern of subsequent contributors will be to lead you into different techniques and aspects of the analysis of such structures. Nevertheless it is important to begin with the problem facing the operating oil company, namely to optimise the economic production of oil or gas from a specific subsea reservoir. During the lifetime of the oilfield, appropriate provision is required for drilling, well completion, hook-up and test, production, oil transfer, monitoring and control, maintenance and repair of the well and of the associated production plant. Any external constraints whether environmental, technological or psychological, will probably be traded off, one against another, to secure the "best" solution. In the present day the solution will most probably specify a platform located above the sea surface to accommodate the equipment to drill and to produce the oil accumulated in the reservoir. In turn that platform will be supported by a structure in steel or concrete, and most probably fixed to, or resting in the seabed. Such structures are visually the most spectacular items of offshore hardware but should not be allowed to distract attention from the existence of many other kinds of structure associated with a developed offshore oil field.

An an abbreviated comparative form, this paper selectively reviews five North Sea oilfields in the UK sector and points out some differences in the engineering configurations adopted. For the purpose of this introductory review, the following fields will be considered:-

FORTIES	(BP)
AUK	(Shell)
BERYL	(Mobil)
BRENT	(Shell), and
ARGYLL	(Hamilton)

The name in brackets is that of the operating company for the named field. The location of these and other North Sea oil producing fields is shown in Figure 1.

FORTIES

The Forties field was discovered in November 1970, was brought into production in November 1975, and now has a daily production rate exceeding 300,000 barrels. There are four platforms from each of which can be drilled 27 deviated wells. The supporting steel structures were constructed at two different UK sites, many of the massive and complex nodes and other elements having been prefabricated elsewhere. The design had to be checked for structural integrity during all stages of construction. Special purpose flotation chambers, which formed the base on which the structure was assembled, conveyed the jackets on the sea journey of at least 200 km to location. Next followed the complicated and remotely-controlled manoeuvre of partially submerging the structure, rotating it to the vertical position, then landing it on the seabed. For the structure the final installation activity was the driving of the necessary 1.5 m diameter open-ended piles. Subject always to suitable weather conditions, it was then possible to instal the platform deck and the associated equipment, accommodation, drilling and other modules. It should be noted that 100 year design maximum wave height and maximum wind speed conditions for the Forties area are 30 m and 240 km/hr respectively. Two steel structures, or jackets, were installed in summer 1974, and the other two in June 1975.

The availability of oil at the production platform is not sufficient - it is essential to convey the product to the onshore market. If the oil reservoir, or a group of neighbouring oil fields is large enough, it may be economic to

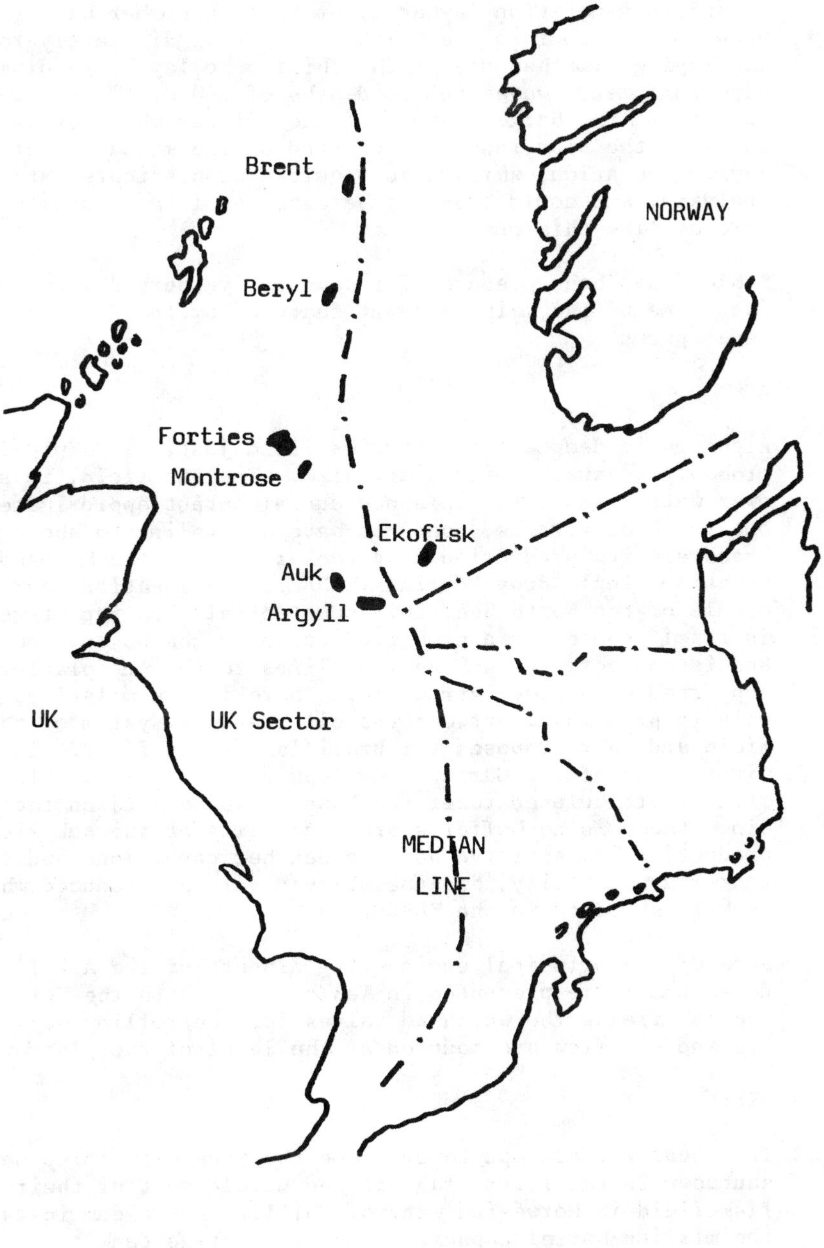

FIGURE 1 Location of North Sea Oil Producing Fields (Aug. 1977)

instal a subsea pipeline. This decision was taken for the Forties field, despite the water depths of 120 m and the necessary pipeline length of 175 km. Where the seabed permits, subsea pipe is normally trenched in and subsequently buried. The mechanics of the pipelaying operation are very complex and the basic problems are made much worse by rough weather. The third generation laybarge, of which a number have recently been introduced into the North Sea, can significantly reduce the laying time because of the ability to lay large diameter pipe in 5 metre waves and to depths of 200 m. There are many aspects of the barge design, of the stinger which assists the entry of the pipe into the sea, and of the mechanics of the laying operation, which also required sophisticated structural analysis, and could properly be explored in an academic conference like this one.

Table 1 has been prepared for comparative purposes and summarises some of the main relevant features of the Forties field development.

AUK

Although in deep waters, Forties field justified a pipeline, probably because of its giant size. The Auk field, in shallower waters and with a planned annual output approximately one tenth of Forties, does not have a pipeline to shore but transfers its production to a tanker through the Exposed Location Single Buoy Mooring (ELSBM). On location near the middle of the North Sea, this substantial floating structure is itself moored by a number of catenary chains and anchors and is connected by subsea flow lines to the Auk platform supported on a steel structure. There is a similarity, but only in principle, between the oil transfer system of the Auk field and that proposed for Brazil's Garoupa field. In the first phase of the Garoupa development a tanker will load oil from an articulated tower the base of which sits on the seabed. Since there is no buffer storage capacity at the Auk field, production is sensitive both to weather conditions and to tanker availability, because oil can only be produced when a tanker is moored to the ELSBM.

Some of the principal engineering aspects of the Auk field development are presented in Table 1. In both the Forties and the Auk fields the wellhead valves for controlling well pressure and oil flow are mounted at the level of the platform deck.

BERYL

In an earlier attempt to overcome the necessity for production shutdown in the first stage of the development of their Ekofisk field in Norwegian waters, Phillips Petroleum installed a one million barrel capacity concrete storage tank to provide a

Field and Operator	Magnitude of Field (Large or Small)	Water Depth	Oil Transfer System (Pipeline or SPM)	Structure Configuration	Structure Material	Wellhead Location (Platform or Seabed)	First Production
FORTIES BP	Large	120 m	Pipeline	Fixed, piled	Steel	Platform	Nov. 1975
AUK Shell	Small	85 m	SPM (ELSBM)	Fixed, piled	Steel	Platform	Feb. 1976
BERYL Mobil	Small	115 m	SPM (Mobere)	Fixed, gravity	Concrete	Platform and Seabed	June 1976
BRENT Shell	Large	140 m	Pipeline and SPM (SPAR)	Fixed, gravity & piled	Concrete and Steel	Platform	Nov. 1976
ARGYLL Hamilton	Small	75 m	SPM	Floating, S-S	Steel	Seabed	June 1975

TABLE I. Comparative information of selected UK offshore oil fields

buffer against interruptions in the offshore loading sequence. This experience led the designers to devise the concrete gravity structure, incorporating oil storage, to replace the steel structure. Another advantage claimed for this concept was the existence of sufficient excess buoyancy to permit towout of the structure to its final location with decks, equipment and accommodation modules already in place. By installing this equipment in sheltered water it was possible to avoid some of the bad weather delays out on location.

The first "Condeep" built at Stavanger, Norway, was located on the Beryl field in mid-1975. The complete structure and platform was towed out in the final vertical orientation and slowly ballasted down into position. With a total dry structure weight of approximately 200,000 tonnes, the concrete gravity structure rests firmly on the seabed without piling. With a view to taking further advantage of the Condeep concept, Mobil developed one step-out well on the Beryl field with a Seal subsea wellhead, designed for remote control, and for remote access and maintenance. In theory the produced oil and gas was quickly available to fuel the installed power plant, to provide power for drilling the main wells. The Beryl field has two additional seabed-mounted structures, the Mobere oil loading tower, and the flare towers for gas combustion.

Table 1 contains comparative information.

BRENT

In August 1975 and in May 1976, two concrete gravity structures were installed on the Brent field, another giant with a scheduled production rate by 1982 of 500,000 bbl/day. In the final configuration Brent will have four platforms, three supported by concrete gravity structures, and the fourth by a steel structure. Separate oil and gas pipelines have been laid but are not yet in use. Early production has been achieved by using the SPAR loading system, a mooring and offshore loading buoy of unique construction, with the capability of oil storage and fitted out with the necessary oily water treatment plant.

A notable feature of recent North Sea operations has been the increasing use of sub-submersible drill barges in the construction support role. For example in the Brent field an S-S barge assisted in the temporary installation of a tower crane thus enabling work to be carried out on the upper sections of a flare tower. Each of these items adds to the range of the offshore structures of possible interest.

Again, Table 1 provides some comparative information on the Brent field.

ARGYLL

In addition to providing the first oil from the UK sector, the Argyll field is distinctive in a number of other ways. The three wells were drilled from floating exploration rigs and are controlled by wet wellheads located at seabed level. The oil flow and control lines from the three separate wells are brought together in a seabed manifold and a novel riser assembly connects with the platform production equipment. In this case the platform is not mounted on a rigid structure resting on the seabed but is the deck of a floating semi-submersible drill barge, modified for this specific duty and carrying basic separation and treatment plant. The treated oil returns to the seabed level through the central pipe in the riser assembly and then a distance of approximately 2 Km to a relatively conventional single point mooring system for tanker loading. The time taken to design, construct, instal and commission the whole system was only slightly longer than two years, and in the succeeding two years of production operations the system appears to have worked to expectations. An extrapolation of the configuration used in the first phase of the Ekofisk development, the Argyll system is likely to have a significant influence on development trends.

FUTURE

Table I provides a simple comparative summary of some features of the various UK oilfields referred to. The various combinations and permutations would appear to provide adequate flexibility to meet most future requirements. Nevertheless we can identify some considerations (listed in Table II) which will lead to changes and to new developments. Firstly there is a need to improve on the performance of installations already in place - for example, much more thought is required at the design stage to overcome the problems of inspection and maintenance which have become evident in existing structures, both steel and concrete.

A big incentive for changes comes through cost. For many North Sea oil developments, costs have escalated to two and three times the original estimates. A number of factors have contributed, namely:-

- (a) cost of raw materials,
- (b) cost of labour,
- (c) design changes during manufacture,
- (d) technical complexity of some manufacture,
- (e) delays due to material, manpower and other deficiencies, and
- (f) difficulties in forecasting costs contained in original estimates.

One reason for using subsea well heads has been to speed up

TABLE II Principal Factors Motivating Change

1. IMPROVED PERFORMANCE

   over existing system.

2. IMPROVED COSTS
   - (a) overall cost
   - (b) earlier return on capital
   - (c) recovery efficiency.

3. IMPROVED SAFETY
   - (a) security of operations and plant
   - (b) personnel
   - (c) environmental protection.

4. IMPROVED ACCESSIBILITY TO RESERVES
   - (a) deeper water
   - (b) weather
   - (c) Artic and Antartic locations
   - (d) earthquake areas
   - (e) marginal fields.

the revenue earning process by developing step-out wells to production in advance of production drilling from fixed platform. In Brazil's Garoupa field there will be nine subsea wellheads, and by installing these in Lockeed dry wellhead cellars in conjunction with the atmosphere pressure personnel transfer system there is a reduction in diver involvement and a corresponding reduction in hazard.

Much thought and considerable speculation has been directed to the possibility of oil production from deeper water. In these circumstances the tension leg platform (TLP) has been offered as one alternative to the fixed structure. This concept is reminiscent of the Argyll semi-submersible production in that buoyancy is derived from flotation chambers at depth sufficiently far below the sea surface to minimise the effects of wave motion. In the case of the TLP the buoyancy is much greater than that required to support the platform equipment and the platform self-weight. The net upward force is resisted by tensioned cables which link the platform to an array of seabed anchors. It is essential to appreciate that the success of such an advanced system will depend on the total package, that is the tension leg platform, the subsea wellheads, the manifold system, the floating production riser, the surface.

It is the performance of the total production system which concerns the operator and therefore in assessing structural innovation he will seek convincing evidence that the innovation will play an effective part in ensuring that the offshore oil is produced economically and safely. Structural analysis will provide an integral and central part of that evidence.

REFERENCE

1. "Development of the oil and gas resources of the United Kingdom 1977" , Department of Energy, April 1977, HMSO London.

the [illegible] gaining process by developing [illegible] to production in advance of production drilling from fixed platform. In Brazil's Garoupa field there will be [illegible] subsea wellheads and by installing these in locked dry wellhead [illegible] cellars [illegible] in conjunction with the atmospheric pressure personnel transfer system there is a reduction in diver involvement and a corresponding reduction in [illegible].

Much thought and considerable speculation has been directed at the possibility of oil production from deeper water [illegible] these [illegible] [illegible] (TLP) has been [illegible] [illegible] [illegible] [illegible] from [illegible] [illegible] [illegible] sufficiently far [illegible] [illegible] [illegible] [illegible] [illegible] [illegible] [illegible] [illegible] [illegible] than [illegible] [illegible] [illegible] [illegible] [illegible] [illegible] [illegible] [illegible] [illegible] [illegible] [illegible] [illegible] [illegible] [illegible] [illegible] advantage [illegible] [illegible] [illegible] [illegible] the [illegible] [illegible] [illegible] system [illegible] [illegible] [illegible]

[illegible] production system [illegible] [illegible] [illegible] [illegible] [illegible] [illegible] [illegible] the [illegible] [illegible] [illegible] [illegible] [illegible] [illegible] [illegible] [illegible] [illegible]

[illegible]

[illegible] 1975 [illegible]

PART II ANALYSIS OF OFFSHORE STRUCTURES

THE ANALYSIS OF OFFSHORE STRUCTURES

C.A. Brebbia

Southampton University
England

1. INTRODUCTION

The construction of offshore structures in ever increasing water depths has necessitated the reassessment of current method of design. Since 1940's the tendency to build them in deeper waters has increased linearly (figure 1). Exceptions to this rule are the steel fixed platform for the Santa Barbara Channel erected in 260 m of water and the one designed for 300 m of water in the Gulf of Mexico. With exploration water depth of around 1000 m, these new designs may indicate an acceleration in the rate of growth of offshore platforms.

It is important to determine how far our current knowledge can be extrapolated in order to analyse deep water fixed bottom structures and look at the shortcomings of the current analysis. Fixed bottom structures may also be unsuited for deeper waters for which new structural forms may be required.

Current design processes start by assuming an estimate for wind and water climate which may be substantially wrong (figure 2). In addition these estimates need to be transformed into loads acting on the structures and in this way new sources of uncertainty are introduced. These loads produce a structural response which for linear systems can nowadays be very accurately determined using computational techniques. The main uncertainties occur when trying to estimate the soil properties and the fatigue life of the structure. Design guidelines on these two problems are scanty or unreliable due to the lack of experimental data and the need for more adequate analytical techniques.

In addition to these uncertainties one should take into consideration the possibility of having substantial construction and material errors. All this casts serious doubts on the possibility of carrying out an accurate analysis of offshore

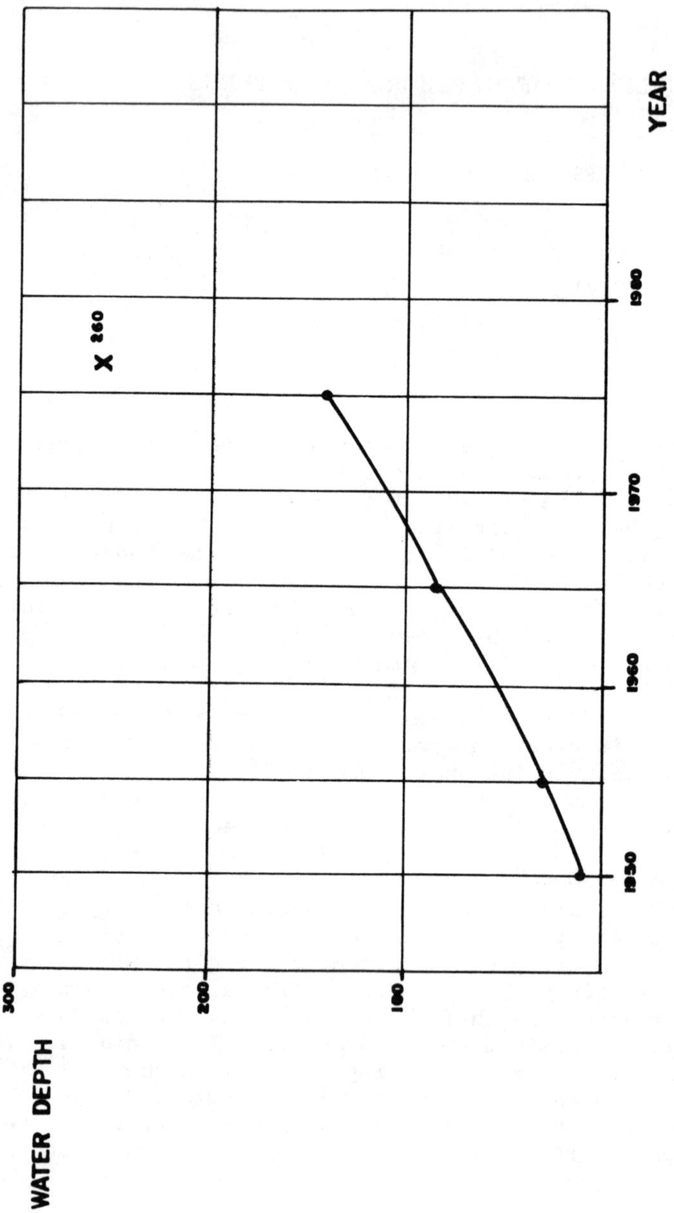

Figure I : WATER DEPTH Vs. YEAR

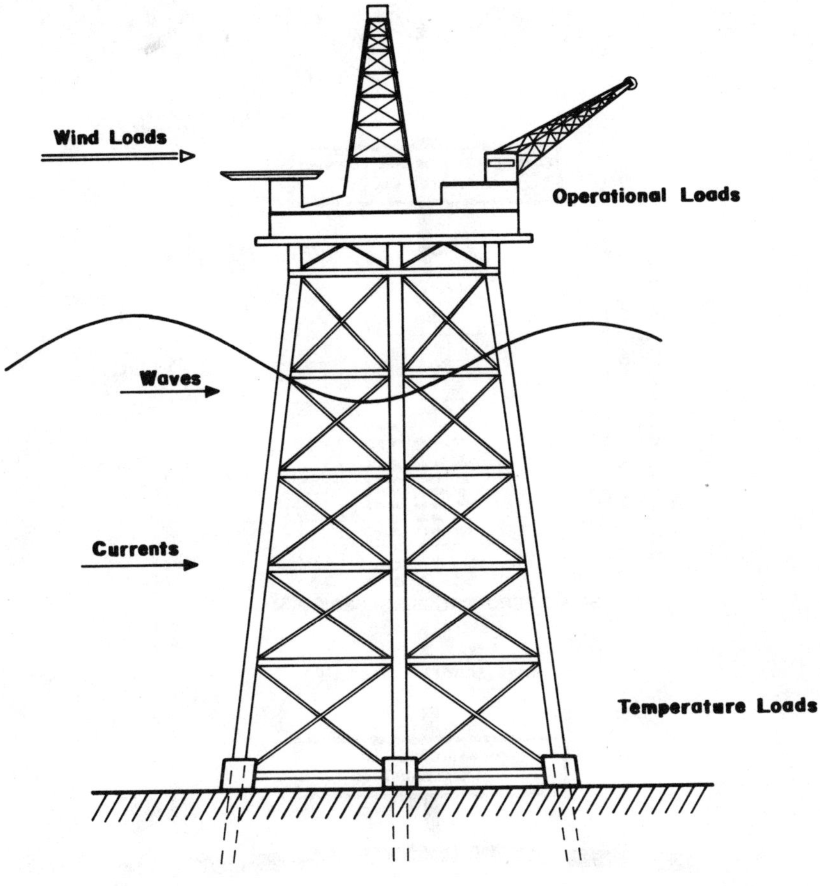

FIGURE 2 - LOADS ACTING ON OFFSHORE STRUCTURE.

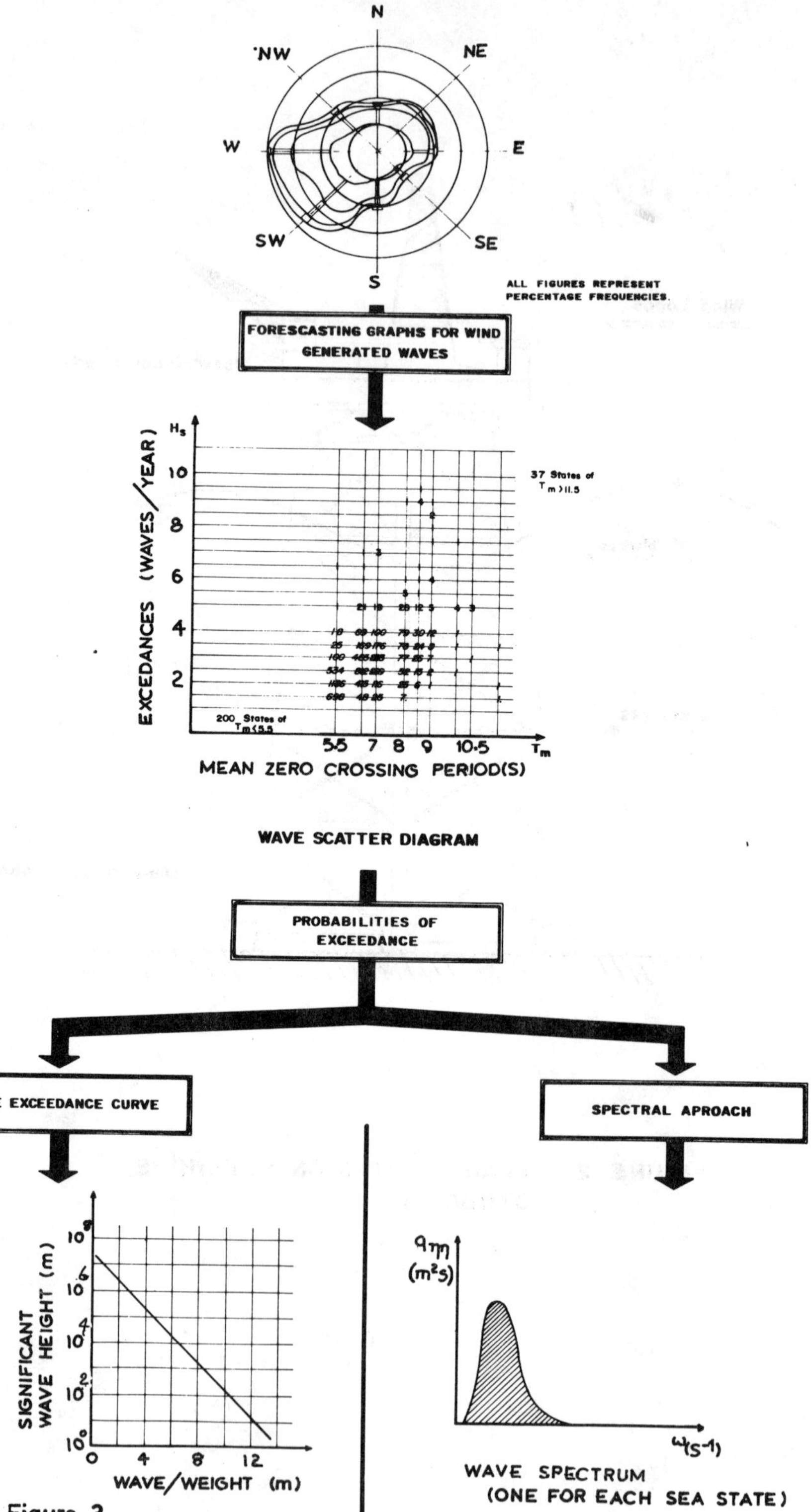
N
NW
NE
W
E
SW
SE
S
ALL FIGURES REPRESENT PERCENTAGE FREQUENCIES.
FORESCASTING GRAPHS FOR WIND GENERATED WAVES
H_s
EXCEDANCES (WAVES/YEAR)
37 States of T_m >11.5
200 States of T_m <5.5
5.5 7 8 9 10.5 T_m
MEAN ZERO CROSSING PERIOD(S)
WAVE SCATTER DIAGRAM
PROBABILITIES OF EXCEEDANCE
WAVE EXCEEDANCE CURVE
SPECTRAL APROACH
SIGNIFICANT WAVE HEIGHT (m)
WAVE/WEIGHT (m)
q_ηη (m²s)
ω(s-1)
WAVE SPECTRUM
(ONE FOR EACH SEA STATE)

Figure 3

structures with our current knowledge but as the structures are urgently needed our research efforts should be directed towards trying to determine the major sources of uncertainties in the design process. In this respect experimental structures, of which few have been built, have a fundamental role to play towards a better understanding of the way in which offshore structures react to environmental forces. Similarly the instrumentation of operating platforms which is now becoming more widespread, will produce more information of direct relevance to the designers.

2. SEA STATES

The response of offshore structures to wave loading is of fundamental importance in the analysis. Waves account for most of the structural loading and because they are time dependent, produce dynamic effects tending to increase the stresses and damage the long term behaviour of the system.

Wave forecasting techniques allow to determine the sea states occurring during the life of the structure and in particular the worst possible of them, usually the sea state corresponding to a storm with a return period of 100 years. While the worst possible storm is necessary to determine the maximum stresses and displacements in the structure, knowledge of all the sea states allows us to calculate long term statistics. These long term statistics are usually more important as the maximum stresses occur only during a very short period in the life of the platform.

Figure 3 summarizes the two main techniques for defining the sea states. In both cases the minimum information that we require is the wind distribution or wind rose. This information which is usually presented in the Beaufort scale, can then be transformed into a wave scatter diagram by relating wind and its duration to the basic parameters at a sea state e.g. significant wave height and mean zero crossing period. This conversion can be carried out using graphics such as these due to Darbyshire, Draper and others [1], which take into account fetch and duration of the wind. From the wave scatter diagram one can pass to a wave exceedance diagram or several wave spectral density curves. The height exceedance diagram is obtained by plotting height versus number of waves on semi-log papers. The results can usually be approximated by a straight line. This diagram can be done for all waves or waves from a particular direction. Directionality of the waves is important for fatigue calculations as it can reduce long term damage by about 40% [2]. The wave scatter and height exceedance diagram shown in figure 3 were deduced from tables prepared by the Institute of Oceanographic Sciences, England and are applicable to the sea region North East of Rio de Janeiro. They are for waves from all directions. The

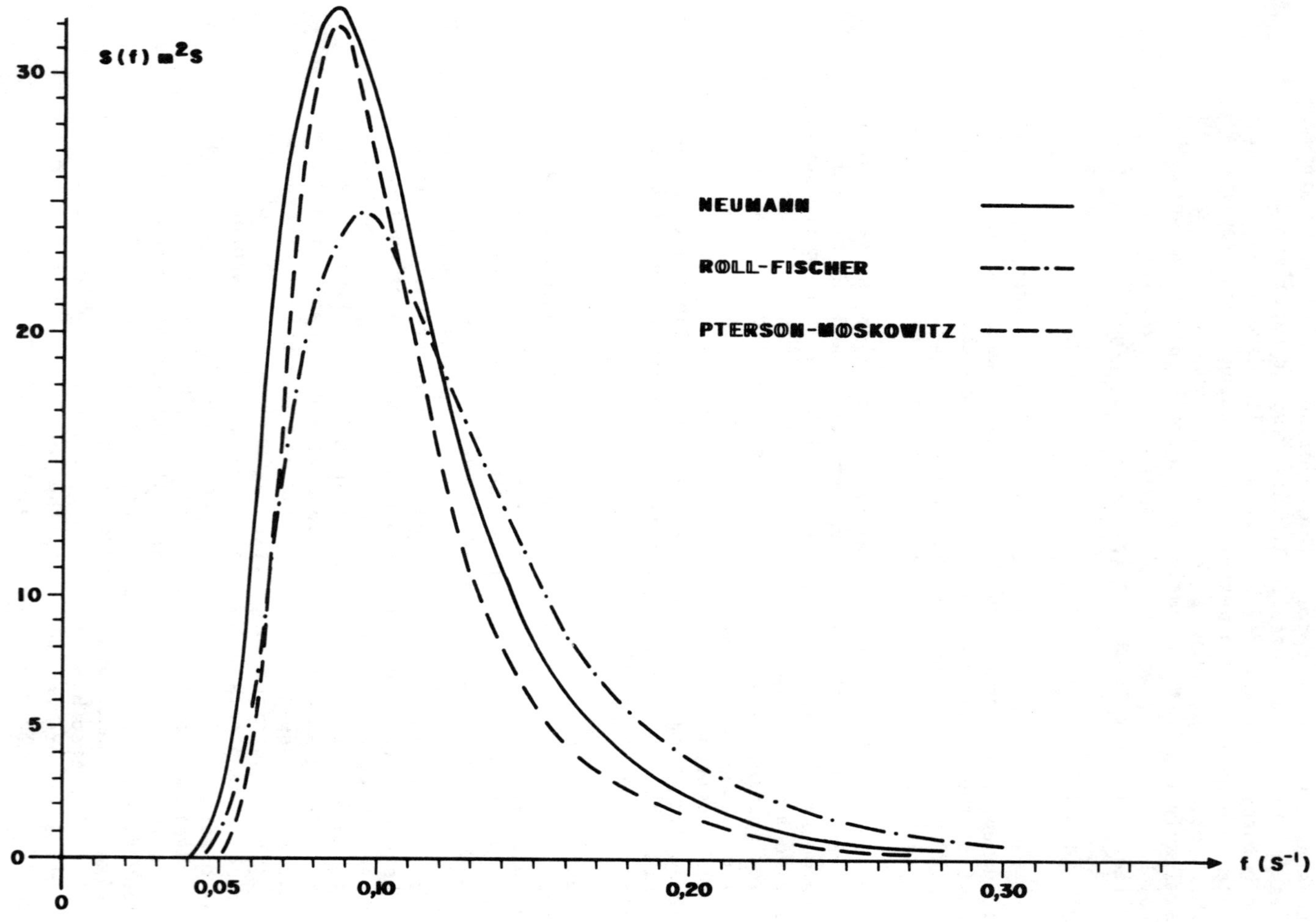

FIGURE 4 - WAVE SPECTRA FOR WIND VELOCITS 15 $\frac{m}{S}$.

results can be extrapolated to determine the maximum wave for a given return period. Plots of wave height versus period can also be obtained from the wave scatter diagram and they are approximated to a straight line when plotted in semi-log paper. From these results the maximum wave for a given return period can be deduced. It gave us H_m = 18 m with T_z = 12 sec. for deep waters but these results are approximate due to the scarcity of the data. The full predictive process requires the use of forecastings graphs for wind generated waves and the assumption that the probability of exceedance can be expressed mathematically (usually as a Rayleigh type probability function).

In the spectral approach exceedance diagrams are not used and instead each of the sea states shown in the wave scatter diagram is transformed into a spectrum. These spectra are then applied to study the response of the system in a probabilistic manner. The spectra are usually applicable to fully developed seas, although some of them are not. A source of uncertainty here is the particular form of the spectrum used. Three different fully developed sea spectra are plotted in figure 4 for a wind velocity of 15 m/sec. [3]. More important than the peak differences are the differences in the higher order frequencies part or tail of the spectrum; i.e. those affecting more the structural response. In addition to this, measured spectra do not have the smooth variation shown in figure 4, but present a series of secondary peaks at higher frequencies, usually multiples of the maximum peak frequency. These peaks have influence in the response but are neglected and a smooth analytical expression used instead.

The spectral approach can be used to determine the maximum response by working with the spectrum of the centenary wave. Once the probabilistic analysis is carried out and the standard deviation of the response found, one can multiply it by a constant value ± λ (usually taken between 3 or 4 for Gaussian process) which gives the maximum response within a certain probability. Another possibility is to take into account the record duration i.e. duration of the storm T, and the expected response is then given by,

$$E\left[x_{max}\right] = \sigma_x \left\{ \left(2 \ln \frac{T}{T_m}\right)^{\frac{1}{2}} + \frac{0.5772}{\left(2 \ln \frac{T}{T_m}\right)^{\frac{1}{2}}} \right\} \qquad (1)$$

where T_m is the mean zero crossing period response

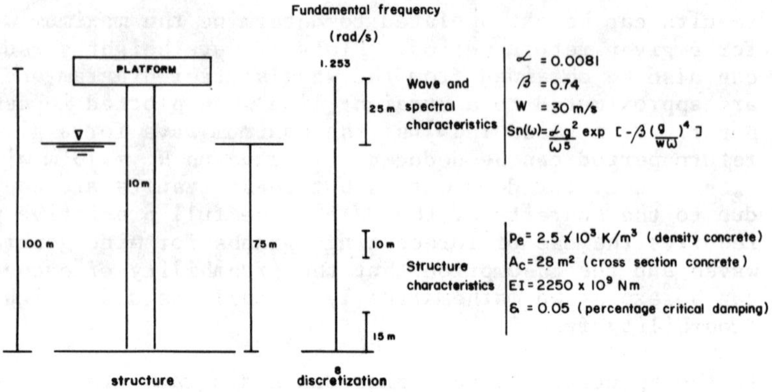

Original Structure and its discretization

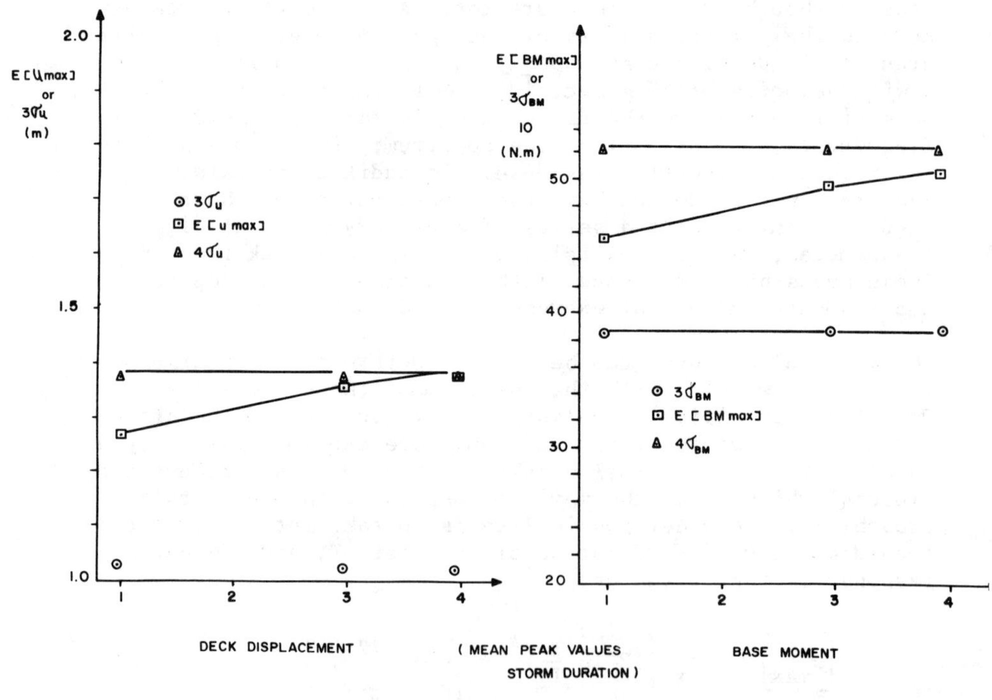

FIGURE 5

$$\frac{1}{T_m} = \frac{1}{2\pi}\left\{\frac{\int_o^\infty \omega^2 S_x(\omega)\,d\omega}{\int_o^\infty S_x(\omega)\,d\omega}\right\}^{\frac{1}{2}} \tag{2}$$

and σ_x is the standard deviation of the response (e.g. displacements, forces, etc.).

If one computes the spectral maximum values of the displacements and base moment for a simple column an example [4] (figure 5) one can see that they increase asymptotically when the storm duration increases. The $\pm$ 3σ values used in the past prove to be inadequate in the practice and the $\pm$ 4σ criterion is recommended.

3. WAVE FORCES

The determination of the forces exerted by waves on structures is a very complex task, even for slender members. Waves can be defined analytically using different theories but in addition to this they produce effects such as slamming and slapping on the structural members that defy simple analytical expressions. The forces are also the result of an interactive process between the structure and the waves with the former modifying the kinematic of the waves.

From the offshore structure designer's point of view it is important to differentiate the various force regimes resulting from the interaction between the structure and the waves.

Hogben [5] has summarised them as follows:

a) for $d/\lambda > 1$ (d: diameter or characteristic dimension and λ wave length) conditions approximate pure reflection;
b) for $d/\lambda > 0.2$ diffraction is increasingly important; c) for $d/w_o > 0.2$ (w_o: orbit width parameter equal to wave length in deep water) inertia is dominant, d) for $d/w_o < 0.2$ drag becomes more important.

The determination of the forces resulting from any of these regimes can be formulated in two basic steps: 1) Computation of the kinematic flow field starting with the water elevations and using a wave theory, 2) Determination of hydrodynamic forces applying Morison's equation, diffraction theory, etc.

In reference [6] Morison et al. developed a formula to represent the total force per unit length on a slender cylinder. The formula takes into account the inertia and the drag components, i.e.

$$P(t) = C_I \ddot{w} + C_D \dot{w} \, |\dot{w}| \tag{3}$$

The forces in the direction of wave advance and the water particle velocity $\dot{w}$ and acceleration $\ddot{w}$ are evaluated at the cylinder axis. C_I is a constant due to inertia consisting of two terms, one due to the 'hydrodynamic' mass contribution and the other to the variation of the pressure gradient within the accelerating fluid.
Hence

$$C_I = C_M + C_A = c_m \frac{\rho \pi D^2}{4} + \rho A \tag{4}$$

c_m: hydrodynamic coefficient for the section, A: cross sectional area.
Thus the inertia force on the body per unit length can be written,

$$P(t)_{inertia} = C_I \ddot{w} = C_M \ddot{w} + C_A \ddot{w} \tag{5}$$

The value of C_I varies with different section shapes. Even for the same shapes i.e. circular, cylinders, Wiegel [7] reported experimental values from 1.0 to 1.4.

When the movement of the cylinders is taken into consideration, u, equation (5) is written as follows

$$P(t) = C_M(\ddot{w} - \ddot{u}) + C_A \ddot{w} + C_D(\dot{w} - \dot{u}) \, |\dot{w} - \dot{u}| \tag{6}$$

The first term is the added or hydrodynamic mass term and it depends on the motion of the member. The second term is the inertial term representing the distortion of the streamlines in the fluid and is independent of the acceleration of the structure in the linear approximation used here.

The drag term is the only non linear term in the above expression. In order to linearize it one can assume that the velocities ($\dot{w}$ or $\dot{w}-\dot{u}$) are distributed as a Gaussian Process with zero mean.

This gives,

$$P(t) = C_M(\ddot{w} - \ddot{u}) + C_A \ddot{w} + \sqrt{\frac{8}{\pi}} \quad C_D(\dot{w} - \dot{u}) \tag{7}$$

where the standard deviation σ can apply for $\dot{w}$ or $(\dot{w}-\dot{u})$. The Gaussian distribution is not justified when working with $(\dot{w}-\dot{u})$ instead of $\dot{w}$ and in this case one generally uses a cyclic procedure to obtain $\sigma_{(\dot{w}-\dot{u})}$. The process is convergent [8].

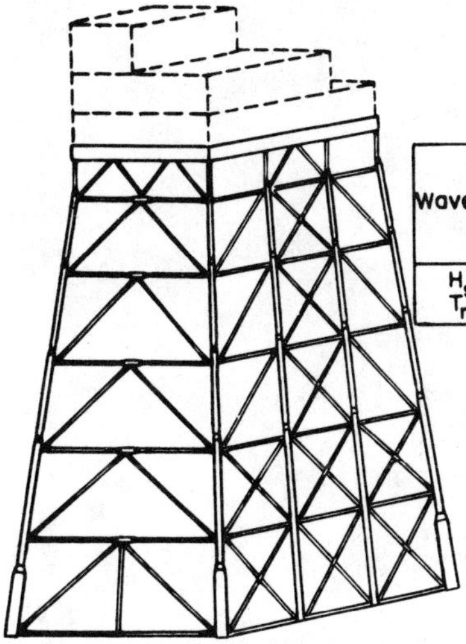

Wave spectrum	Displacement standard Deviation at top σ_u (m)		Fundamental natural frequency ω_1 (rad/s)	Second natural frequency ω_2 (rad/s)
	Inc. drag.	No drag.		
H_s=18.5 m T_m=11.35s	0.1695	0.0626	38.054	10.714

Steel lattice structure

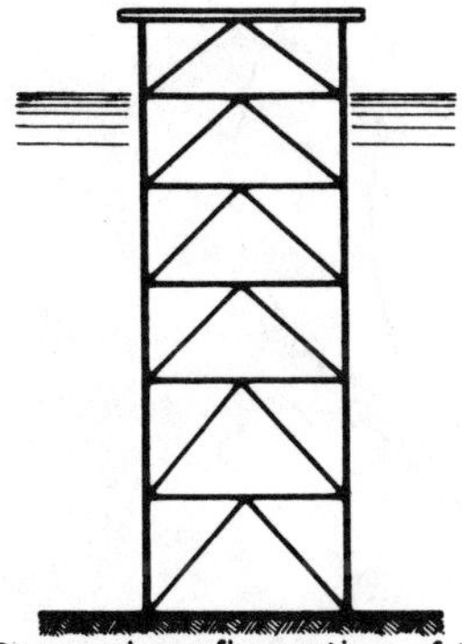

General configuration of the steel platform

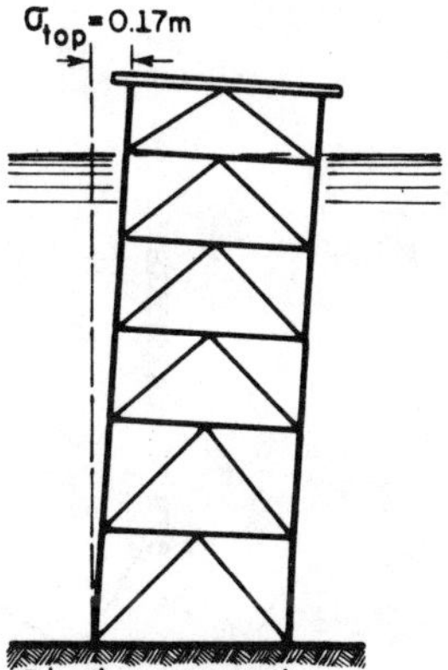

Displacement shape steel platform

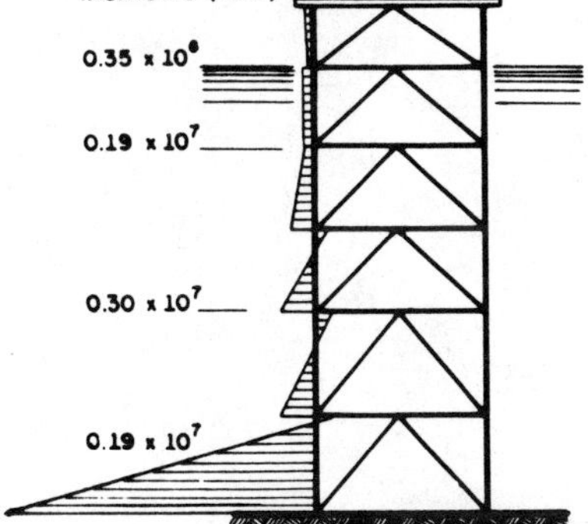

Bending moments in legs steel platform (including drag)

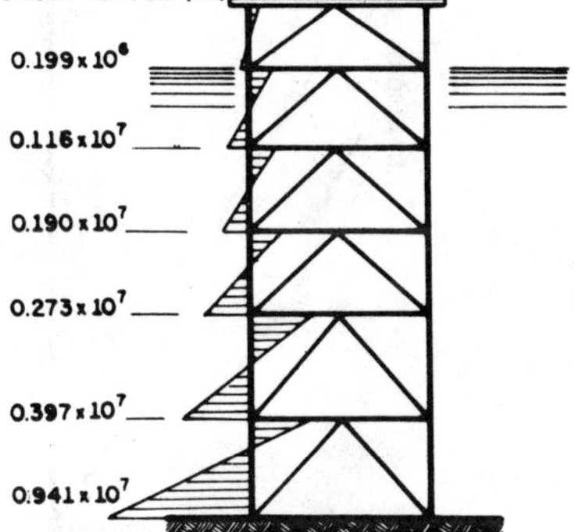

Shear forces steel platform (including drag)

FIGURE 6

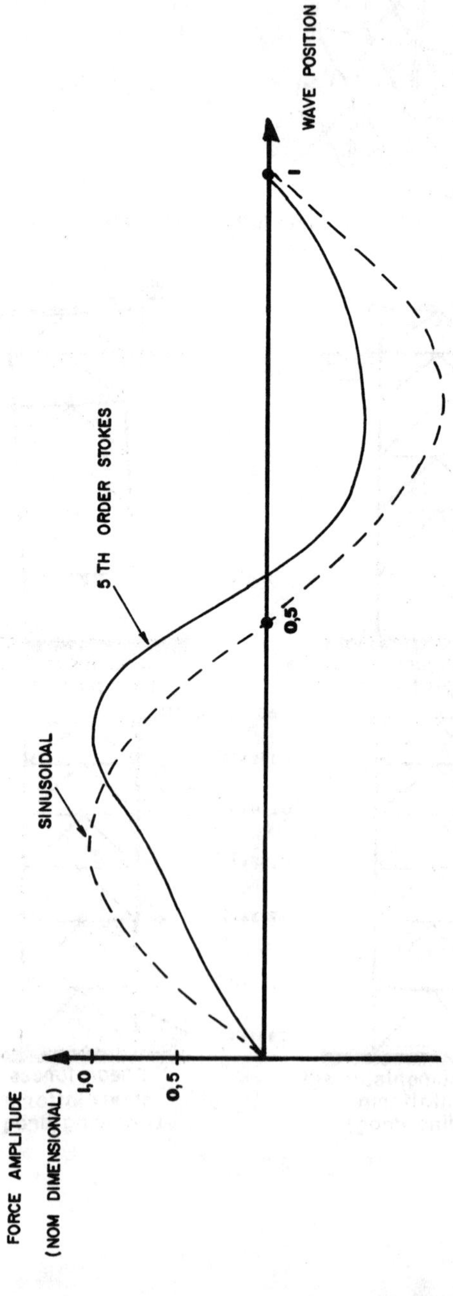

FORCE SHAPE FOR A H = 22,5 m , T = 12 s WAVE

FIGURE 7

The use of the initial $\sigma_{\dot{w}}$ in the calculations without any cyclic improvement can give place to errors in the analysis of offshore structures for which the drag effects are significant.

Figure 6 shows a typical steel platform for 150 m of water in which the displacements and forces have been calculated using the spectral approach [9]. The results shown in the table indicate the relative importance of drag and inertia. Inertia accounts for only 37% of the deck displacement. Figure 6 corresponds to a platform built in the North sea and although the analysis is only two dimensional it is useful for the preliminary design of the system before a full three dimensional finite element analysis is carried out. Here the drag and inertia coefficient were taken to be both equal to 1.

The situation is completely the opposite for structures of larger diameter. For the 10 m diameter column shown in figure 5 for instance, the non inclusion of drag produces an error of less than 5% in the force and displacement response.

A source of uncertainty when applying Morison's equation are the values to be taken for c_I and c_D coefficients. Measures of water particle velocity against the wave forces 10 gave mean values of $c_d \simeq 1.0$ with standard deviation of 24%. They also indicate an average coefficient of inertia $c_a + c_m = 1.2$, with a standard deviation of 22%. Even assuming these comparative low deviations one will obtain large variations of response. For instance taking $c_I \pm 3\sigma_I$ for an inertia only regime the response can vary by as much as 66% and the same variation will apply to stresses.

A further source of uncertainty is the effect of different wave theories to calculate the velocities and accelerations used in Morison's equation [4]. Different theories do not seem to have large influence in the magnitude of the resulting forces but they do alter the shape of the force time curve. Figure 7 shows the normalized forces to be expected when using the linear Airy wave theory and the 5th order Stokes theory. The shape corresponding to the latter is quite different from a sinusoid and its asymmetric shape produces secondary peaks in the dynamic response which tend to amplify the effect of smaller waves. For a fuller discussion see reference [11].

Diffraction is an important effect for large members. It is caused by the disturbance of the flow field due to the presence of the structure. The structure produces a scattering of the waves as these members are in conditions of pure inertia the diffraction theories are based on potential flow. The total velocity potential is given as the sum of the incident and reflected potentials. The two fields are then made to satisfy the boundary conditions. The net effect is a different force distribution than the one obtained if the wave was undisturbed. Figure 8 shows the effect in terms of the force spectral density. The effect is important as most large diameter structures in 100 to 150 m water depth have natural frequencies in the range $\omega = 1.5$ (gravity) to 3.0(steel). Member diameter near the surface may be as large as 10 m for gravity structures and 2 m for steel structures. To give an idea of the importance of the problem a 12 m diameter column in 10 m of water considering or not diffraction will show a difference for the deck displacements of 100%. The values of the displacements are smaller considering diffraction and this is the reason for neglecting it in many cases. Diffraction is however of fundamental importance for gravity structure bases and in other applications such as group members, etc. It is generally considered as a linear problem but is strongly frequency dependent.

4. METHODS OF ANALYSIS

The methods of analysis are probably the more reliable part of the design process due to the many recent advances in computational mechanics [12]. But even here a variety of options are opened to the analyst. The most important of these possibilities are summarized in figure 9.

The first option is to use static analysis. A rule of thumb indicates that when the period of the system is larger than 2 seconds dynamic amplification becomes important. (Note that the period of the structure-water-soil system is generally quite different from the period of the structure only). It is also important to point out that even structures analysed statically will eventually require some dynamic consideration to analyse the effect of fatigue.

If the inertia forces are comparatively important one uses dynamic analysis and two possibilities exist. One can work in the frequency domain, in which case transient effects are neglected and one concentrates in the steady state. The method implies having a linear system. The other possibility is to analyse the system in the time domain by some step technique, in which case transient effects may be considered as well as non linearities.

Due to computer time savings the frequency domain analysis is usually preferred, but if one wants to study effects such as drag, an earthquake, or non linear wave theories the time integration method is needed.

Frequency domain methods can be divided into deterministic and probabilistic methods. The deterministic analysis applies a series of design waves and uses the long term exceedance diagram for fatigue. Probabilistic methods have become increas-

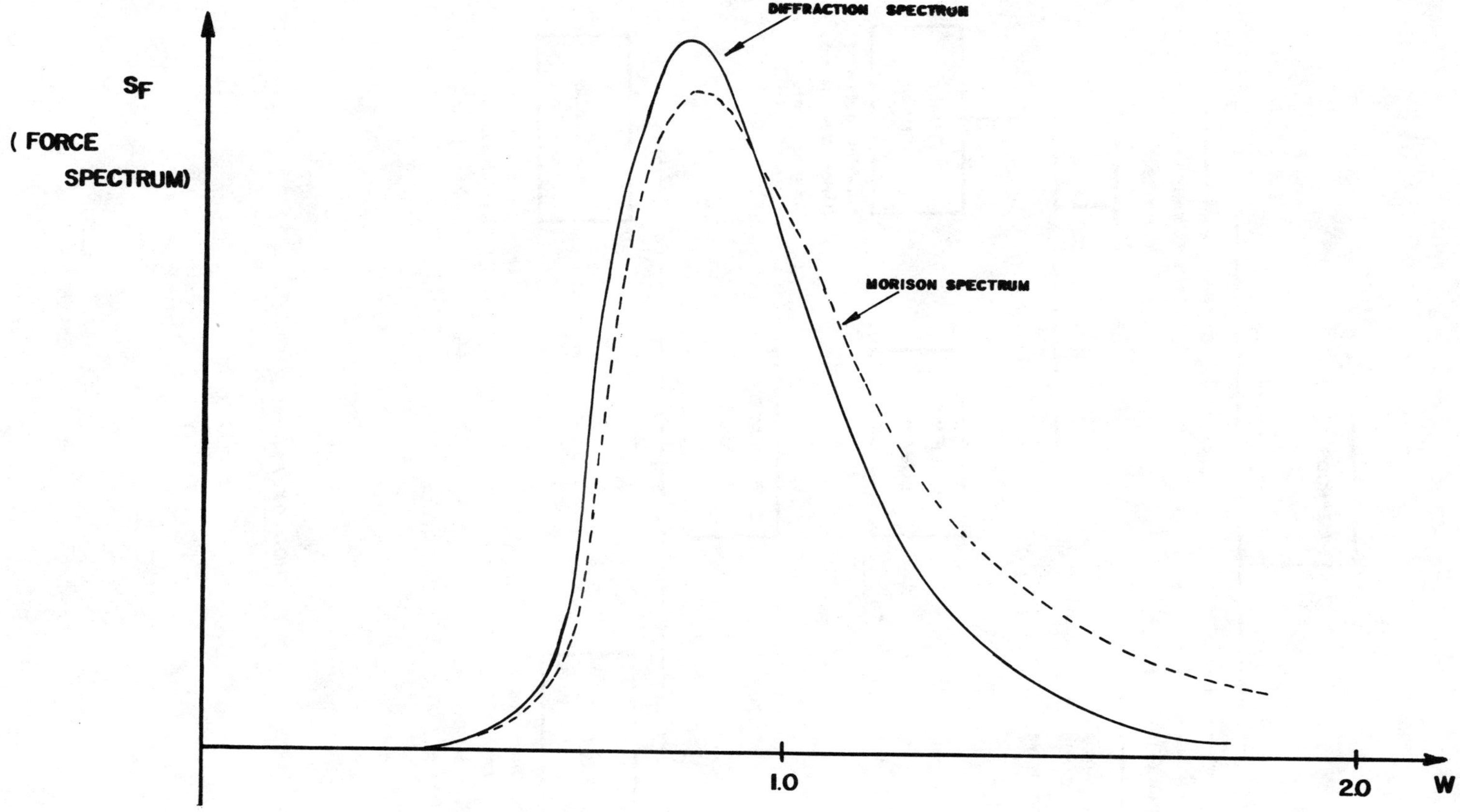

Figure 8 : DIFFRACTION Vs. MORISON FORCE SPECTRUM

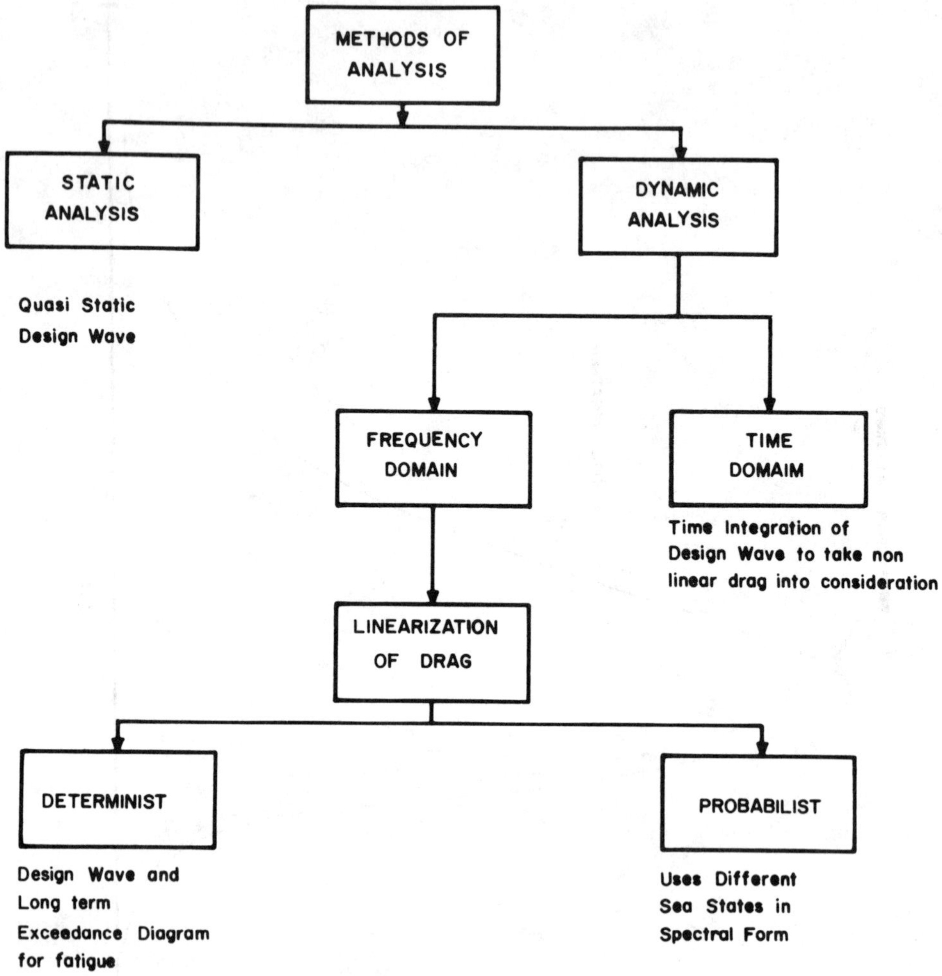

METHOD OF ANALYSIS

FIGURE 9

ingly popular with analysts during recent years. They can be used to study the behaviour of the structure during the extreme design storm and its long term behaviour by analysing it for a range of sea states.

Spectral curves of wave height are the information needed to start the probabilistic analysis. They are then converted into spectral estimates of forces by applying a transfer function which is simply the square of the resulting wave force per unit wave height. The new spectrum (see figure 10) is then multiplied by the ordinates of the structural transfer function. This transfer function is the square of the dynamic response per unit wave force. The final response curve can then be integrated and it produces the variance σ^2 , for displacements or stresses. The procedure can be extended to obtain other statistical information about the system such as long term behaviour. The drawback is that one analysis is required for each sea state in order to study cumulative damage.

It is nowadays accepted that the probabilistic analysis gives good result and is the type of analysis recommended for large structures.

5. FATIGUE ANALYSIS

With our present knowledge the fatigue analysis of offshore structures only gives a rough estimate of the behaviour of the system. This is because of the many approximations involved in the design process. A fatigue analysis needs the following three parts,

a) The statistics of the sea, presented in the form of wave exceedance diagrams or spectral curves.
b) The relationship between the sea state and the stresses at the points under consideration (called 'hot-spots').
c) The determination of the stress-number of cycles fatigue curves (S-N curves).

In addition the stress concentration factors may be needed (This depends on the type of fatigue curve used). The S-N curves obtained for the hot spot are compared with the S-N curves to failure given by the norms. This is done by using Miner's cumulative damage rule, i.e.

$$(D) = \sum_i \frac{n_i}{N_i}$$

i = stress range, n_i number of cycles occurring at stress range i, N_i number of cycles needed for failure at the stress range i.

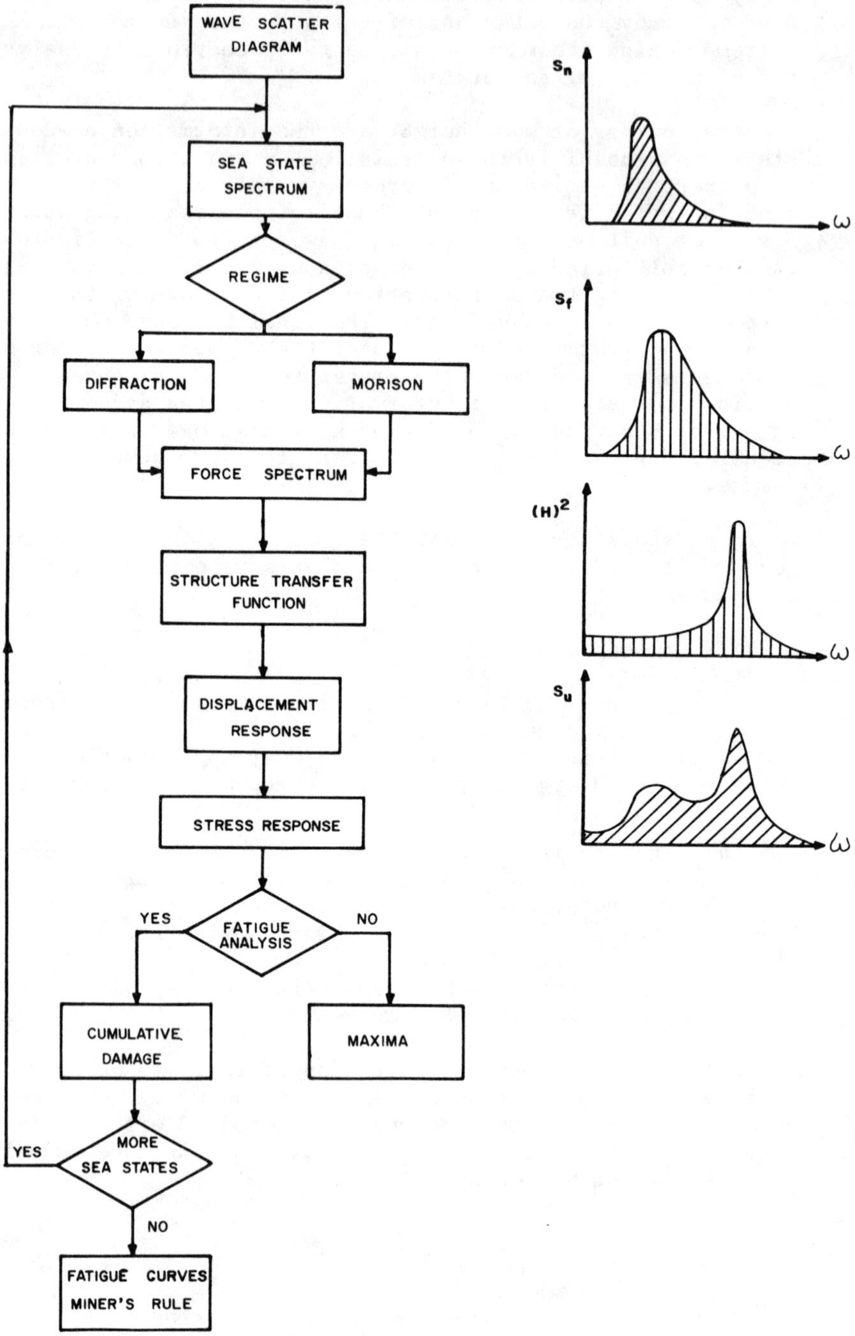

FIGURE 10 **PROBABILISTIC ANALYSIS**

If $(D)_1$ is the accumulated damage occurring during one year, the total life of the structure is,

$$\text{Life (in years)} = \frac{1}{(D)_1} \quad (8)$$

The methods for fatigue analysis follow a deterministic or a probabilistic approach (figure 11). The deterministic approach is based in knowing the height exceedance diagram, usually for one year. Then the relationship between stresses at the hot spot and wave height is found and finally an accumulative stress damage curve is computed by plotting number of cycles versus stress range. The number of the cycles can be obtained from the wave exceedance curve. It is generally recommended that the dynamic amplification factor is taken into account and sometimes this is obtained by computing the dynamic response of a simplified system instead of the response of the multi-degree of freedom structure. The accumulative stress history curve thus obtained is compared against curves such as tht AWS or Lloyds failure curves using Miner's rule.

The process is relatively simple and straight forward. By fitting closed form expressions for the curves, some authors [13] [14] have deduced formulae for the accumulated fatigue damage. Reference [13] deduced two formulae; one for the fatigue damage resulting from a single sea state (storm) and the other for that occurring during the service life of the structure. The difference is that the maxima for a single sea state can be approximated by a Rayleigh distribution while a more general Weibull distribution is applied for long term damage. Several types of deterministic fatigue analysis can be proposed but they are all related to the scheme shown in figure 12, their differences being in the degree and type of approximation used. It is important to keep in mind that each simplification implies another approximation in our results and and that they can easily introduce large errors due to the exponential character of the curves.

Another shortcoming of the analysis is the difficulty of knowing the zero crossing period for the stress histories, from which the number of stress cycles can be calculated. This period depends both on the structure period and the period of the exciting waves. The period is computed for the probabilistic fatigue analysis by taking spectral moments (equation 9) but has to be approximated for the deterministic analysis. The rule is to use the same period as the period of the waves for structures with negligible dynamic amplification. When the dynamic effects can not be neglected (i.e. for those waves which frequencies are within a certain interval of the fundamental frequency of the structure) the period is taken as that of the structure.

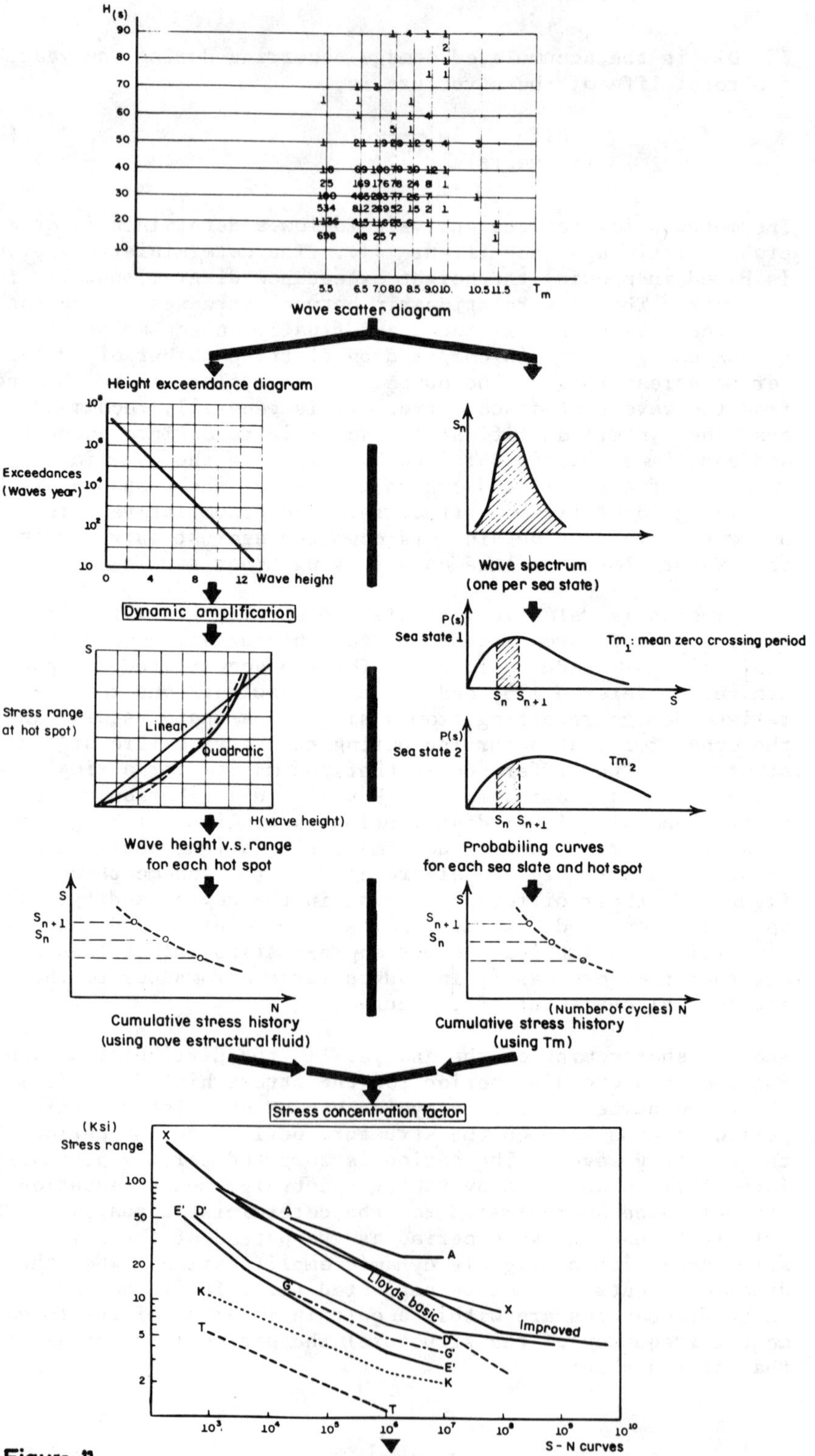
Wave scatter diagram
Height exceedance diagram
Exceedances (Waves year)
Wave height
Dynamic amplification
(Stress range at hot spot)
Linear
Quadratic
H(wave height)
Wave height v.s. range for each hot spot
Cumulative stress history (using nove estructural fluid)
Wave spectrum (one per sea state)
Sea state 1
Tm1: mean zero crossing period
Sea state 2
Probabiling curves for each sea slate and hot spot
(Numberof cycles) N
Cumulative stress history (using Tm)
Stress concentration factor
(Ksi) Stress range
Lloyds basic
Improved
S - N curves
Accumulate damage (mine's rule)
Fatigue analysis

Figure 11

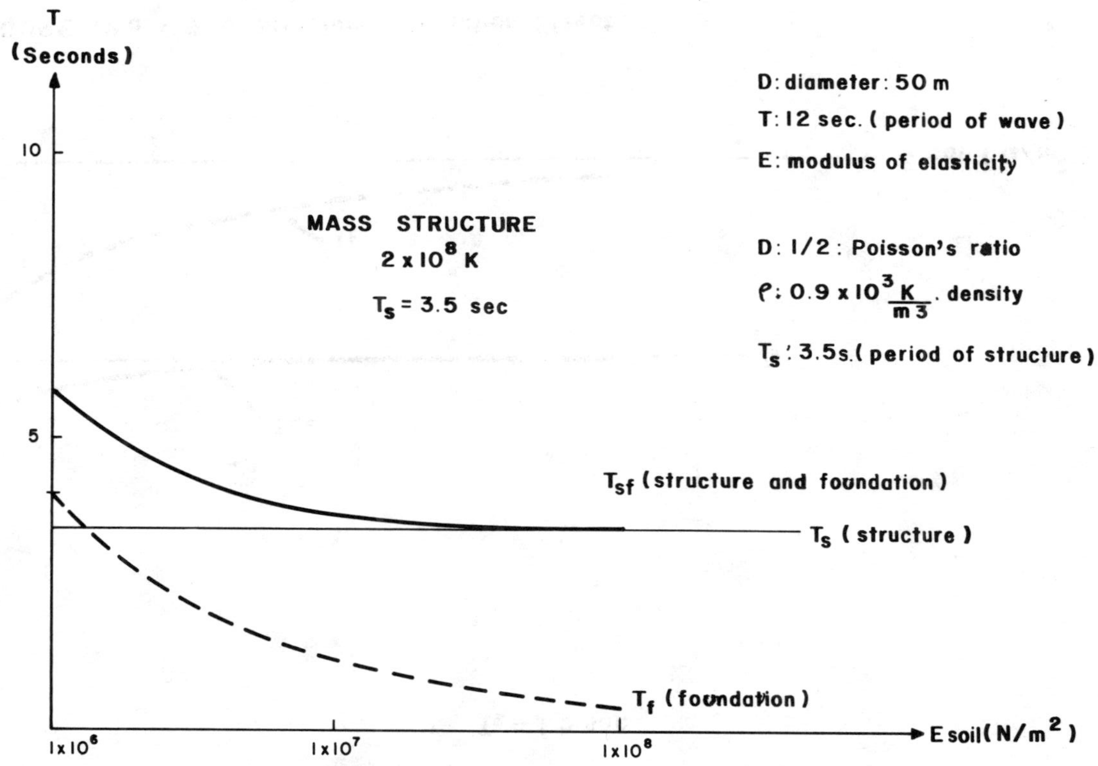

FIGURE 12a - Soil Structure Interaction Effect
(MASS 2×10^8 K, period 3.5 s)

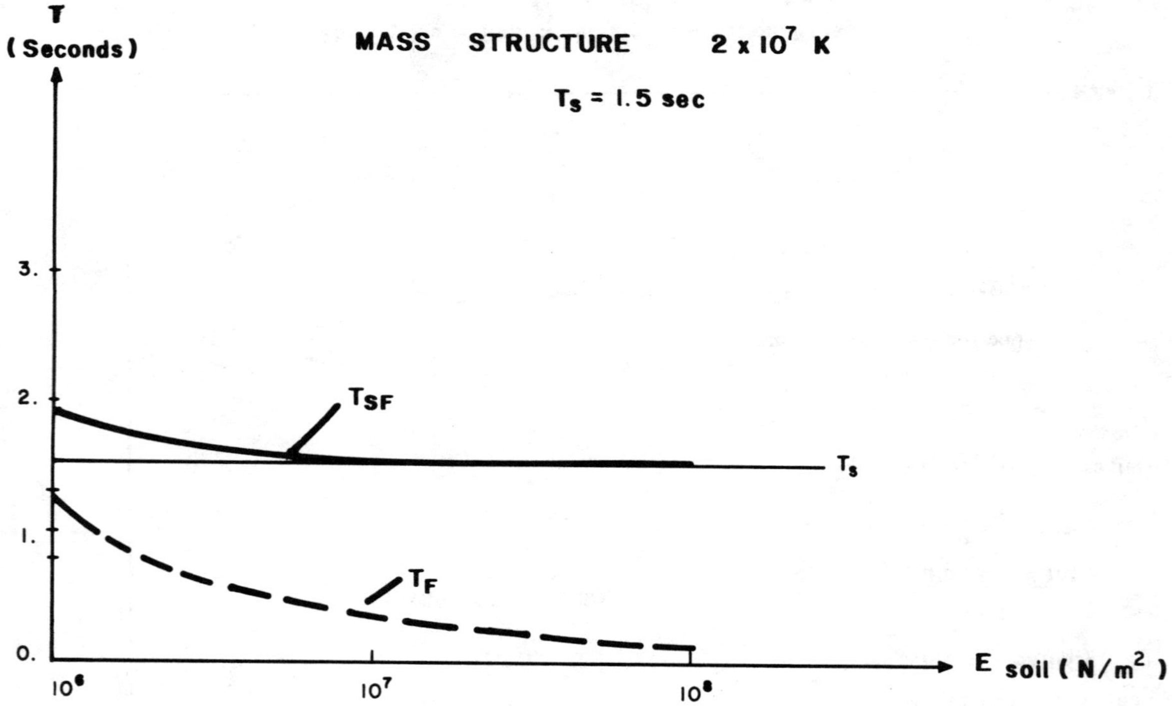

FIGURE 12 b - Soil Structure Interaction Effect
(MASS 2×10^7 K, period 1.55s)

For a probabilistic fatigue analysis one starts by considering the different sea states for one year period (figure 11) and the percentage of time during which they act. The number of times a sea state produces peaks that exceed a particular stress level can be computed from the distribution of stress peaks. For each sea state this distribution is fully defined by knowing σ_S (stress deviation at the point or 'hot spot' under consideration). The probability curve can then be plotted for each sea state against the stress level. The curve is a Rayleigh distribution if the process is narrow banded.

The number of cycles per sea state is calculated by first determining the mean stress period.

$$T_{ms} = 2\pi \left\{ \frac{\int_0^\infty S_s \, d\omega}{\int_0^\infty S_s \, \omega^2 d\omega} \right\}^{\frac{1}{2}} \tag{9}$$

Hence the number of the cycles results,

$$N_s = \frac{T}{T_{ms}} \tag{10}$$

T: duration of the sea state under consideration.

Note that the total area under probability curves is always equal to 1. The area A for instance describes for each sea state the proportion of cycles in a particular stress range. This proportion is then multiplied by the total number of cycles for the particular stress range under consideration. The same is done for all the cycles and one obtains a cumulative stress history curve for a one year period. One can then multiply the results by a stress concentration factor and compare them against the S-N curves using Miner's rule. Hence the fatigue damage for the particular point or 'hot spot' under consideration is obtained. The results are influenced by the inaccuracies in the stress concentration factors and fatigue design curves. In reference [15] fatigue life in several 'hot spots' was computed using different S-N curves and different concentration factors.

The ratio between two different curves for the same hot-spot (American Welding Society category X and British Welding Institute which is substantially Lloyd's basic but continue as shown by dash line in figure 11) oscillated between 7 and 30. It is concluded that until more reliable high cycle data become available the dilemma of which SN curve to use will be difficult to resolve.

Stress concentration factors are impossible to determine in a simple form and although some empirical formulae exist the differene between applying one or the other or carrying out a full finite element analysis can be considerable. Reference [16] reports differences of up to 100%. This factor has great influence in the results due to the exponential character of the S-N curves. In other words relationship between number of cycles and stress is of the type

$$N = m\, S^{n} \tag{11}$$

where n is the slope of the curve approximately between -2 to -4. Assuming n = -4 for instance a difference of 2. in the stress concentration factors produces a ratio of 16 i.e. fatigue life is 16 times longer for the case with the smaller concentration factors.

Another important point is the directionality of the waves. Using directional spectra yields a substantial reduction in computed forces. Directional spreading is essential for the lower waves of primary interest on fatigue analysis and Marshall [2] concluded that taking it into account the stress may be influenced by a factor of 2 and the fatigue damage ratio by a factor of 10.

6. SOIL-STRUCTURE INTERACTION

An area of still greater uncertainty than fatigue is that of soil-structure interaction. The main difficulty is the characterization of the soil, which presents a complex non-linear behaviour. In some cases this behaviour is approximated by an equivalent histeretic damping (5 to 15%). In addition the loss of energy through the boundaries extending to infinite is taken into account by radiation damping and this is usually estimated using half space theories. The half space results [17] used have the disadvantage of being valid for homogeneous elastic soils for which no material damping occurs. Recent studies allow the extension of the theory to account for damping but the results are not in tabulated form. In addition it has been found that layered medium behave in a different way than the homogeneous one, resulting in a stronger frequency dependency for the radiation damping [18].

All these difficulties occurring with half space theories have led some researchers to use finite element representation for the soil. Although finite elements allow for a better variation of the soil properties, they tend to represent inaccurately the radiation damping and have problems connected with mesh resolution. Radiation damping is very significant for soils but finite element discretizations are unable to represent it properly due to the approximations made in the boundary conditions. Special elements for the boundaries have been developed but there is still controversy regarding their performance. Care should also be taken to discretize the region with a mesh fine enough as not to produce artificial filtering of frequencies. The mesh size is function of the velocity of wave propagation or celerity. The wave length can be defined as

$$\lambda = T\ V_s \qquad (12)$$

where T is the period of the exciting frequency for steady state oscillations and V_s is the velocity of propagation of the wave. Typical wave lengths are,

	Exciting frequency	
	T = 10s	T = 4s
$E_{soil} = 3 \times 10^7\ N/m^2$	1000 m	400 m
$E_{soil} = 3 \times 10^5\ N/m^2$	100 m	40 m

This table gives an idea of the finite element grid needed. Typical element size ought to be around $\lambda/10$ to $\lambda/4$ depending on the element used (linear or quadratic). Discretization problems increase when one considers processes with very short periods, such as earthquakes.

It is now generally accepted that soil foundation interaction is important and should be taken into account when calculating the response of the system. How important it is will depend on the particular soil and structure under consideration. Figure 12 shows a structure idealized as a block on an elastic foundation that can only have rock movements. The period of the structure is assumed to be T_S = 3.5 seconds and the frequency has a period of 12 seconds (or ω = 0.52 cycles/hours). To evaluate the importance of interaction, one first determines the period system assuming the structure to be rigid and then the period of the structure-soil system is found. These periods were determined using the stiffness coefficients for rocking presented in reference [17] which are frequency dependent. The final soil-structure interaction period was found using Dunkerley's laws, i.e. $T_{SF} = T_S{}^2 + T_F{}^2$. Figure 20 shows how the period varies when the soil modulus increases. For hard soils the interaction effects become less important. Similarly if the mass of the structure is reduced the interaction becomes less marked (figure 12b). This simple example serves to indicate the complexity of the soil-structure interaction problems. The coefficients used here correspond to a rigid circular base, flexible foundations will behave somewhat differently.

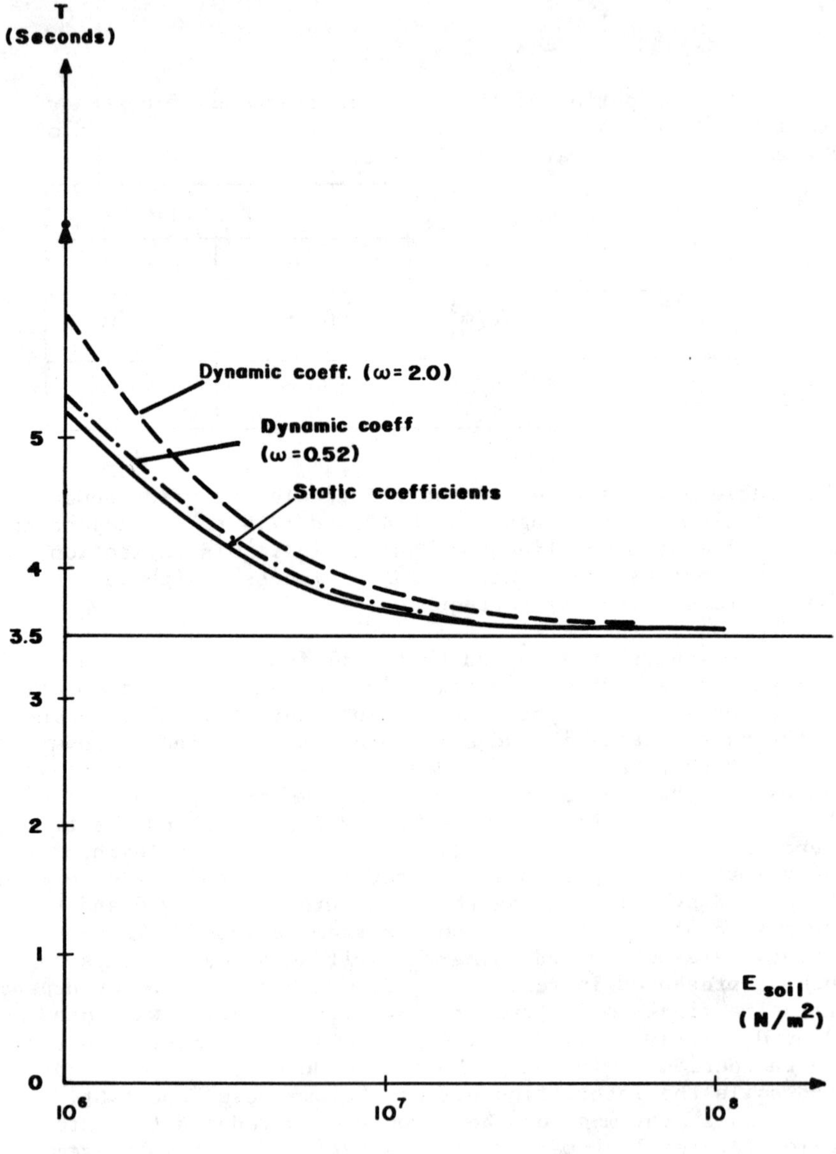

T
(Seconds)
Dynamic coeff. (ω= 2.0)
Dynamic coeff
(ω =0.52)
Static coefficients
5
4
3.5
3
2
1
0
E soil
(N/m²)
10⁶
10⁷
10⁸

FIGURE 13

Another important question is how important is the frequency dependency for the interaction coefficients. Figure 13 compares the case of using always the static coefficients against two cases (for different ω for which the variation with respect to frequency is taken into consideration. The graph shows that good results can be obtained in offshore-structure using the static (i.e. constant frequencies) coefficients for low frequency excitations or hard soils but that for higher order ω or soft foundations the dynamic coefficients are required.

Approximate procedures to determine the damping are generally used when working with modal analysis. Modal damping coefficients are expressed as weighted average of the soil and structural damping in proportion to the deformation energy U. In such a way the weighted coefficient results [19].

$$\beta = \frac{(\beta U)_{soil} + (\beta U)_{structure}}{U_{structure} + U_{soil}} \tag{13}$$

This simple procedure has two disadvantages. First that the soil damping is histeretic in character but structural damping is viscous. In the above formulation a viscous type damping is for the global system. The second disadvantage is that radiation damping is usually of a higher order of magnitude than structural and hydrodynamic damping. The validity of the above weighting procedure is limited to damping coefficients of similar order of magnitude.

These disadvantages cast doubts on the validity of the weighted modal damping coefficients for soil-structure interaction problems. In our opinion the more adequate treatment is to use the eigen-modes in order to reduce the order of the system producing then an uncoupled but small system of equations (the uncoupling is due to the non-diagonal damping matrix). One can work with the new matrix system without trying to force force it to become an independent system of equations. This procedure has the main advantage of classical modal analysis but not the more important disadvantage, i.e. the need to diagonalize the damping matrix.

Methods such as aemi-infinite space give better results than finite elements but they are rather limited in their applications. Because of this a new and more flexible method related to semi-infinite space solutions is now becoming widespread. This is the boundary element technique 20 21 and can be interpreted as a way of making a known solution (called the fundamental solution) satisfy the boundary conditions of the problem. If one considers a type of solution such as a point load in an infinite domain, obviously the conditions of infinity are well represented and radiation is adequately

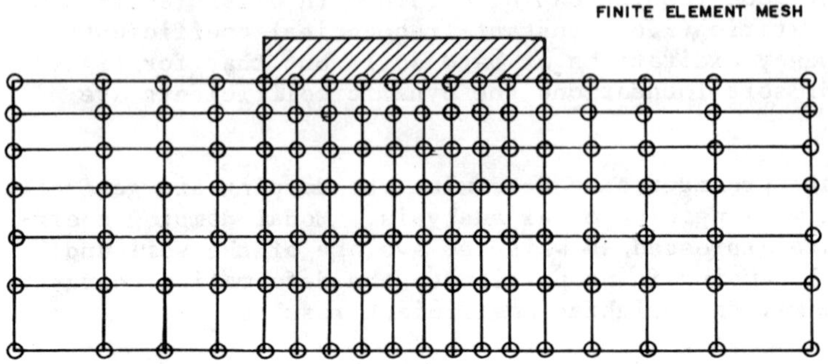

a) FINITE ELEMENT – Representation of the soil.

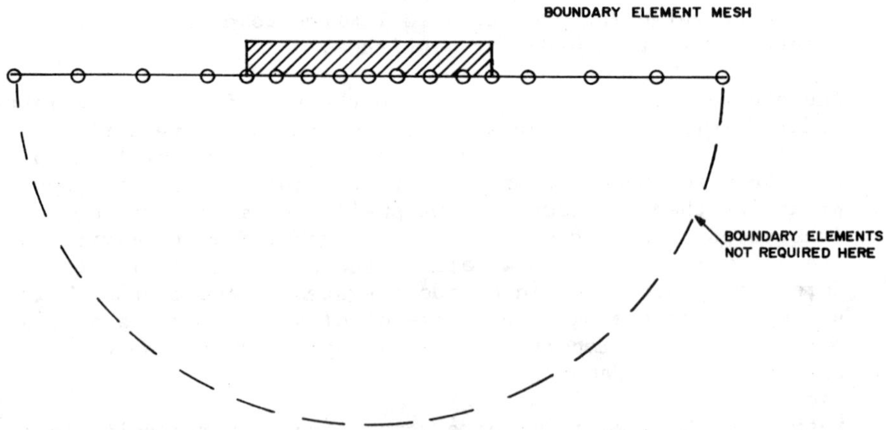

b) BOUNDARY ELEMENT REPRESENTATION OF THE SOIL.

FIGURE 14 – FINITE ELEMENT VS. BOUNDARY ELEMENT MESH.

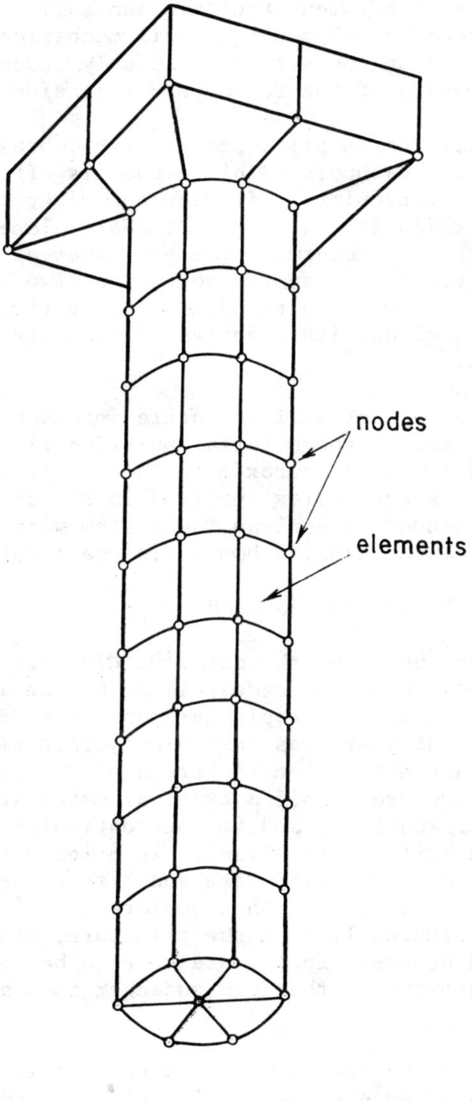

Pile divided boundary elements

Figure 15

taken into account. The fundamental solution needs to be applied only on the boundaries instead of over all the domain (figure 14). Furthermore if the fundamental solution is not the one for an infinite space but for a semi-infinite one the mesh becomes very simple as one only needs elements in the contact region between structure and soil. The advantages of using boundary elements in soil mechanics are obvious as the number of unknowns reduces to only those required at the finite boundary of the region under consideration.

The soil effects on piles are usually represented by a series of spring and dashpots. This method is often open to criticism due to the difficulty in finding the right coefficients and a better approach is required. Boundary elements provide also a good method to determine the behaviour of piles or group of piles (figure 15). Layered soils can also be taken into consideration using boundary elements by dividing the domain into different regions with a series of elements on the internal boundaries.

The main problem of soil structure interaction analysis is the representation of the material non-linearities. Histeretic damping attempts to approximate the non linear behaviour but due to the uncertainties involved in the process it is generally recommended to analyse the system with a range of soil properties and establish bounds in the results.

7. NEED FOR FURTHER RESEARCH

It seems obvious from the previous discussions that there is a lack of data and experience related to the analysis of offshore platforms. The rapid development of deep water structures in recent years was possible by gathering data from different and until then uncorrelated disciplines (oceanography, structures, soil mechanics, material sciences, etc.). Gaps in our knowledge and in the continuity of the different topics are still to be filled. In principle the analyst could start with a simple wind rose and little more information such as soils, and determine the complete dynamic behaviour of the system, including life of the structure, soil structure interaction and others. In the practice to be able to do so he has to make assumptions that can endanger the safety of the structure.

Nevertheless the structures have to be built and future developments may even require further extrapolation of our already scanty knowledge. It is then fundamental to try to gather as much data as possible and that new structures be instrumented. In addition smaller platforms that may serve as oceanographic stations as well as help to understand the structural behaviour of the platforms need to be built.

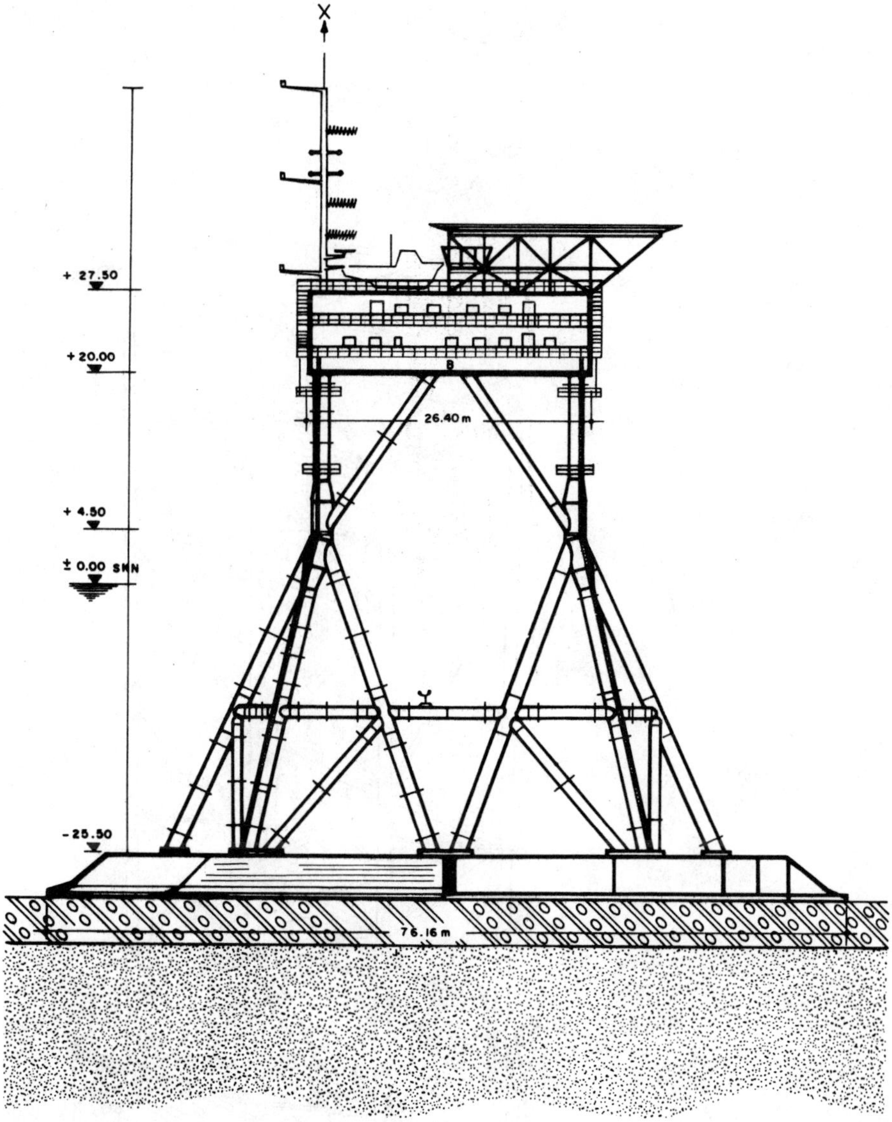

FIGURE 16

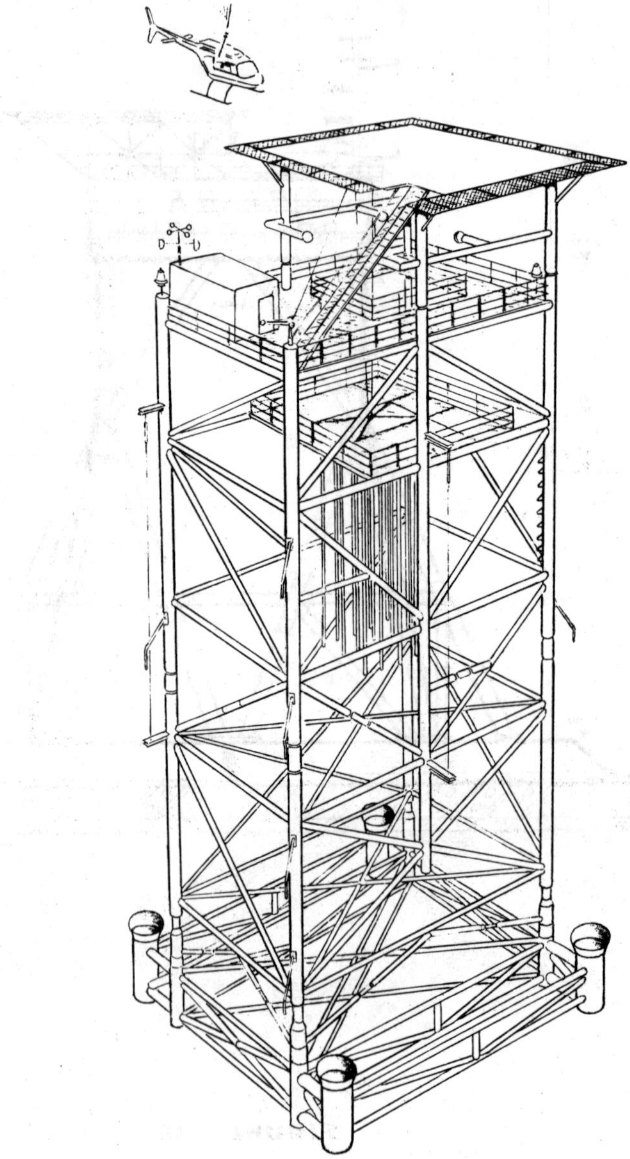

Ocean test structure

Figure 17

There are presently few research platforms although most companies understand now the need for instrumentation. The most well known of those are the Nordsee platform, the Ocean test platforms of Exon, the Guyed tower and the Christ Church experimental platform.

The German Research platform Nordsee [22] started to gather data from the beginning of 1976. It is a fixed bottom gravity type structure with the dimensions shown in figure 16. It measures meteorological and oceanographic data, serves as control station for a marine test field and as a research platform to study structural behaviour and the relationship between structure and the environment. It records wind, wave data, wave forces and stresses in the members and at hot spots, the vibration of the structure and the soil behaviour. The unusualydesign of the platform reflects the requirement that the behaviour of the sea be affected as little as possible.

The Exxon platform [23] instead is mounted on piles and is primarily designed to measure environmental conditions and associated forces. Total bases shear and overturning moment will be directly registered. It is being operated in the Gulf of Mexico in 20 m of water (figure 17).

Exxon has also initiated [23] the installation of a Guyed tower structure in the Gulf of Mexico in 90 metres of water (figure 18). This is a 1/5 scaled version of the prototype designed for 460 m of water. The guyed tower is permitted to comply with the waves rather than resist them as a conventional fixed bottom structure. The tower is held upright by a combined system of dump weights and drag anchors. The experiment is mainly designed to provide data directly relevant to the design of this new type of structure in much the same way as the tension leg platform tests carried out by Deep Oil Technology.

The Christchurch experiment in England was specifically set up to study wave forces and correlate them to measured velocities and sea states. The experiment was designed in such a way that it would produce information on the different waves regimes (such as diffraction, inertia and drag) according to the sea state acting on the platform. The structure basically consists of a large diameter cylinder on a circular basis with a wave staff in front. The wave staff is a slender cylinder instrumented to measure pressures, forces and particle velocities. Soil measurements under the base were also carried out. The tower collapsed due to soil failure and is now being redesigned with a larger base. The tower is not designed to act as a reduced scale model of a platform but rather to collect information about wave forces and soil behaviour under dynamic loads.

Experimental structures such as these give invaluable informa-

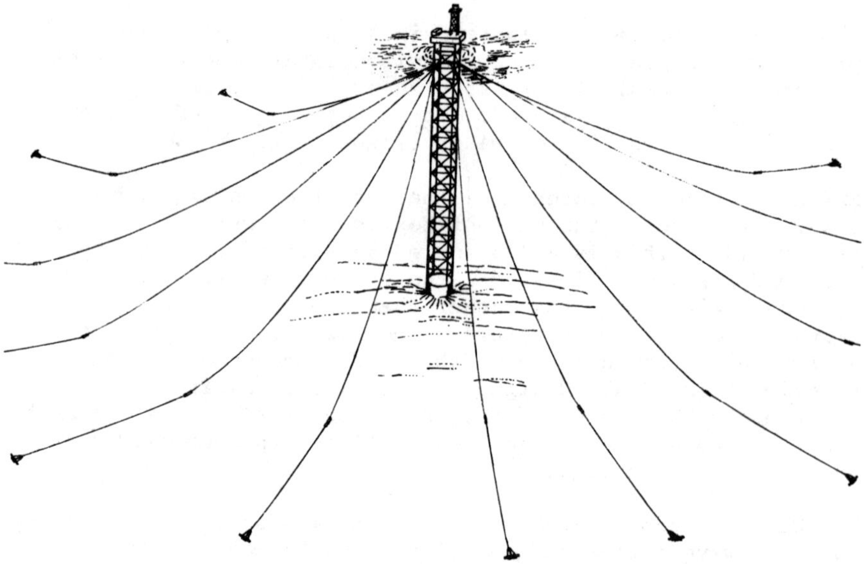

The guyed tower

Figure 18

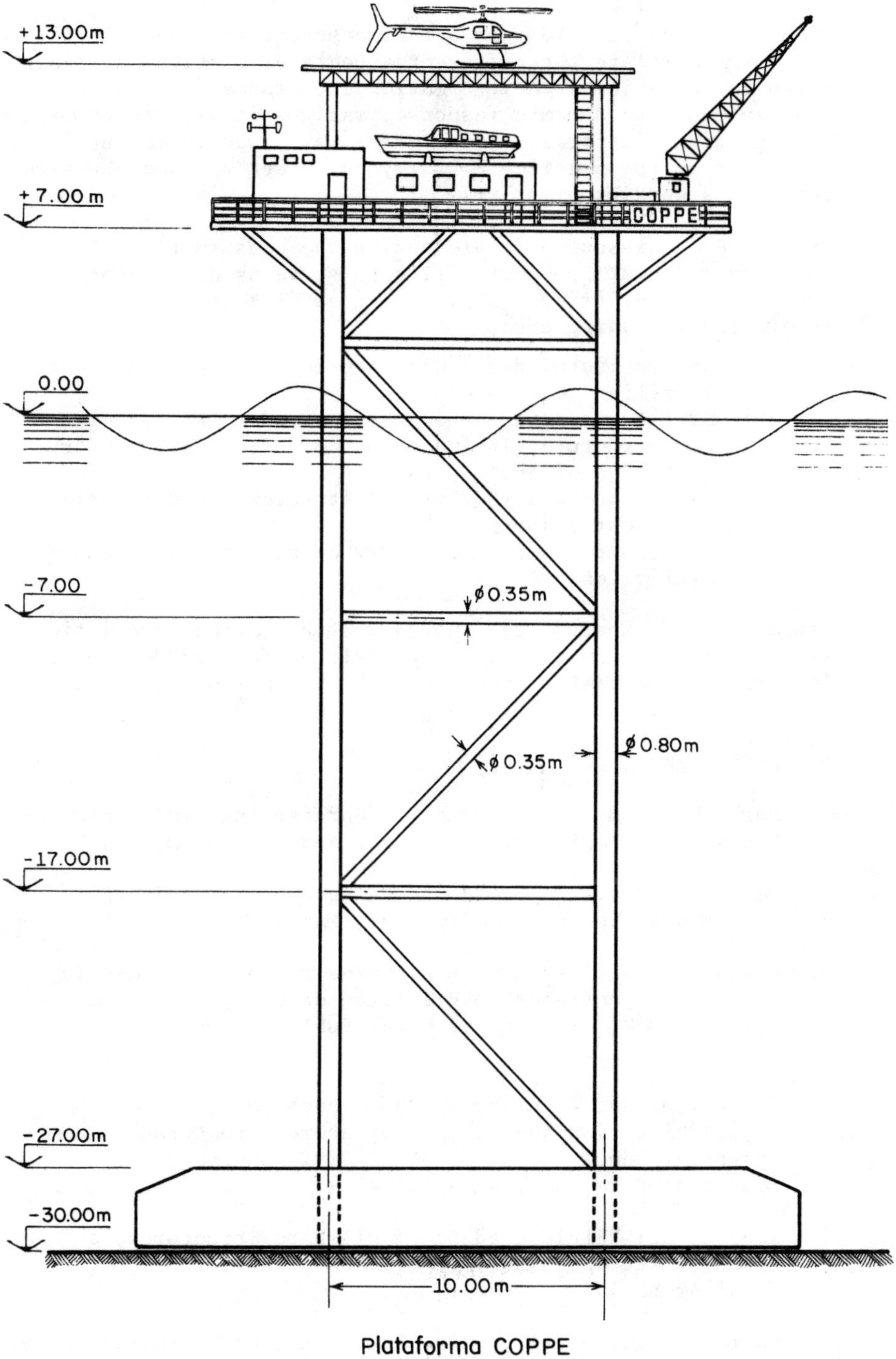

Plataforma COPPE

Figure 19

tion about the behaviour of offshore structures and help the designer to relate the different aspects of platform analysis (such as rendering wave information into forces, or forces on the members into dynamic response, fatigue etc.). Instrumenting existing platforms is also of primary importance but it is difficult in the practice to carry out a set of comprehensive tests in a working platform.

The COPPE is presently interested in the construction of an experimental platform that will also serve as an oceanographic station, (figure 19). The idea is to be able to record the following data,

a) Environmental data (wind and fofces) relevant to the Brazilian coast.
b) Stresses and forces in members and total forces on the structure, including number of cycles. Instrumentation of 'hot spots'.
c) Vibration behaviour of the structure (natural frequencies; damping).
d) Behaviour of the soil under bearing stresses and cyclic loads.

Comparison of experimental values and theoretical predictions will result in a better set of guidelines for the designer. In addition the platform would be able to perform a large series of oceanographic tests.

8. REFERENCES

1. Darbyshire, M. and L. Draper "Forecasting Wind Generated Sea Waves" Engineering, 5, 195, April 1963, p.482.

2. Marshall, P.W. "Dynamic and Fatigue Analysis using Directional Spectra". OTC 2537, May 1976.

3. Losada, M., L. Tejidor, M.A. Corneiro and J.L. Tejería "Metodos Espectrales de Previsión de Oleaje. Estudio Comparativo". Revista de Obras Públicas, Madrid, Junio 1977.

4. Chavez, M. and C. Brebbia "Some considerations on the Random Vibration Analysis of Offshore Structures". Technical Report, Department of Civil Engineering, Southampton University, April 1977.

5. Hogben, N. "Fluid Loading of Offshore Structures, a State of Art Appraisal: Wave Loads". Rev. Inst. of Naval Arch. 1974.

6. Morison, J.R. et al., "The Force exerted by Surface Waves on Piles" Petrol. Trans., Am. Inst. Mining Metal Engineering, 1950.

7. Wiegel, R.L. "Oceanographic Engineering". Prentice Hall, Englewood Cliffs, 1964.

8. Malhotra, A.K. and J. Penzien, "Nondeterministic Analysis of Offshore Structures". Proc. ASCE (EM6), 1970.

9. Ebert, M. and C. Brebbia. "Some Considerations on Random Vibration Analysis of Offshore Structures". Technical Report, Department of Civil Engineering, Southampton University, March 1977.

10. Kim, Y.Y.and H.C.Hubbard. Analysis of Simultaneous Wave Force and Water Particle Velocity Measurements". OTC. Paper 2102, Houston, May 1975.

11. Ralph, J.A. "Offshore Platforms; Observed Behaviour and Comparison with Theory" OTC Paper 2553, May 1976.

12. Brebbia, C.A. and J.J. Connor, "Fundamentals of Finite Elements Techniques". Butterworth, London, 1973.

13. Nolte, K.G. and J.E. Hansford. "Closed-form Expression for Determining the Fatigue Damage of Structures Due to Ocean Waves". OTC 2606, May 1976.

14. Williams, A.K. and J.E. Rime "Fatigue Analysis of Steel Offshore Structures". Inst. of Civil Eng., Proceeding, Part I, 60, Nov. 1976, 635-654.

15. Vughts, J.H. "Probabilistic Fatigue Analysis of Fixed Offshore Structures". OTC 2608, May 1976.

16. Kallaby, J. and J.B. Price "Evaluation of Fatigue Considerations in the Design of Framed Offshore Structures". OTC. 2609, May 1976.

17. Veletsos, A.S. and Y.T. Wei "Lateral and Rocking Vibration of Footings". Proceedings, Journal of Soil Mechanics, September 1971.

18. Luco, J.E. "Independence Function for a Rigid Foundation in a Medium Nuclear Eng. and Design, 31, 1974, 204-207.

19. Nataraja, R. and C.L. Kirk "Dynamic Response of a Gravity Platform under Random Wave Forces". DTC 2904, May 1977.

20. Brebbia, C.A. "The Boundary Element Method for Engineers" Pentech Press, London, 1978.

21. Brebbia, C.A. and J. Dominguez "Boundary Element Methods for Potential Problems". Applied Mathematical Modelling, Vol.2 no. 3, December 1977.

22. Longree, W.D. "Aspects of the Instrumentation and Measurement Performance of the Research Platform Nodesee". NOSS/1976, The Norwegian Institute of Technology, Trondheim, Norway.

23. Kerry Koone, K. "Technology needs for Deepwater Operations". BOSS/76, The Norwegian Institute of Technology, Trondheim, Norway, 1976.

THE STRUCTURAL DESIGN APPRAISAL OF DEEPWATER STEEL PLATFORMS

W. Schum

Lloyds Register of Shipping, England.

1.0 INTRODUCTION

The steady increase of offshore activity in deep water locations is resulting in platforms which have to withstand appreciable dynamic loading. Consequently the technology used in the design and design appraisal of offshore structures has had to be developed to meet the new environmental conditions.

Two of the main aspects of structural design and design appraisal involve the analysis of platform strength for a particular load condition and the effect of long term fatigue damage.

This paper describes the three basic stages required for both these analyses, which are:

- Wave loading
- Dynamic analysis
- Post-processing of structural responses

with special emphasis on the methods of dynamic analysis and of the stochastic nature of the wave loading.

A more general description of the approach to the certification of offshore structures is given in reference 1.

2.0 OVERALL STRENGTH ANALYSIS

In general the overall strength of a particular platform is found by first calculating internal member forces caused by a particular loading condition and then carrying out a checking procedure using standard design codes.

The usual approach to the problem of random excitations in structural dynamics is to use a spectral analysis technique mainly because it offers an efficient solution method to what can be a complex problem. Its main disadvanage is the requirement for linearity both in the spectral techniques and the frequency domain dynamic analysis.

As this particular problem does not require an enormous amount of data processing the advantages to be gained in a time domain solution outweigh the main disadvantage of an increase in the computing time.

The main advantages to be gained in a time domain approach are:

(i) The correct structure submergence for members in the region of the water surface;
(ii) The drag term can be included in the force equation without having to linearise the velocity squared term.
(iii) The relative velocity between structure and water particles can be correctly accounted for in the solution of the equations of motion.
(iv) Stress combinations (i.e. Von Mises) can be carried out at each time step in a deterministic manner, i.e. the structure is in equilibrium at each step.
(v) It allows the structural responses to be calculated as the sea conditions build up to the maximum wave condition, provided sufficient time is allowed for the initial transients to disappear, whereas using a regular wave approach in a dynamic analysis implies that the '100 year wave' is a steady state condition.

2.1 Wave Analysis

The stochastic nature of the forcing function is introduced using a method of simulating one possible realisation of a time history representing the water surface elevation and corresponding structure loading.

There are several methods of simulating a continuous random process as a time series [3, 10] all of which are essentially the same and each having advantages and disadvantages. The random process to be simulated is described in terms of its power spectrum together with some form of random input to generate phase differences.

The technique used for the analysis is based on the principle of superimposing regular waves of varying periods and phases. The water surface elevation can be considered as being the combination of a large number of regular waves having different periods, heights and phase angle.

This can be expressed in the integral form:

$$\eta(t) = \int_{0}^{\infty} [S_{\eta\eta}(f)\,df]^{\frac{1}{2}} \cos(kx - 2\pi ft + \psi)\,df \qquad (2.1)$$

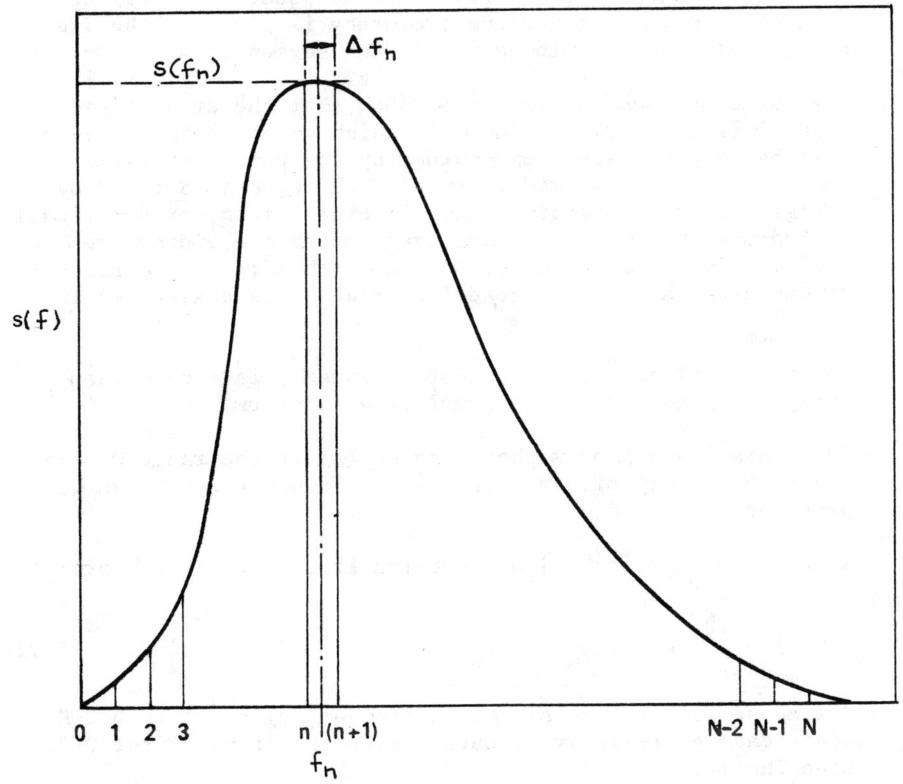

TYPICAL SEA STATE SPECTRUM

FIGURE 1

where $S_{\eta\eta}$ is the sea state power spectrum and ψ is a random phase shift between 0 and 2π .

The method [9] is to divide the required spectrum into N segments of variable width Δf_n (see Fig. 1). The amplitude of each component wave is given by the square root of the area of each segment and the frequency is given by the frequency value at the mid-point of each segment.

The value of each Δf_n is chosen such that the area of each segment is a constant value (A^2) which is equal to the total area beneath the spectrum divided by the number of segments. The frequency at the mid-point of each segment is found by integrating the spectrum either in closed form, or numerically to produce the cumulative spectrum and then dividing the ordinate into N equal divisions of width A^2; the required frequencies can then be read from the abscissa as shown in Fig. 2.

For practical purposes a cut-off frequency is chosen when integrating to obtain the cumulative spectrum.

By generating N random phase angles (ψ) in the range $0 \quad 2\pi$ the time history of wave elevation at a point $x = 0$ can be produced.

Re-writing equation 2.1 and putting it in a discrete form:

$$\eta(m\Delta t) = \sum_{n=1}^{N} A \cos(\psi_n - 2\pi f_n m\Delta t) \qquad (2.2)$$

A time series of particle kinematics at any point in a 2-D space can be similarly produced using the linear First Order Wave Theory.

The force/unit length on a cylindrical member can then be produced at each time step using a discrete form of the Morison Equation.

$$F/\ell = \tfrac{1}{2}\rho\, C_D\, D\bar{u}(m\Delta t) + \rho C_{M1} \frac{\pi D^2}{4} \dot{\bar{u}}(m\Delta t) + \rho C_{M2} \frac{\pi D^2}{4} \times \{ \dot{\bar{u}}(m\Delta t) - \ddot{x}(m\Delta t) \} \qquad (2.3)$$

To make the system compatible with the deterministic approach a random wave history equivalent to the 100 year design wave has to be generated. This is achieved by simulating a sea state representing the 100 year condition and inspecting the water surface elevation record until a section of record is found which contains the required wave condition. The simulation is then re-run to generate wave loadings on the structure

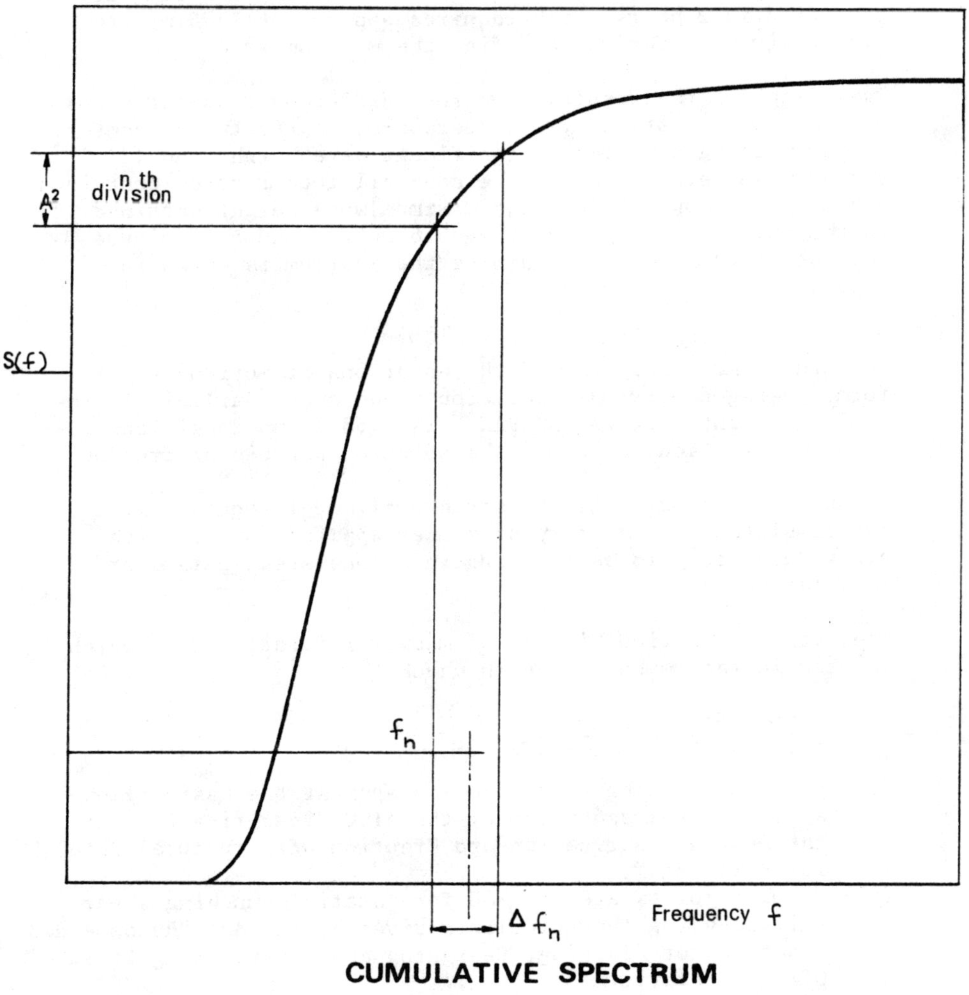

FIGURE 2

for this particular section of record. These records will usually be of the order of 40 seconds in length with sampling at 0.5 second intervals.

The average record length required to find the maximum wave condition is 3 hours which requires approximately 5 minutes c.p.u. time to simulate and find the maximum wave.

Table 1 gives the results from four different 3 hour records of the same sea state i.e. a Pierson-Moskowitz two parameter spectrum using a 15 metre significant wave height and 12 second mean period. As can be seen all four records contained maximum wave conditions. The maximum wave height obtained in the full 12 hour period was 33.6 metres and a 17 minute section of the record containing the maximum is shown in Fig. 3.

2.2 Dynamic Analysis

The time domain solution of the equations of motion is performed using the dynamic solution techniques available in the finite element package NASTRAN. The modal superposition technique is used to reduce the solution degrees of freedom.

Input to the program is the structural model together with the simulated time history of member applied forces which are reduced to grid point loads in a consistent parameter formulation.

The output is a time history of member internal forces which is used in the member strength checks.

2.3 Member Strength Checks

Two types of check are performed.

(i) Structural members are checked against the basic allowable stresses specified by the AISC "Specification for the Design, Fabrication and Erection of Structural Steel for Buildings".

(ii) Tubular joints are checked for 'static' punching shear failure using the guidelines given by the API "Recommended Practice for Planning, Designing and Constructing Fixed Offshore Platforms".

A flow diagram of the computer system is shown in Fig. 4.

3.0 FATIGUE LIFE ESTIMATION

The most widely used method for the evaluation of the fatigue life of an offshore structure is the Palmgren-Miner Law of Accumulative Damage. Using this law the fatigue life is said to be expended when the total damage equals the accumulative damage ratio, normally taken as 1.0.

RECORD NO.	TIME (MINS)	NO. OF READINGS	ZERO CROSSING PERIOD (SECS)	H_{SIG}	H_{MAX}	PROB-ABLE	E
1	180	5401	12.35	14.55	26.7	28.8	0.57
2	180	5401	12.12	14.59	26.5	28.9	0.54
3	180	5401	12.0	16.25	33.6	32.2	0.55
4	180	5401	12.16	14.84	29.4	29.4	0.55

Table 1 Results from 12 hour record.

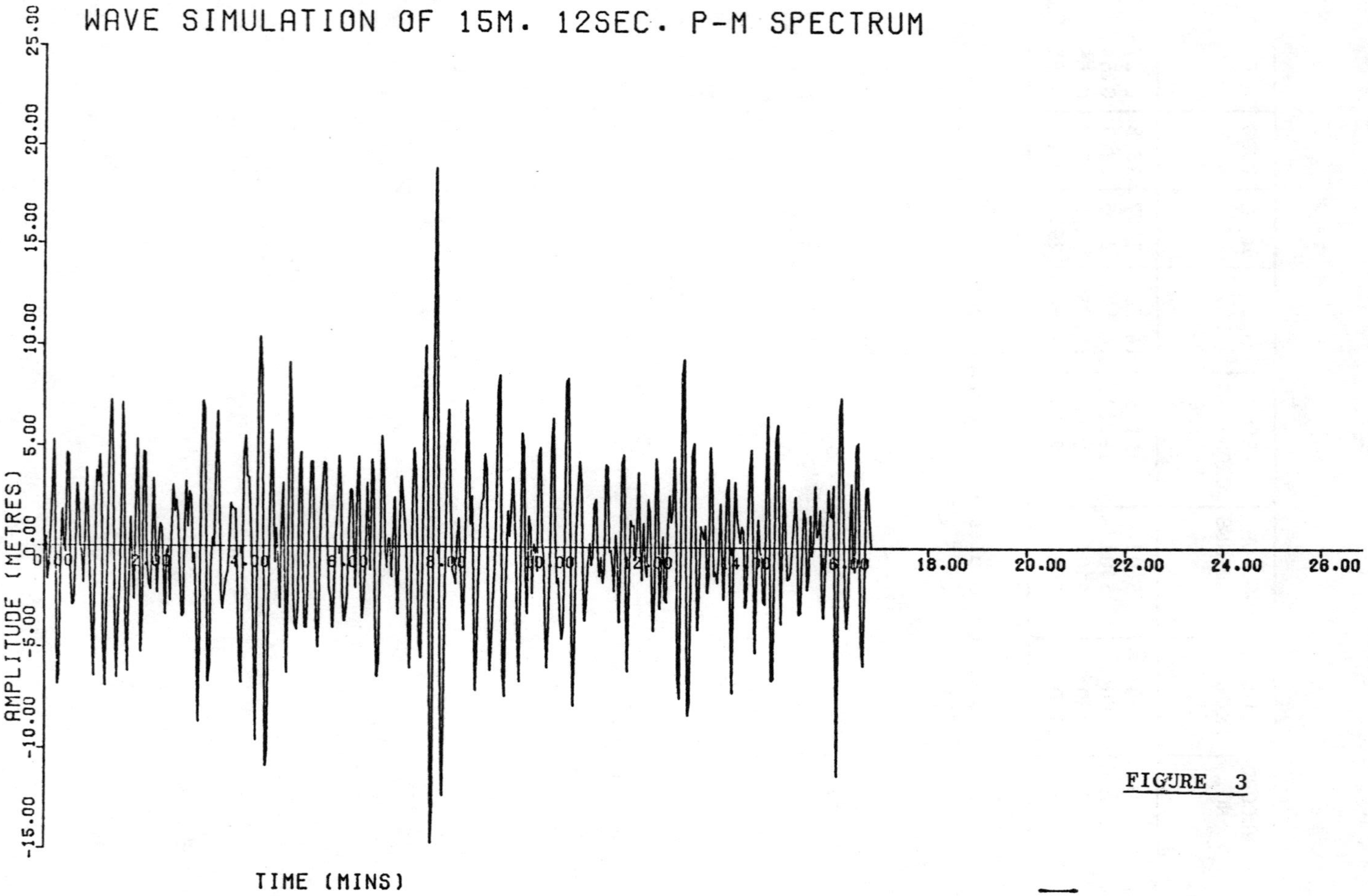
WAVE SIMULATION OF 15M. 12SEC. P-M SPECTRUM
AMPLITUDE (METRES)
25.00
20.00
15.00
10.00
5.00
0.00
-5.00
-10.00
-15.00
18.00
20.00
22.00
24.00
26.00
TIME (MINS)

FIGURE 3

The method requires the determination of the number of cycles of every stress range experienced by the structure which, because of the randomness of the process, must be based to some degree on statistics. Either a long term wave exceedance diagram, as is used in the deterministic approach, or some form of long term sea state occurrence formulation is required.

Despite the advantages of a time domain analysis, its application to a long term fatigue calculation would produce an unmanageable amount of data as well as requiring an enormous amount of computing time. Consequently in this case a spectral analysis is essential from the point of view of data handling, and at the same time the assumptions which have to be introduced into the spectral approach (Section 2.0) are less significant in a fatigue analysis where the majority of the damage is caused by the smaller, short period waves.

3.1 Wave Analysis

The mathematical representation of water particle kinematics is defined using the linear First Order Wave Theory.

Again, as in Section 2.1, the wave loads are calculated using the Morison Equation.

$$F/\ell = \tfrac{1}{2}\rho C_D D(u-\dot{x})\,|u-\dot{x}| + \rho C_{M_1} \frac{\pi D^2}{4}\dot{u} + \rho C_{M_2} \frac{\pi D^2}{4}(\dot{u}-\ddot{x}) \qquad (3.1)$$

As mentioned in section 2.0 the frequency domain approach requires the non-linearities in the system to be accounted for.

In considering the relative velocity term between water particles and structure a non-linearity is introduced into the differential equations of motion. Ignoring the structural velocity in the generalised force equation is a conservative assumption that maximises the wave force and becomes increasingly penalising with an increase in the dynamic response of the structure, as can be seen from equation 3.1.

A further non-linearity in the spectral analysis procedure is caused by the velocity squared term in the drag force expression. This is removed by linearising the squared term and incorporating the drag coefficient to give an equivalent linear coefficient.

i.e. neglecting the structural velocity.

$$C_D\, u|u| = C_{D_e}\, u + e \qquad (3.2)$$

The coefficient C_{D_e} can be optimised so that the error e will

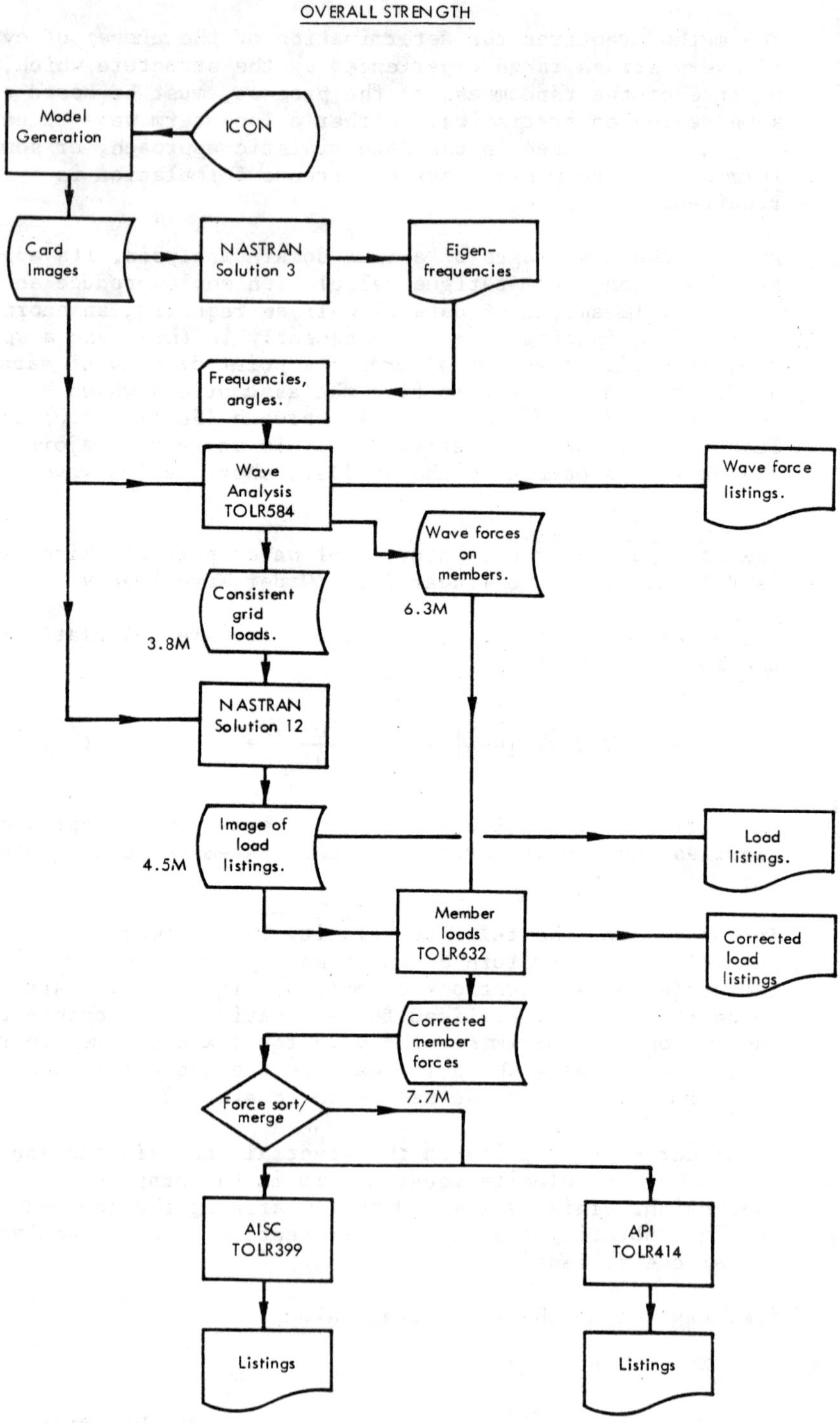
OVERALL STRENGTH
Model Generation
ICON
Card Images
NASTRAN Solution 3
Eigen-frequencies
Frequencies, angles.
Wave Analysis TOLR584
Wave force listings.
Wave forces on members.
6.3M
Consistent grid loads.
3.8M
NASTRAN Solution 12
Image of load listings.
4.5M
Load listings.
Member loads TOLR632
Corrected load listings
Corrected member forces
7.7M
Force sort/ merge
AISC TOLR399
API TOLR414
Listings
Listings

FIGURE 4

be minimised in the mean square sense, giving [2]:-

$$C_{D_e} = C_D \sqrt{\frac{8}{\pi}}\, u_{r.m.s.} \qquad (3.3)$$

The wave forcing function is calculated for frequencies upto 1 Hz and perhaps even higher. At these frequencies the wavelengths become comparable with the diameter of members found on steel platforms and consequently diffraction effects become significant. In an attempt to account for this effect the inertia coefficient is made a function of frequency according to the equation [6]

$$C_M = \frac{4A(ka)}{\pi (ka)^2} \qquad (3.4)$$

where k is the wave number, a the member radius and

$$A(ka) = (J_1'^2(ka) + Y_1'^2(ka))^{-\frac{1}{2}}$$

where J and Y are the appropriate forms of the Bessel Function.

The function is plotted in Fig. 5.

The distributed loads are converted to equivalent grid loads using a consistent parameter formulation.

3.2 Dynamic Analysis

The modal superposition technique is used to solve the equations of motion in the frequency domain.

The output is in the form of frequency response functions of member internal forces $H_F(f)$.

3.3 Fatigue Analysis

Fatigue damage is calculated for the welded tubular joints of offshore jacket structures.

It is normally satisfactory to consider two types of failure at these joints:

(i) Failure through the brace adjacent to the weld (brace to weld failure)

(ii) Punching type failure of the chord wall adjacent to the weld (chord to weld failure)

Both types of failure are shown diagrammatically in Fig. 8.

Both these types of failure are dependent on the stress range at a point in the member some distance from the joint, i.e. the brace nominal stress range, multiplied by a suitable stress concentration factor.

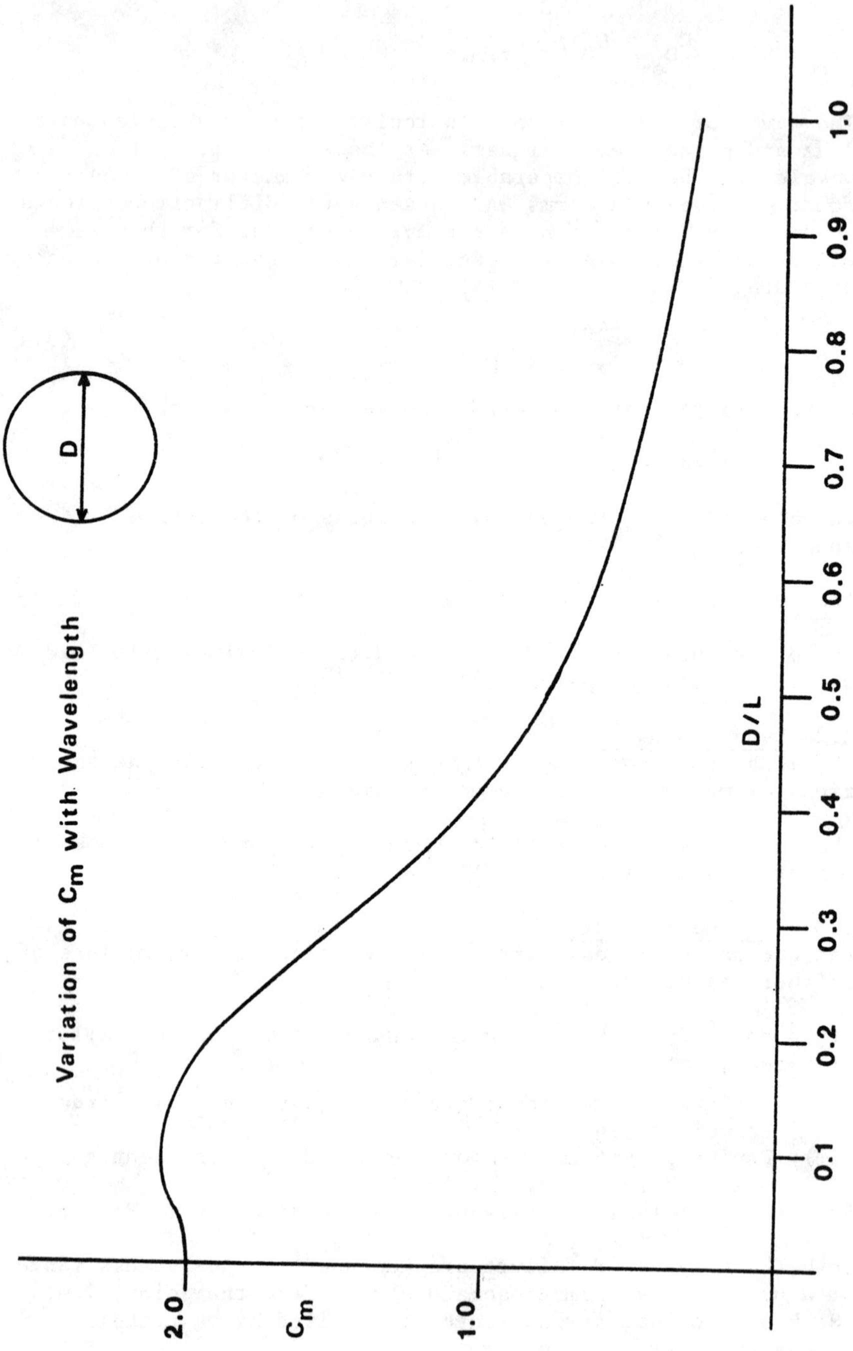
Variation of C_m with Wavelength
D
2.0
C_m
1.0
0.1
0.2
0.3
0.4
0.5
0.6
0.7
0.8
0.9
1.0
D/L

FIGURE 5

As the exact spot on the circumference of each brace at which failure is going to occur is unknown, the damage calculation has to be performed at eight positions around the circumference of selected brace ends to cover all possible outcomes.

The calculation of fatigue damage at each of the eight positions can be considered in two parts. Firstly the calculation of damage caused by a single, short term, sea state, and secondly the estimation of the probability of obtaining the sea state, in the long term.

<u>Short term damage</u> If one adopts the Palmgren-Miner Law, the short term accumulative damage due to a single storm can be expressed in the form: [8]

$$AD_{ST} \doteq \nu T \sum_{z=o}^{z=\sigma_{MAX}} \frac{p(Z)}{N(Z)} \Delta Z \qquad (3.5)$$

where ν is the predominent frequency of the stress response given by

$$\nu = \left\{ \frac{\int_o^\infty f^2 \, S_{ZZ}(f)\, df}{\int_o^\infty S_{ZZ}(f)} \right\}^{\frac{1}{2}} \qquad (3.6)$$

$S_{ZZ}(f)$ is the stress response spectra.

T is the duration of the sea state i.e. νT is the number of cycles occurring in time T.

p(Z) is the probability density function of stress range, and N(Z) is the number of cycles to failure for a particular stress range. This information is taken directly from S-N curves.

The problems involved in the calculation are:

A. to choose a suitable form for p(Z)
B. establish the stress response spectra
C. use a suitable S-N curve and stress concentration factors.

A. The majority of the work that has been carried out on stochastic fatigue life estimation uses the Rayleigh Distribution to define the form of p(Z) the stress range p.d.f.

$$p(Z) = \frac{Z}{2m_o} e^{-Z^2/8m_o} \qquad (3.7)$$

where m_o is the variance of the process i.e. zeroth moment of the response spectra of stress amplitude.

This is based on the original work carried out by Rice and later extended to marine applications by Cartwright and Longuett-Higgins [4], on the probability density function of the maxima of a random process.

Other formulations have been proposed and Söding [11] has used simulations of a broad band process to develop a joint p.d.f. of stress range and mean stress. The form of the function depends on the spectral width parameter ε however, the final equations, particularly for the case of ε less than 0.5, become quite involved. So, despite the fact that the Rayleigh Distribution is for narrowbanded process it still serves as a good first approximation for the p.d.f.

B. The stress spectra for a particular sea state is calculated for the frequency response function of stress, Hz(f), which is obtained from $H_F(f)$ using a simple coefficient matrix based on engineers bending theory and the cross-sectional properties of the member end.

$$H_Z(f) = [G]\ H_F(f) \tag{3.8}$$

To take account of the short crested sea condition the input power spectrum should be defined in terms of a directional sea spectrum, i.e.

$$S_{\eta\eta}(f,\alpha) = S_{\eta\eta}(f).f(\alpha) \tag{3.9}$$

where $S_{\eta\eta}(f)$ is the unidirectional spectral definition.

α is the angle between the wave component direction and the predominant direction of the wave system

and a suitable formulation of $f(\alpha)$, the spreading function, would be

$$f(\alpha) = \frac{2}{\pi}\cos^2\alpha \qquad -\frac{\pi}{2} < \alpha < \frac{\pi}{2}$$

$$f(\alpha) = 0 \qquad \text{otherwise}$$

The stress spectra for a particular spread angle and circumferential position k, then becomes

$$S_{ZZ_k}(f,\alpha) = \{|H_{Z_k}(f)|\}^2.S_{\eta\eta}(f).f(\alpha) \tag{3.10}$$

and hence the variance of stress amplitude

$$m_{o_k} = \int_{o}^{\infty} S_{ZZ_k}(f,\alpha)\,df.d\alpha \qquad (3.11)$$

C. The fatigue life on both the brace and chord side of the weld are calculated using the S-N curve 'Q' shown in Fig. 7 with the following stress concentration factors.

(i) Brace to weld failure

S.C.F. = 5.0 nominally
= 6.0 for an attached overlapping brace

(ii) Chord to weld failure (single brace)

$$S.C.F. = \tau \frac{\sin\theta}{ka} \frac{1}{Q_\beta} .(4 + 0.67\gamma) \qquad (3.12)$$

$$Q_\beta = 1.0 \qquad \beta \leq 0.6$$

$$Q_\beta = \frac{0.3}{\beta(1.0 - 0.833\beta)} \qquad \beta > 0.6$$

ka is a geometrical factor allowing for change in contact area of brace on the chord, and is given in the API document.

Chord to weld failure (multi brace, no significant out of plane bending)

$$S.C.F. = \frac{\tau\sin\theta}{ka} . \frac{1}{Q_\beta} .(4+0.2\gamma) \qquad (3.13)$$

where significant out of plane bending exists the single brace formula is used.
The joint parameters are shown in Fig. 8.

The S-N curve 'Q' (fig. 7) is the latest result of several years development and production work calculating fatigue failures in offshore structures. It should be used in conjunction with the above stress concentration factors.

The stress amplitude variance at the joint is then $m_o' =$ S.C.F. × brace nominal stress variance (m_o) substituting into equation 3.7 gives the p.d.f. of stress range at the joint.

Long Term Damage Having calculated the damage for a particular sea state the data has to be accumulated in some way depending on the long term statistics of sea occurrences.

Several methods are available for the prediction of long term trends.

(i) Measured scatter diagrams

FATIGUE FAILURE TYPES

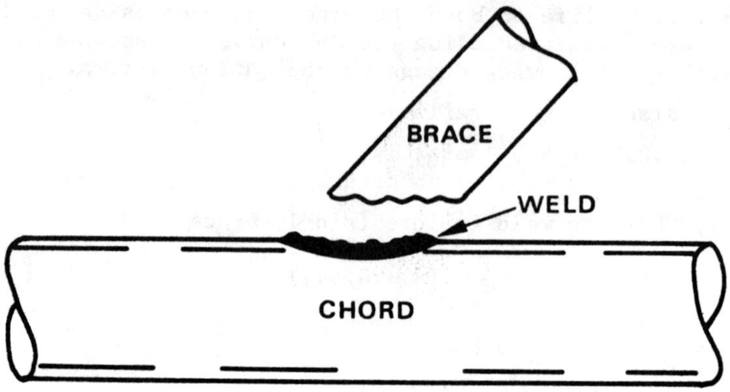

BRACE TO WELD FAILURE

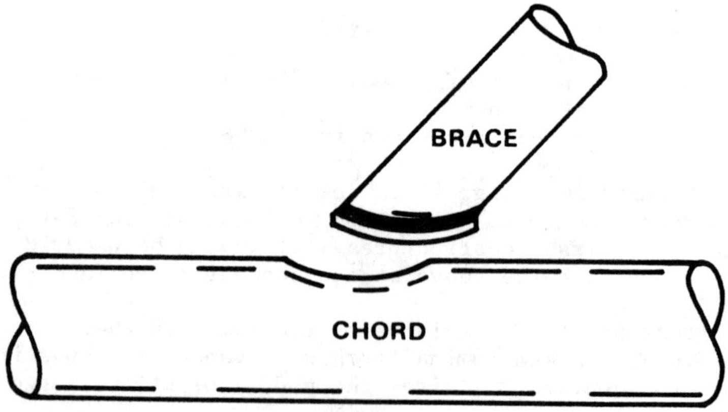

CHORD TO WELD FAILURE

FIGURE 6

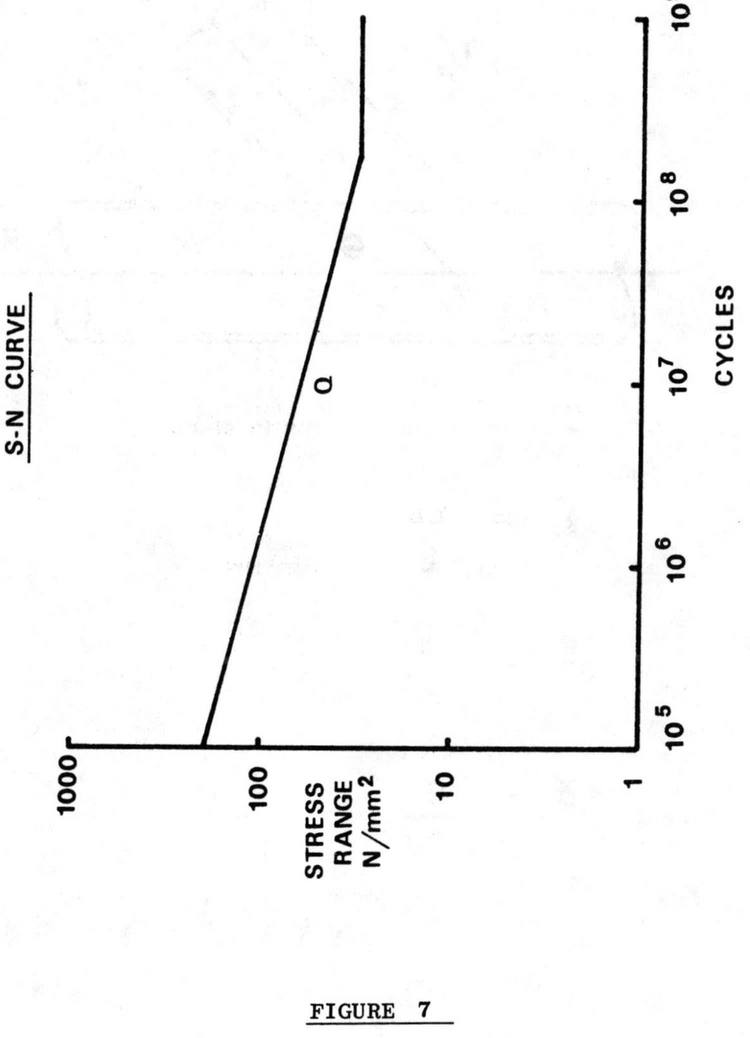
S-N CURVE
1000
100
10
1
STRESS RANGE N/mm2
Q
10 5
10 6
10 7
10 8
10 9
CYCLES

FIGURE 7

JOINT PARAMETERS

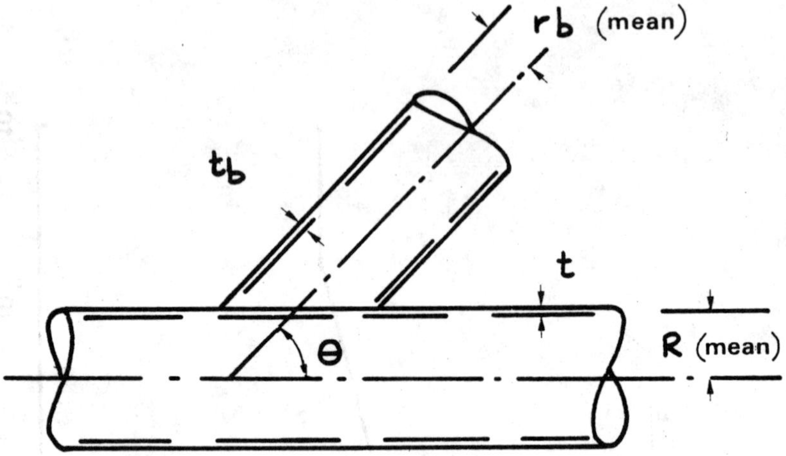

θ = angle of brace to chord

$$\tau = \frac{tb}{t}$$

$$\beta = \frac{rb}{R}$$

$$\gamma = \frac{R}{t}$$

FIGURE 8

(ii) Normal distribution of Beaufort Number
(iii) Wave height and period distribution

Obviously a measured scatter diagram for the area of ocean being considered is preferable to either of the other methods.

The amount of data of this sort being made available is continually increasing, particularly for North Sea locations.

When the sea state is defined in terms of Beaufort Number the long term probability of response Q(Z) is given by the summation

$$Q(Z) = \sum_{B_n=o}^{B_n=\alpha} p(B_n)\; p(Z)\; \Delta Z\; \Delta B_n \tag{3.14}$$

where $p(B_n)$ is the probability density of B_n and p(Z) is the short term probability density of response 'Z' relating to a particular value of B_n.

The Beaufort description assumes one height and one period relates to one specific Beaufort Number i.e. for each B_n only one short term response is necessary. The advantage of this simple approach is that it relies on wind data only, which is readily available for locations throughout the world.

In the case of the sea state being defined by a joint distribution of wave height and period, the long term probability is given by

$$Q(Z) = \sum_{T=T_{MIN}}^{T=T_{MAX}} p(T) \left\{ \sum_{H=H_{MIN}}^{H=H_{MAX}} p(T|H).p(Z)\Delta Z\Delta H \right\} \Delta T \tag{3.15}$$

where p(T) is the probability of obtaining a particular period group and is input from smoothed actual probability densities obtained from wave statistics for the sea area under consideration. $p(T|H)$ is the conditional probability density of obtaining wave height H within the period group under consideration and is obtained using either a Weibull distribution or a Log-Normal distribution of wave heights.

p(Z) is the short term probability of the response for a particular T,H combination.

The use of this method is similar to the scatter diagram approach where a whole range of short term responses have to be calculated to cover all possible T,H combinations.

Scatter diagrams are usually constructed from data collected for a one year period. Reducing the information to equivalent Weibull parameters can be used as a method of smoothing the

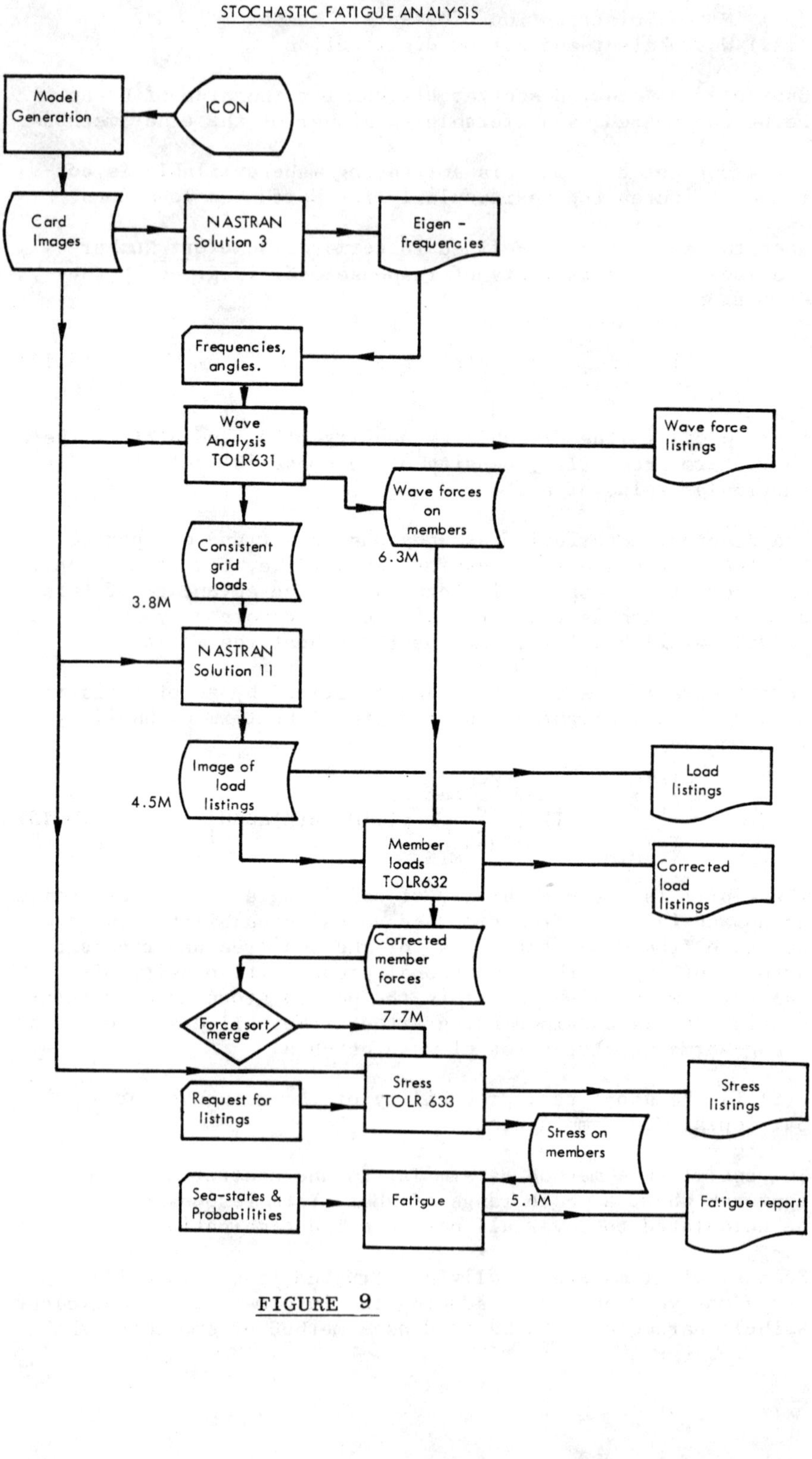
STOCHASTIC FATIGUE ANALYSIS
Model Generation
ICON
Card Images
NASTRAN Solution 3
Eigen - frequencies
Frequencies, angles.
Wave Analysis TOLR631
Wave force listings
Wave forces on members
6.3M
Consistent grid loads
3.8M
NASTRAN Solution 11
Image of load listings
4.5M
Load listings
Member loads TOLR632
Corrected load listings
Corrected member forces
Force sort/ merge
7.7M
Stress TOLR 633
Request for listings
Stress listings
Stress on members
Sea-states & Probabilities
Fatigue
5.1M
Fatigue report

FIGURE 9

measured data to take account of the lack of results for the longer period seas.

Therefore, combining the long term distribution of environmental conditions with short term damage due to particular sea states an expression is obtained for the total damage in the life of the structure.

Hence, if p(s) is the probability of obtaining the sea state 's' and p(ϕ) is the probability of a wave approach angle 'ϕ' then the total damage in the life of the structure T_L is

$$AD_{TL} = T_L \sum p(\phi) \sum p(s) \left\{ \nu \sum \frac{p(Z)}{N(Z)} \Delta Z \right\} \Delta s \; \Delta\phi \qquad (3.16)$$

More commonly the fatigue life is quoted, which is the reciprocal of the damage per year.

A flow diagram of the system is shown in Fig. 9.

4.0 DYNAMIC ANALYSIS

In the past dynamic excitation has been accounted for by multiplying the results of a static analysis by a dynamic amplification factor applicable to the structural natural frequency.

Fig. 10 is a typical curve used in this type of analysis. Area 'A' represents a platform with a fundamental period of 2.5 seconds for wave periods between 7 and 11 seconds. The amplification for this type of structure is slightly greater than 1.0.

However, as the fundamental period increases to 5 seconds, as shown by area 'B', the amplification in this case can rise to a factor of 3.0. Preliminary investigations have shown that dynamic amplification can vary significantly throughout a particular structure, consequently for this level of amplification, factoring the static analysis could lead to either high over or under stressing.

For this reason the analysis systems are based on the solution of the dynamic equation of motion.

$$M\ddot{x} + C\dot{x} + Kx = P$$

In order to produce a viable system the solution technique should be capable of handling and processing large amounts of data efficiently.

The program used to perform this task is the finite element analysis package NASTRAN [7] which is mounted on the Society's

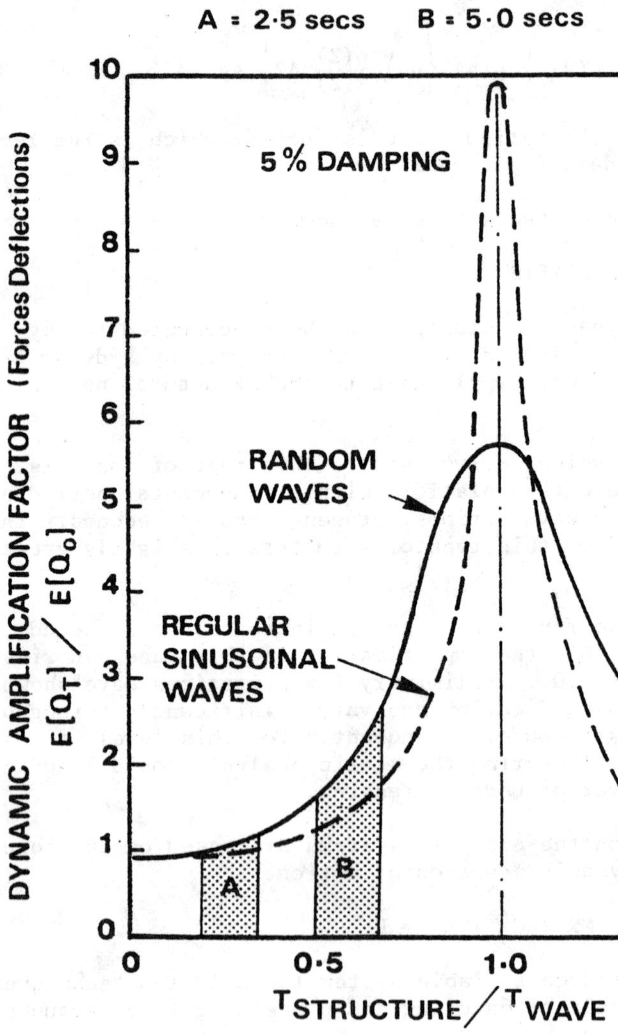
A = 2·5 secs B = 5·0 secs
5% DAMPING
RANDOM WAVES
REGULAR SINUSOINAL WAVES
A
B
DYNAMIC AMPLIFICATION FACTOR (Forces Deflections)
E[Q_T] / E[Q_o]
0
1
2
3
4
5
6
7
8
9
10
0
0·5
1·0
1·5
T STRUCTURE / T WAVE

FIGURE 10

IBM 370/158 computer.

The program was specifically designed to treat large problems with many degrees of freedom, the limitations being those imposed by practical considerations of running time and auxilliary storage.

In order to maximise efficiency for the solution of various structural configurations both direct and modal methods of dynamic formulation are incorporated into the program.

In general the dynamic response of steel platforms is best suited to the modal superposition technique, in which case the major response of the structure can be adequately represented by the first ten mode shapes, without a significant loss of accuracy.

4.1 The Modal Technique

The modal technique involves initially solving the eigenvalue equation:

$$[K - \lambda M] \, x = 0 \tag{4.1}$$

There are numerous methods of performing the eigenvalue extraction and NASTRAN provides three possible methods in order to cover all possible structural configurations. From experience the order of magnitude of the eigenvalues of steel platforms is well known and so the search methods of extraction offer a fast efficient technique.

The technique which has been found to be most useful is the inverse power method with shifts, which is a version of the more familiar power method but with the advantage that the eigenvalues are found in order of closeness to an arbitrarily selected shift point rather than in order of closeness to the origin. Difficulties with convergence, and with rigid body modes, are thereby avoided.

The eigenvalue analysis produces the vibration mode shape matrix ϕ and the frequency vector ω.
The modal coordinates Y are defined as $x = \phi Y$.

The modal mass and load is then calculated for each mode using ϕ .

i.e., for the kth mode,

$$m_k = \phi_k^T M \phi_k \tag{4.2}$$

$$p_k = \phi_k^T P \tag{4.3}$$

the modal damping and stiffness terms are then

$$c_k = \omega_k \cdot g(\omega_k) m_k \qquad (4.4)$$

$$k_k = \omega_k^2 \, m_k \qquad (4.5)$$

(note the damping factor is twice the damping ratio.)

By transforming to modal coordinates the problem is simplified to solving N independent equations of motion, where N is the number of modes [6].

The time domain form of the equation for a particular mode k is:

$$\ddot{Y}_k + g_k \, \omega_k \, \dot{Y}_k + \omega_k^2 \, Y = \frac{p_k(t)}{m_k} \qquad (4.6)$$

the general response expression is given by the Duhamel integral for each mode.

$$Y_k(t) = \frac{1}{m_k \, \omega_{D_k}} \int_0^t p_k(\tau) e^{\frac{-g_k \omega_k}{2}(t-\tau)} \sin \omega_{D_k} (t-\tau) d\tau \qquad (4.7)$$

ω_{D_k} is the damped frequency.

The solution in the frequency domain is even simpler

$$Y_k(\omega) = \left\{ \frac{p_k(\omega)}{-m_k \omega + i c_k \omega + k_k} \right\} \qquad (4.8)$$

The displacements in the physical set are now obtained from

$$x = \phi Y \qquad (4.9)$$

To improve accuracy in the data recovery, the mode acceleration technique is used. When modal analysis is used the selected modal coordinates will usually have the effects of higher modes omitted, however, the influence of the higher modes can be approximated by observing that their response to low frequency dynamic excitation is almost purely static. Thus the inertia and damping forces on the structure contain very little contribution from the higher modes.

In the mode acceleration method the inertia and damping forces are computed from the modal solution. These forces are then added to the applied forces and are used to obtain a more accurate displacement vector for the structure.

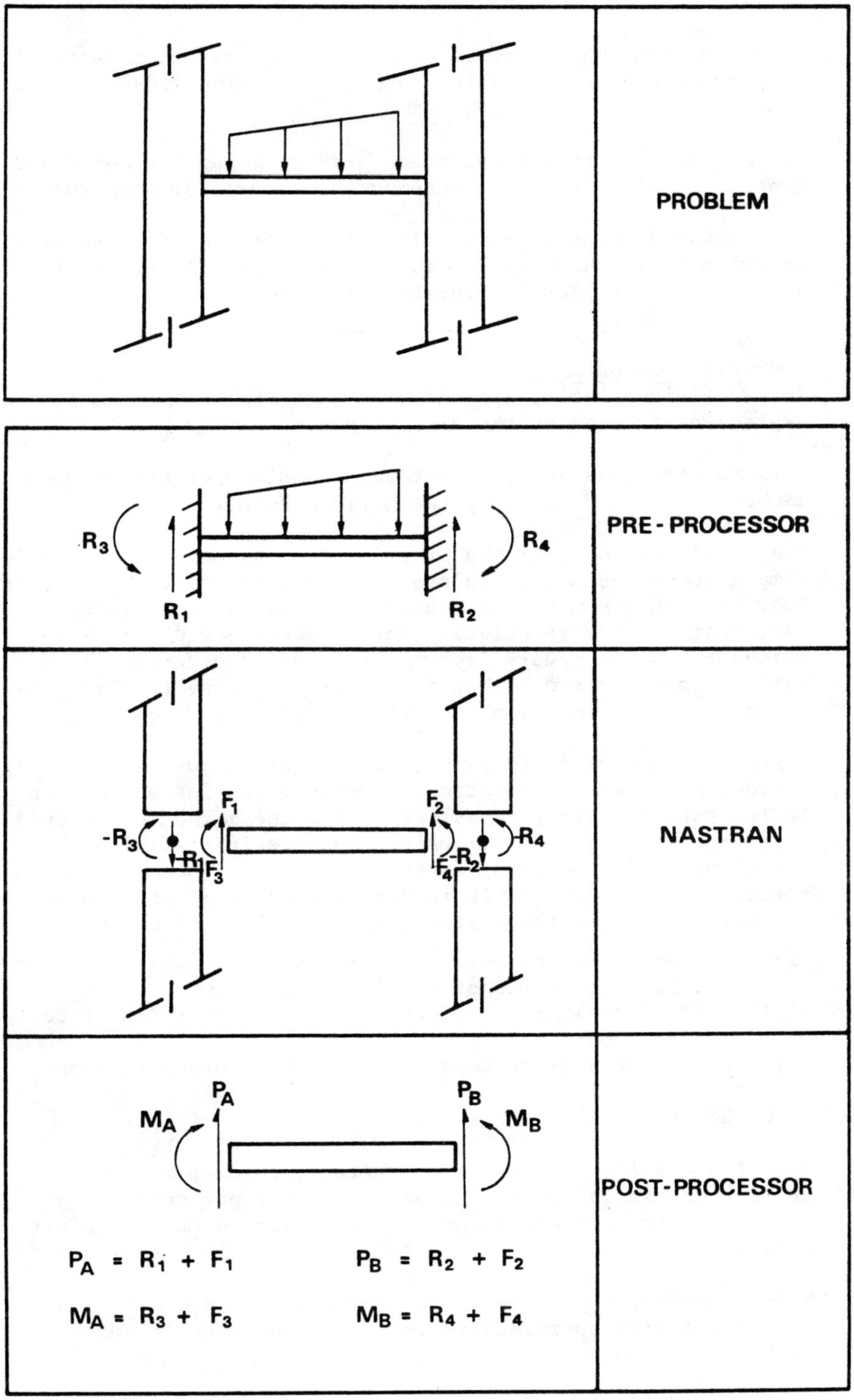
PROBLEM
PRE-PROCESSOR
R3
R4
R1
R2
NASTRAN
F1
F2
-R3
R4
R1
F3
F4
-R2
POST-PROCESSOR
PA
PB
MA
MB
PA = R1 + F1
PB = R2 + F2
MA = R3 + F3
MB = R4 + F4

FIGURE 11

The modal damping factor is an overall equivalent structural damping term which takes into account contributions from all sources.

Overall damping is generally accepted as being between 2 and 5 per cent of critical, depending on the type of construction.

Foundation damping is less well defined for piled structures. Unfortunately, there is no equivalent to the elastic half space equations used in gravity structures.

4.2 Distributed Loads

In any finite element approach the analysis is reduced to the degrees of freedom at the grid points.

This is particularly important when considering distributed member loads as defined by the Morison Equation.

The simple approach to the problem is to assume the member is simple supported and calculate the end reactions. The applied load is then the negative reaction at each end. This method is acceptable for relatively short members and when member forces are of secondary importance to displacements. In a fatigue analysis the member forces are the important responses and hence a more sophisticated approach is required.

A suitable method is to calculate the grid loads using a consistent parameter formulation. The term consistent is used to indicate that it is consistent with the method of calculating the stiffness matrix using the principle of virtual work. As it is dependent on the overall stiffness of the structure the technique is incorporated into a finite element analysis using the three steps, shown in Fig. 11 which are:

(i) Pre-processor to calculate the end reactions using the initial assumption of a fully fixed beam;
(ii) NASTRAN analysis to determine the distribution of forces due to global stiffness;
(iii) Post-processor to correct the member forces as shown.

CONCLUSIONS

With the placing of offshore structures in deep and more hostile waters it has become necessary for not only the design but the design appraisal techniques to become more sophisticated.

This paper has presented two alternative procedures to the existing design appraisal systems based on regular quasi-static analyses.

These existing systems have been established over several

years and are generally accepted as giving meaningful results. Consequently before any new method is introduced, into the appraisal procedure, the assumptions used and results obtained have to be verified.

It is intended to carry out a programme of correlation studies using the information presently being collected and stored in a data bank at Lloyds Register of Shipping. The information consists of both structural and environmental recordings for four offshore platforms, two of which are of the piled jacket type.

The proposed studies are:

1. Correlation between measured and simulated wave forces and particle kinematics for a particular wave elevation record. This will be using the simulation program LR584 which has an option of simulating a history of particle kinematics and wave forces for a given wave elevation history. Wheeler [12] has carried out a similar study using results from single piles and has obtained encouraging results.

2. An investigation of skewness in the Gaussian Distribution of wave elevation.
   Steep waves in nature have skewed surface elevations. That is, the crests tend to be higher above mean sea level than the troughs are below.
   Borgman has suggested a method of incorporating skewness into the simulation of a random sea using factors obtained from actual wave records.
   If enough records are investigated it should be possible to formulate some empirical formulae which could be applicable to all simulations.

3. The present technique in stochastic fatigue analysis is to assume that the member stress is a narrow band process and hence its p.d.f. can be represented by the Rayleigh Distribution.
   As more data becomes available it is hoped to investigate the true form of the stress range p.d.f. and its variation from the Rayleigh Distribution.

4. As well as the data from instrumented platforms Lloyds Register also has a large amount of information on actual fatigue failures on offshore platforms. Correlation studies have already been carried out using the deterministic fatigue analysis system and extremely good results were obtained.
   After carrying out the previous investigation it is intended to repeat the correlation exercise using the stochastic fatigue system.

ACKNOWLEDGEMENTS

The author wishes to thank the Committee of Lloyds Register of Shipping for permission to publish this paper.

NOMENCLATURE

A Constant amplitude of component waves

C_D Drag coefficient

C_{M_1} Froude-Kryloff force coefficient

C_{M_2} Inertia force coefficient

D Member diameter

d Water depth

k_n Wave number of nth component

$\underline{u}$ Particle velocity resolved normal to member axis

$\dot{\underline{u}}$ Particle acceleration resolved normal to member axis

ρ Mass density of water

REFERENCES

1. Bainbridge, C.A. "Certification of Fixed Steel Platforms in the North Sea", OTC (1975) Paper OTC 2410.

2. Berge, B., J. Penzien. "Three-Dimensional Stochastic Response of Offshore Towers to Wave Action", University of California.

3. Borgman, L.E., "Ocean Wave Simulation for Engineering Design", Proc. Conf. Civil Engineering in the Oceans, 1968.

4. Cartwright, D.E., M.S. Longuett-Higgins. "The Statistical Distribution of the Maxima of a Random Function", Proceedings Royal Society, Vol.237, 1956.

5. Chakrabarti, S.K., W.A. Tam. "Gross and Local Wave Loads on a Large Vertical Cylinder - Theory and Experiment", OTC (1973) Paper OTC 1818.

6. Clough, R.W., J. Penzien. "Dynamics of Structures" McGraw-Hill, 1975.

7. MacNeal, R.H., C.W. McCormick. "The NASTRAN Computer Program for Structural Analysis", Computers and Structures, 1971, Vol.1, pp.389-412.

8. Penzien, J., M.K. Kaul and B. Berge, "Stochastic Response of Offshore Towers to Random Sea Waves and Strong Motion Earthquakes", Computers and Structures 1972, Vol.2, pp.733-756.

9. Shaw, P., A.D. Coates, R. Hobbs and W. Schumm, "Analytical and Field Data Studies of the Dynamic Behaviour ot Gravity Structures and Foundations", OTC (1977) Paper OTC 3008.

10. Shinozuka, M. "Monte Carlo Solution of Structural Dynamics", Computers and Structures, 1972, Vol.2, pp.855-874.

11. Soding, H., "Calculation of Long-term Extreme Loads and Fatigue Loads of Marine Structures", The Dynamics of Marine Vehicles and Structures in Waves, 1974, Paper 35.

12. Wheeler, J.D., "Method for Calculating Forces Produced by Irregular Waves", OTC (1969) Paper OTC 1006.

SOME THOUGHTS ON THE DYNAMICS OF OFFSHORE GRAVITY PLATFORMS

R. Eatock Taylor

University College London, England.

SUMMARY

This note provides a précis of some work aimed at understanding the response of gravity platforms. Their dynamic behaviour in waves, as described by transfer functions and magnification factors, is considerably more complex than a simple resonance phenomenon at the fundamental natural frequency.

The dynamic analysis of gravity platforms has been motivated by the need to quantify both the maximum stresses anticipated (with a given probability) in the life of the structure, and its probable fatigue life (e.g. the steel deck). More recently, attempts to carry out in service monitoring of structural integrity, based on measured dynamic responses, have emphasized the need for sophisticated methods of quantifying the dynamic behaviour of gravity platforms.

The dynamic response of these platforms is strongly influenced by the foundation characteristics, about which accurate data is seldom available from direct measurements. It becomes necessary to perform a series of analyses of the same structural idealisation on a range of different seabed representations. An efficient method of achieving this is the method of substructuring, in which the structure is first analysed on a fixed base. Its dynamic response is then defined in terms of the first few principal coordinates of the fixed base structure, and the rigid body motions of the base on a deformable foundation. [1]

There are two principal advantages to this approach in the linear frequency domain analysis that is appropriate to gravity platforms excited by waves. One is the possibility of reducing a finite element system with a very large number of degrees of freedom, having frequency dependent coefficient matrices, to a small system which may be efficiently solved at a large number of frequencies. And the second advantage is that the very different magnitudes of damping in structure and soil may be accurately represented. It has been found that the alternative procedure, a direct modal analysis of the complete system (structure + foundation) using the best weighted modal damping ratio [2], can lead to large errors in the calculated responses (of the order of 100% for certain resultants). Fig. 1 shows a comparison between results from such a modal analysis with weighted damping, and results obtained from an exact solution of the damped equations of motion for a simple representation of a gravity platform.

Results from the analyses may be obtained in various forms such as transfer functions, but an instructive illustration of the behaviour is given by curves of dynamic magnification factor. Results for deck displacement, and shear and overtuning moment at the foundation for typical platforms, are much as might be expected: each is a smooth curve displaying a maximum corresponding to the first system resonance (Fig. 1a). If the second resonance is close to the first there will be some interaction (Fig. 1b), but again the curves are smooth. However the curves of shear force and bending moment magnification in one tower of a multi-towered gravity platform are much less straightforward. Superimposed on the resonant hump of each of these is a series of peaks or troughs, ranging from minor ripples in some cases to sharp spikes and inverted spikes in others (Fig. 2). The frequencies at which these irregularities occur are very close to those frequencies where the distance between the towers equals an odd number of half wave-lengths. But the characteristics of the curves for a three towered platform are found to be quite different from those of four towered platforms.

In order to provide an understanding of these phenomena, very simple two and three degree of freedom models have been studied [3]: one generalised coordinate represents distortion of the superstructure, and the others represent the motions of the base (rocking and swaying). It has been found that by a rational choice of parameters for the simple models, the characteristics of the results from the more complex finite element idealisations may be closely approximated. In particular the magnification factors for the tower stress resultants oscillate violently with frequency in a similar manner for both the simple and complex representations. On this basis it has proved possible to clarify the physical mechanisms accounting for these somewhat unexpected results.

Two consequences of this effect are especially noteworthy. First the magnitication for one stress resultant at a particular frequency may be radically different from that for

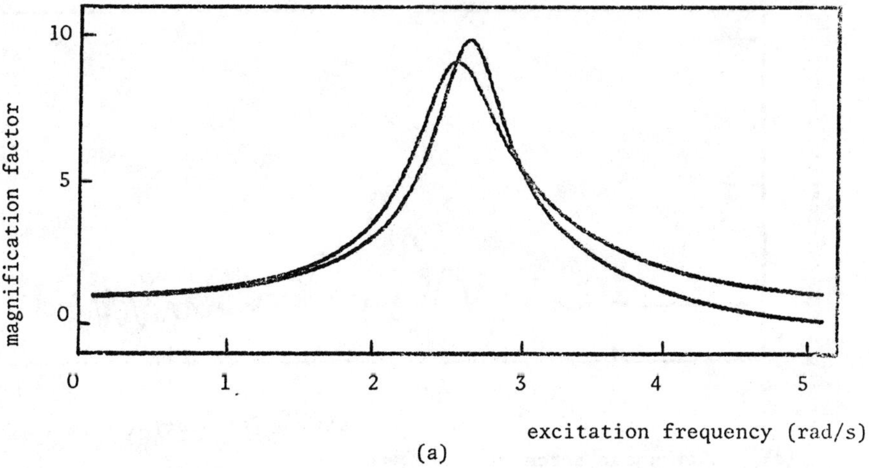

(a)

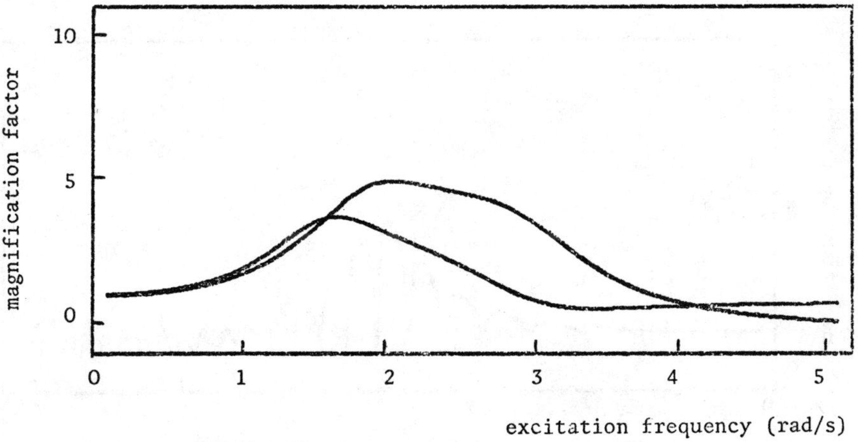

(b)

Figure 1. Dynamic magnification factors for the wave induced shear force at the foundation of a four-towered gravity platform, computed using weighted modal damping ratios and "exact" coupled damping. (a) Hard soil conditions; (b) soft soil conditions.

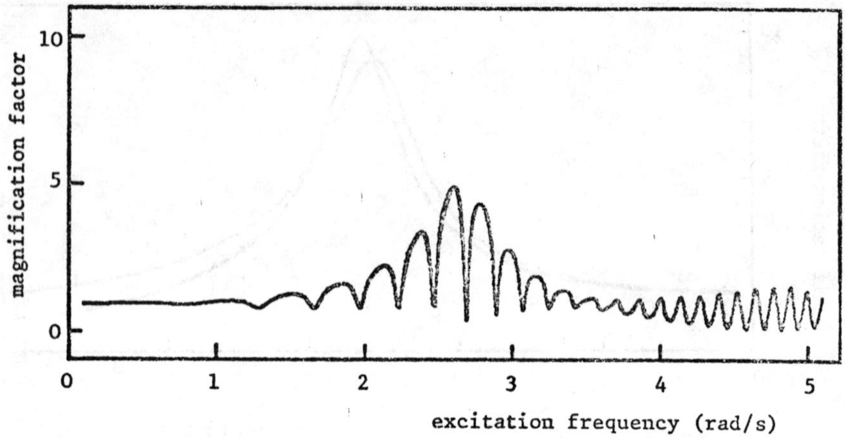

(a) tower shear force

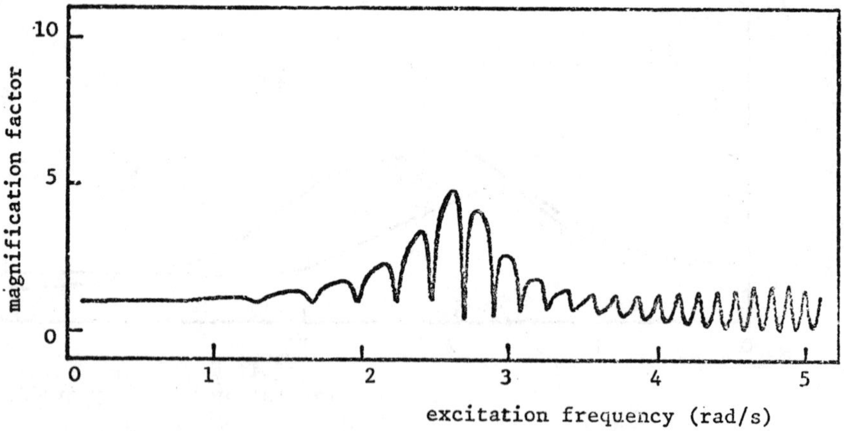

(b) tower bending moment

Figure 2. Dynamic magnification factors for the wave induced stress resultants in one tower of a four-towered gravity platform, computed from the fully coupled equations of motion.
(a) Shear force; (b) bending moment.

another stress resultant. And secondly, it is possible for the magnification factor of some stress resultants to be less than unity at frequencies below the fundamental resonance.

REFERENCES

1. Eatock Taylor, R. "A Preliminary Study of the Structural Dynamics of Gravity Platforms", Offshore Technology Conference, 2406, III, 695-706, 1975.

2. Roesset, J.M., R.V. Whitman and R. Dobry. "Modal Analysis for Structures with Foundation Interaction", ASCE J. Struct. Div. 99 ST3, 399-416, 1973.

3. Eatock Taylor, R. and P.E. Duncan. "The Dynamics of Offshore Gravity Platforms: Some Insights afforded by a Two Degree of Freedom Model", Int. J. Struct. Dyn. and Earthquake Eng., 6, 1978

FATIGUE ANALYSIS OF STEEL-JACKET PRODUCTION PLATFORMS INCLUDING THE EFFECT OF DYNAMIC RESPONSE

M.G. Hallam and A. Hooper

Atkins Research and Development, England.

ABSTRACT

Up to the present time steel jacket structures have been designed on the basis of an extreme static design load. Subsequently, the completed design has been checked for dynamic amplification and fatigue life, often only by the certification authority.

As jackets are designed for deeper waters, from 150 m depth, their natural periods become lower and the dynamic excitation begins to have a profound effect on the stress levels achieved in all wave conditions.

In order to meet the design requirements for dynamic and dynamic fatigue analysis a new suite of computer programs called FATJACK has been developed, which embodies several innovations in the analysis of steel-jacket structures.

The whole system is centred on the Atkins Stress Analysis System (ASAS). It uses the sub-space iteration technique to handle very large eigenvector extractions which are necessary if the structure is to be modelled with sufficient accuracy to be able to find the dynamic stress levels in the smaller members.

This is followed by a spectral approach using internally generated transfer functions to relate the climatic conditions to the dynamic response and hence fatigue life duration.

The application of this program to large jacket structures is described in detail.

In addition, a suite of computer programs has been developed which can use conventional wind records to predict the wave climate at specific locations. The predicted waves are fetch

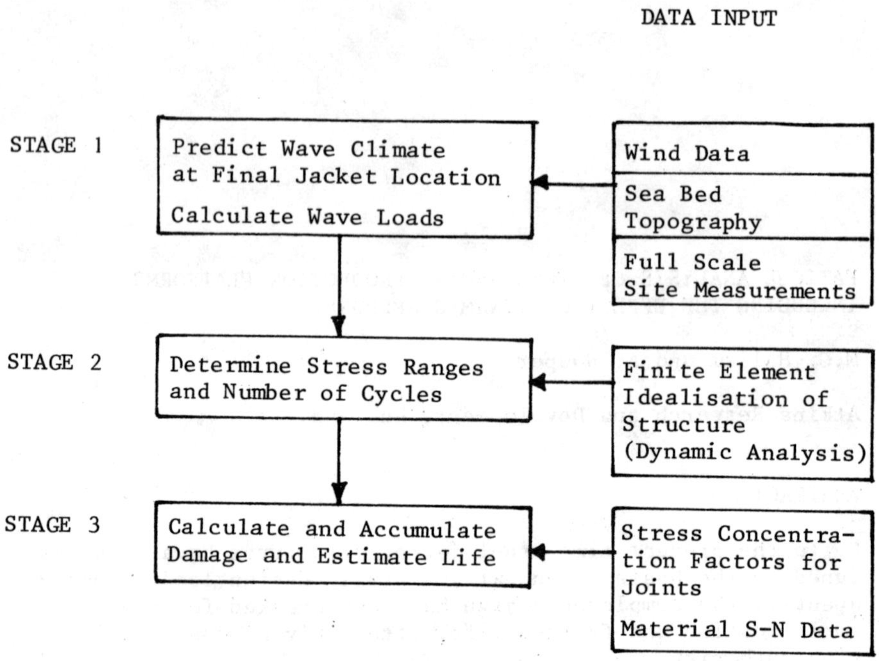

FIGURE 1 FLOW CHART OF TYPICAL FATIGUE ANALYSIS

limited and can be calculated from either the Jonswap or Bretschreider formulations. The effects of refraction are then computed for each wave, where applicable.

INTRODUCTION

The escalating demand for hydrocarbon fuels has necessitated the exploration and development of offshore fields in ever increasing water depths. The enormous size of the jacket structures required for these depths has lead to a new breed of problem and, in particular, the fatigue and dynamic response of such structures has become increasingly dominant.

Experience in the North Sea has shown that some structures are more prone to fatigue damage from small, low period waves rather than overload failures due to the long period design waves.

Fatigue analysis is by no means new and has been used for many decades, particularly in the aero-space industry. Many civil engineering structures, most notably bridges, have also been investigated for fatigue loading. However, there are differences in the way in which fatigue calculation may be presented for offshore structures.

The relative smallness of aircraft structures permits closely controlled manufacture and stress relieving heat treatment so that due account can be taken of both the mean and fluctuating stresses to assess the fatigue life. Additionally, full scale component testing leads to useful feedback to "calibrate" the theoretical analysis. The combination of accurate analytical analysis backed up by experimental component testing has resulted in fatigue analysis reaching a very advanced stage of reliability.

For civil engineering structures, the sheer size and unknown in-built stresses predetermine that the initial stresses and therefore the mean stresses cannot be accounted for. Moreover, full scale testing is out of the question, so that fatigue analysis has to be based solely on the fluctuating stresses (stress range) and much more reliance has to be placed on an analytical approach.

Also, for most civil engineering structures, the live or fluctuating loads are small compared with the dead weight. This is not so with offshore structures where the fluctuating stresses caused by waves are very high indeed.

For the very large structures now being considered, fatigue analysis is further complicated by the need to consider the dynamic magnification caused by the closeness of the structural natural frequencies with the exciting wave frequencies.

The ASAS Jacket Fatigue Suite

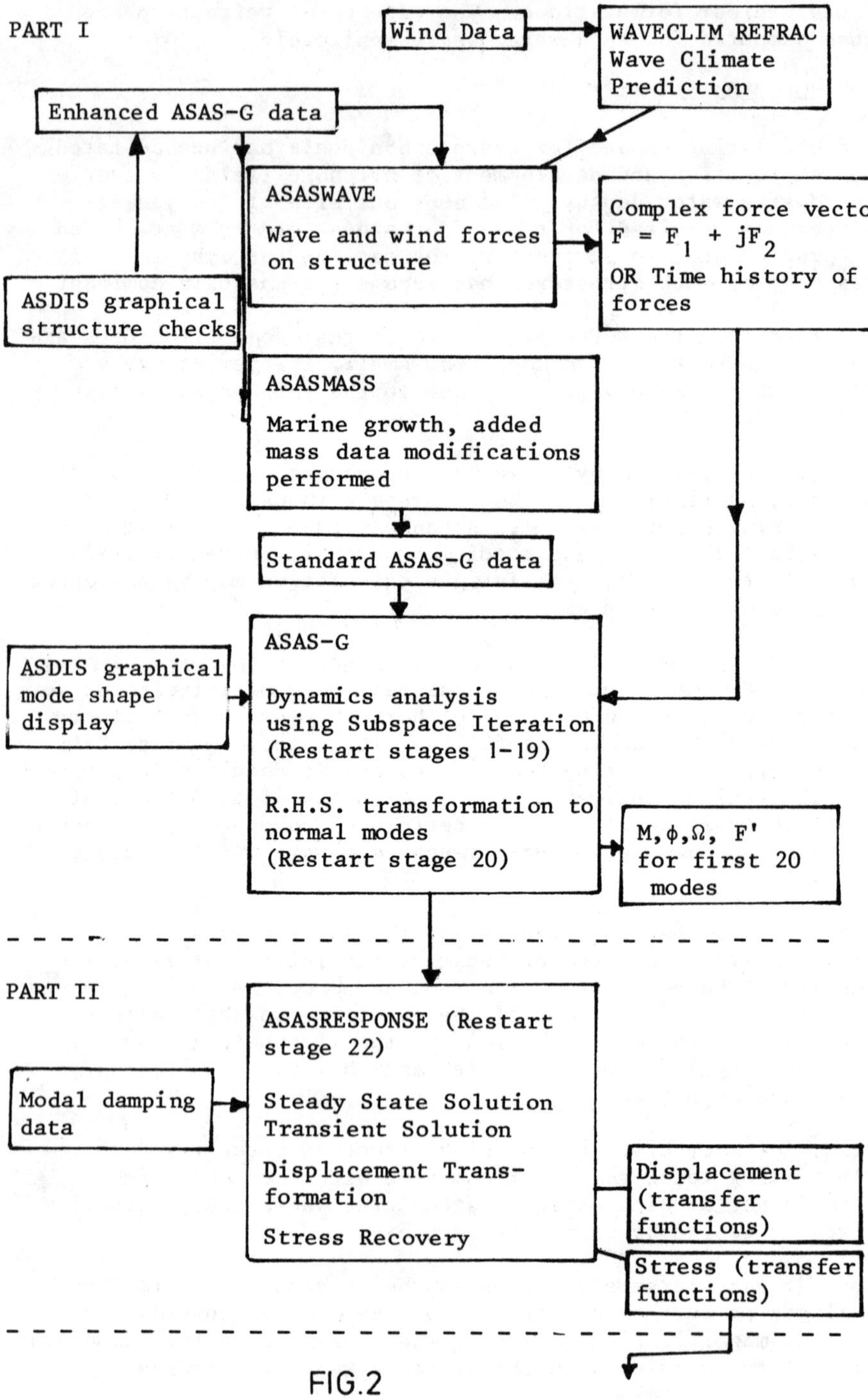

FIG.2

PART III

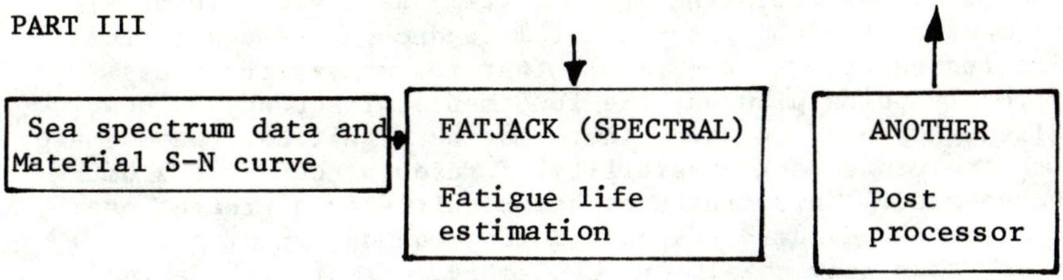

FIGURE 2.

With stiffness factoring approximately as the square of size and mass as the cube, the natural frequency reduces as the size increases with the result that for many structures currently being planned, the fundamental frequency is now in a frequency band in which there may be significant wave energy with the consequent possibility of resonant conditions being encountered. This, in turn, has resulted in a greater need to allow for dynamic response during the design process rather than rely solely on the "equivalent static" design methods hitherto used.

Putting the problem into numerical terms, for the smaller structures with fundamental periods less than 3 sec., static design methods appear adequate. At 3-4 sec. periods they become increasingly unreliable and at natural periods greater than 4 sec., totally misleading.

Dynamic fatigue analysis for any structure follows the same basic pattern outlined in figure 1, although the detailed methods within each stage may vary considerably.

All of the calculations for the analysis presented in this paper may be carried out within the FATJACK suite of programs which is built around the Atkins Stress Analysis System Finite Element Program ASAS. A detailed flow chart of the FATJACK suite of programs is shown in figure 2.

STAGE 1 - WAVE LOADS

As far as offshore platforms are concerned, the first stage in determining the loads is to estimate the wave climate at the particular location of the platform.

A realistic estimate of this can be achieved either by field measurement or prediction techniques.

The remote measurement of waves at sea is now a relatively routine operation, but requires a considerable length of time to gather an adequate volume of data. Normally, observations are taken for a maximum period of 12 months and almost invariably this includes periods of missing data due to instrument failure, maintenance, loss, etc. Records of this nature provide a valuable basis for wave climate analysis, but should not be used in isolation as they do not indicate annual variations in wave climate.

The measurement of waves for a sufficient number of years to include these long term variations is usually impractical and an alternative approach is needed.

Such an approach is provided by the program WAVECLIM, which is designed to provide wave climate data predictions at specific locations from conventional wind records. Wind

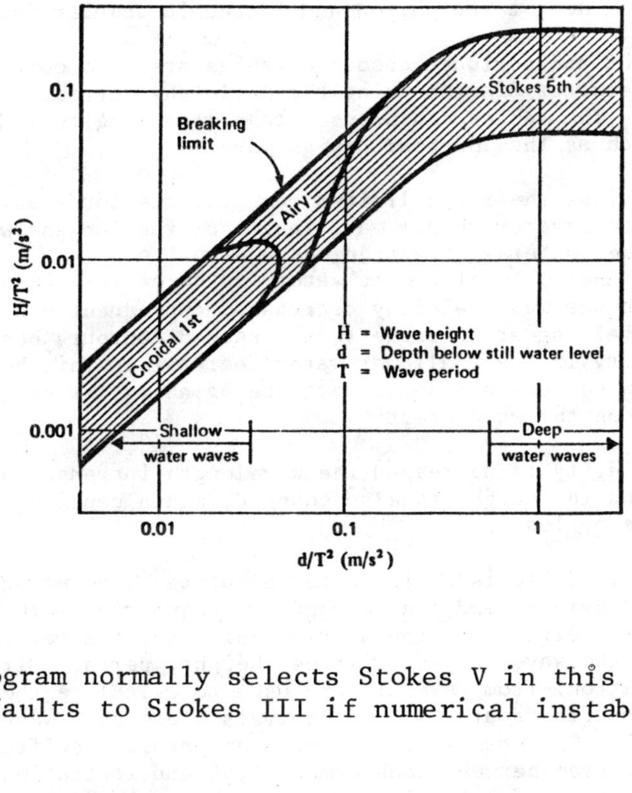

* The program normally selects Stokes V in this regime but defaults to Stokes III if numerical instability occurs

FIGURE 3 The range of validity of various wave theories

information for a period of several years can be used to obtain a realistic deep water wave climate which takes account of annual variations in wind characteristics. Alternatively, data for individual months enables the program to provide seasonal variations and likely periods of relative calm suitable for the more delicate offshore operations.

The probability of occurrence of wind for discrete speed and direction ranges are used together with the fetch in each direction sector to calculate a scatter diagram of significant wave height and mean zero crossing period using either the Bretschneider [1] or Jonswap [2] formulations. The scatter diagram probabilities are summed for intervals of significant height and period and listed in tabular form.

Wave height and period exceedance tables are also computed from the scatter diagram using the Rayleigh distribution function. The tables can be used to derive long term design values such as the 100 year design wave.

WAVECLIM gives the fetch limited wave climate for areas of water depth greater than ½ wave length of the longest wave. In shallower waters refraction has a significant effect on energy levels. At water depths of less than ½ wavelength the wave celerity decreases, consequently a wave crest travelling at an angle to the seabed contours bends as the celerity in the shallower water decreases. This bending results in convergence and divergence of energy at different points along the wave crest.

As the celerity is decreased the wavelength increases proportionally and the height is also changed, again causing changes in the energy levels.

The program REFRAC is designed to calculate these effects of varying bathymetry and can be used in conjunction with WAVECLIM. Bathymetric data (on a grid basis over the relevant area) and the wave characteristics (height, period, direction of propagation) from WAVECLIM are used to calculate the refraction effects at regular intervals along each wave orthogonal. Options within the program permit specification of the interval between each computation and initiation of the calculation of lateral energy flow between adjacent orthogonals. At each step the water depth of the four surrounding grid points are examined; if any are less than ½ the wavelength of the wave in question the refraction effects are computed and the wave of new height, celerity and wavelength propagated in the new direction until the next calculation point is reached. The process is repeated until the wave reaches the specified location. The results can be output in graphical and tabular form showing the wave orthogonal positions and wave characteristics of each step.

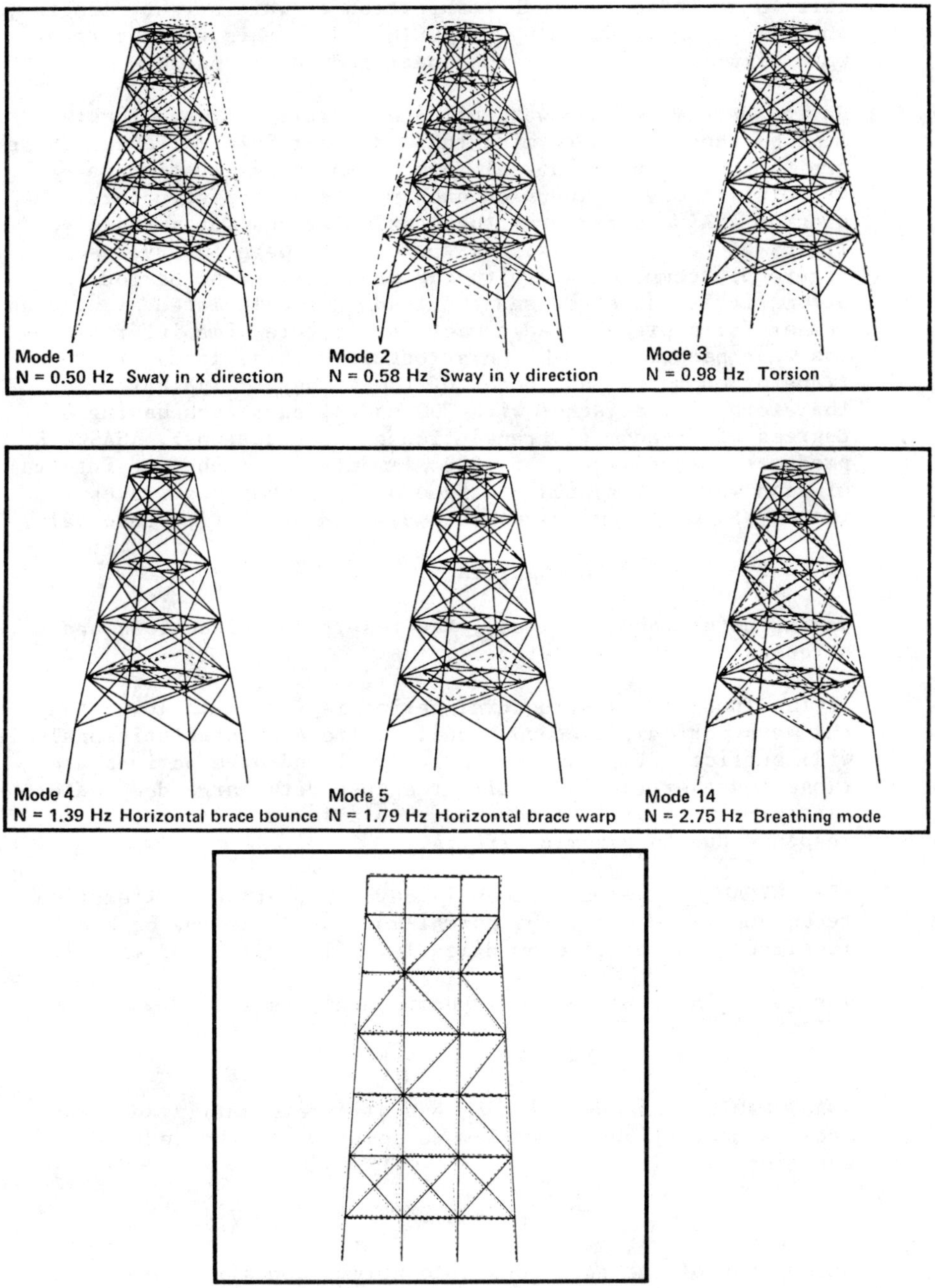

FIGURE 4 Global and "Local" Modes of Two Typical Structures

This is repeated for each combination of wave height, length and direction produced by WAVECLIM and in this way the true wave climate derived for the specified location.

Having determined the wave climate, a suitable wave theory must be chosen to determine the water particle motions. These are then used in the normal form of Morisons equation to calculate the wave loads on each member of the structure. The program ASASWAVE performs these calculations for successive trains of regular waves. An appropriate wave theory (see figure 3), commensurate with the wave height and period is automatically selected and the loads are calculated along each member for a preselected number of discrete time intervals as the wave passes through the structure. These loads are then transformed to the end nodes of the member. Typically, therefore, for a jacket with 300 nodes, say, each having 6 degrees of freedom (3 translation and 3 rotations), ASASWAVE produces a load vector of 1800 elements for each time interval of each wave. A typical fatigue analysis may require approximately 50 different waves each defined at 10 time intervals.

STAGE 2 - STRUCTURAL ANALYSIS

Having determined the loads, the resulting deflections and stresses may be calculated.

If the fundamental structural period is very much lower than the wave periods, a conventional finite element static analysis will suffice. If, however, structural and wave periods are close together (as is usually the case with large deep water platforms), it will be necessary to allow for increased response due to dynamic effects.

ASAS DYNAMICS uses the recently developed Subspace Iteration technique [4] whereby the normal mode analysis may be handled in all degrees of freedom directly. (Typical 2000, or so).

The governing equation for dynamic analysis is

$$M\ddot{x} + C\dot{x} + Kx = F(t)$$

ASASDYNAMICS assembles a mass and stiffness matrix and produces a conventional normal mode solution to the reduced equation.

$$M\ddot{x} + Kx = 0$$

Decoupling of the equations into normal modes is achieved by the operation,

$$\phi_x^T M \phi_x \ddot{u} + \phi_x^T K \phi_x u = \phi_x^T F(t)$$

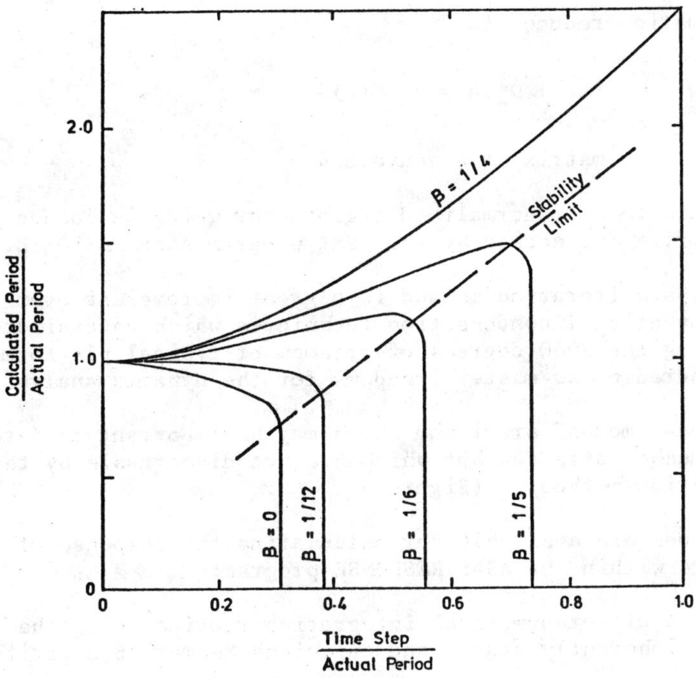

FIGURE 5 Distortion and Stability of the Newmark-β Integration Scheme

where $x = \phi_x u$, and

ϕ_x are the eigenvectors

With ϕ_x normalised such that:

$$\phi_x^T M \phi_x = I$$

this equation reduces to

$$I \ddot{u} + \Omega_n^2 u = \phi_x^T F(t)$$

where Ω_n^2 is a matrix of eigenvalues.

It is this form of normalised eigenvector which is loaded onto backing file for access by the next program ASAS RESPONSE.

The subspace iteration method is a great improvement over the more conventional condensation techniques which necessitate condensing the 2000 degrees of freedom of typical platforms to a hundred or so master freedoms for the dynamic analysis.

Many "local modes" are found which may be important in determining member stresses but which are not discernable by the condensation methods. (Figure 4).

Two methods are available for calculating the response of the structure within the ASAS RESPONSE program:-

(i) A direct numerical integration routine using the inherently stable and efficienc Newmark B algorithm

$$U_{m+1} = \frac{\{2(1+h^2\Omega_n^2(\beta-\frac{1}{2})\ U_m-(1-\eta\Omega h+h^2\Omega^2\beta)U_{m-1}\ +h^2F_m\ \}}{(1+\eta\Omega h+h^2\Omega^2\beta\ )}$$

where U_{m-1}, U_m, U_{m+1} are the nth normal mode coordinates at time intervals m-1, m and m+1, respectively.
Ω_n is the nth natural frequency
h is the time step
F_m is the value of the generalised force at time m
η is the damping ratio
β is the constant

The value of β is chosen automatically within the program such that minimum distortion occurs during the integration process whilst numerical stability is always ensured [5] (figure 5).

(ii) A steady state complex response routine of the form

$$U_1 + U_2 = \frac{F_1 + jF_2}{\Omega_n^2 \left| \left(1 - \frac{\Omega^2}{\Omega_n^2}\right)^2 + \left(2j\eta \frac{\Omega}{\Omega_n}\right)^2 \right|}$$

where Ω_n = natural frequency

Ω = wave frequency

η = damping ratio

$j = \sqrt{-1}$

The first solution routine is useful for looking at the transient behaviour of the platform as the waves are applied and may also be used for fatigue life calculations based on deterministic methods (see later).

The second method may be used for either deterministic or spectral based fatigue calculations and is capable of handling larger numbers of wave cases with greater economy. For this method the time histories of wave forces at each node of the structures are "fitted" with a Fourier series of harmonics. (From 1 to 5 Fourier terms may be taken.)

For each harmonic, displacement transfer functions are then computed by the transformation back to original coordinates.

$$X_1 + jX_2 = \phi_x(U_1 + jU_2)$$

Stress transfer functions are recovered from the displacement functions using the same ASAS-G techniques as in a standard linear stress analysis:

$$\phi_1 + j\phi_2 = f(X_1 + jX_2)$$

These transfer functions are stored on file for use by the spectral post processor program.

For both response methods, the user inputs damping as a percentage of critical for each mode.

It should be noted that the Subspace Iteration method produces reliable stress information. Condensation methods, whilst giving reasonable results for the overall jacket deflections, are somewhat prone to error in recovering back to stress. (The stresses essentially relate both to jacket deflection and curvature for which the computed accuracy is not so good. Moreover, condensation methods require that the distributed wave loads over the jacket to be represented by relatively few master freedom point loads and that the coarse mass modelling

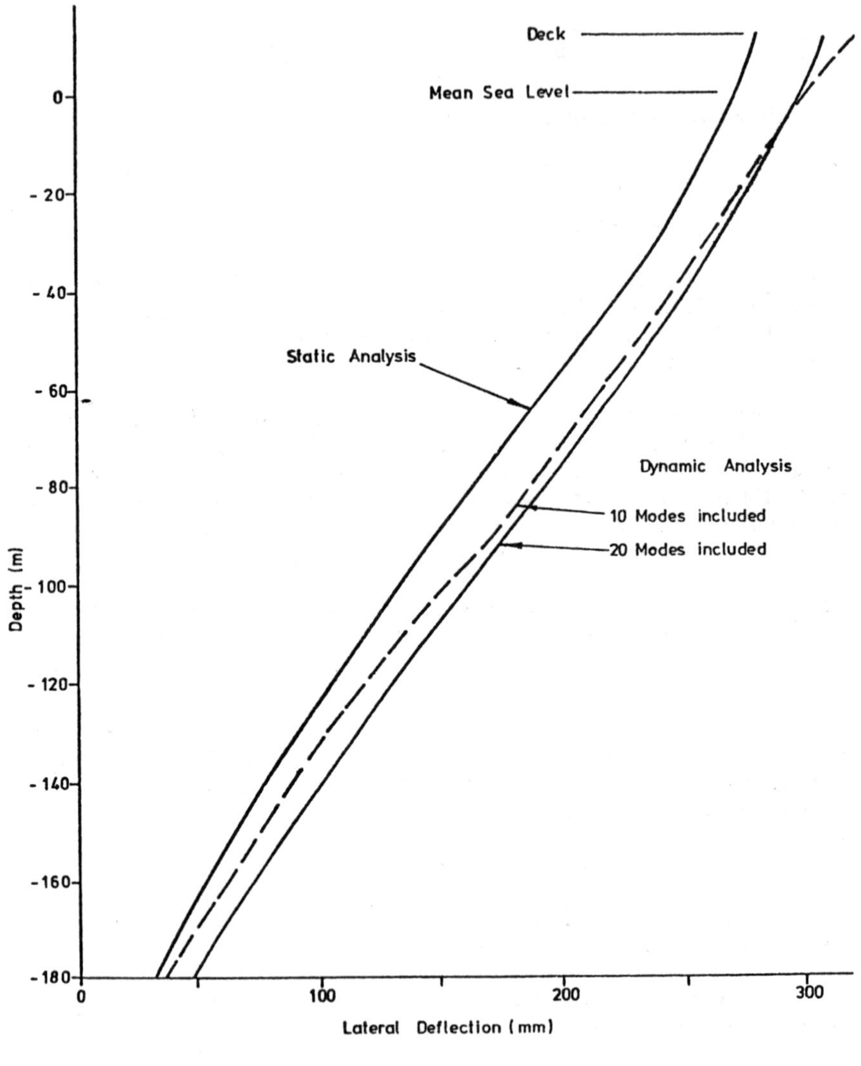

FIGURE 6 Effect on Accuracy of Incorporating too few Nodes in the Analysis. The Errors are more Pronounced on Stress (curvature) than Overall Deflection.

will result in many "local" modes being missed altogether (figure 4). Both of the shortcomings significantly affect the reliability of the stress calculation.) One final point, in order to achieve sufficiently accurate stress information, it may be necessary to incorporate a reasonable number of modes in the final summation. Inaccuracies may result if too few modes are taken (see figure 6). Alternatively, "static" inclusion of higher modes often improves the accuracy to a satisfactory degree but care must be exercised.

An example of the type of information output from the program is given in figure 7, which shows some typical time histories of response as regular waves pass through the structure |3|.

STAGES 3 AND 4 - DAMAGE CALCULATION

At this stage the fatigue damage can be assessed directly by assigning the number of occurrences for each of the regular waves and performing a conventional "cycle count" within each stress range. The assumption here, of course, is that regular waves reasonably represent the true sea.

More realistically the true random nature of the sea can be included by interpreting the steady state response due to regular seas as points on a transfer function. A complete transfer function, therefore, will be assembled from a number of base case regular waves. A wave spectrum may then be multiplied into this transfer function to produce the response spectrum and by assigning a probability of occurrence to this spectrum the damage may be assessed. (Figure 8).

If the material SN data is a straight line on log-log scales with slope B and a point (S_1, N_1), one formula which may be applied is:

$$D = T\left(\frac{M_2}{M_o}\right)^{\frac{1}{2}} M_o^{B/2} \left(\frac{2^{B/2}}{S_1^B N_1}\right) \int_o^{\infty} X^{B/2} e^{-x} \, dx$$

where $X = S^2/2M_o$

D is the damage

M_o and M_2 are the zeroth and second spectral moments of the stress

$$(M_n = \int_o^{\infty} f^n S_{\sigma\sigma}(f)\,df)$$

f = frequency

and T is the duration of the sea state.

The integral term simply reduces to a constant and is probably best evaluated using a Gamma function.

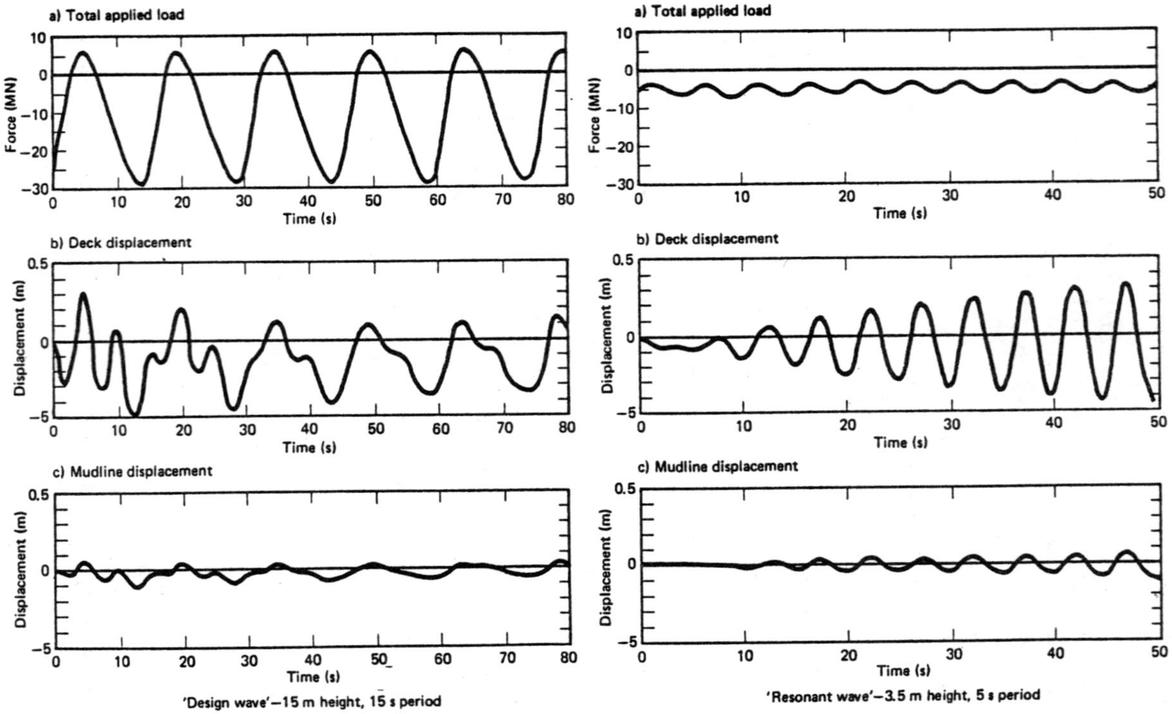

FIGURE 7 Typical Results from Transient (Newmark-β) Analysis

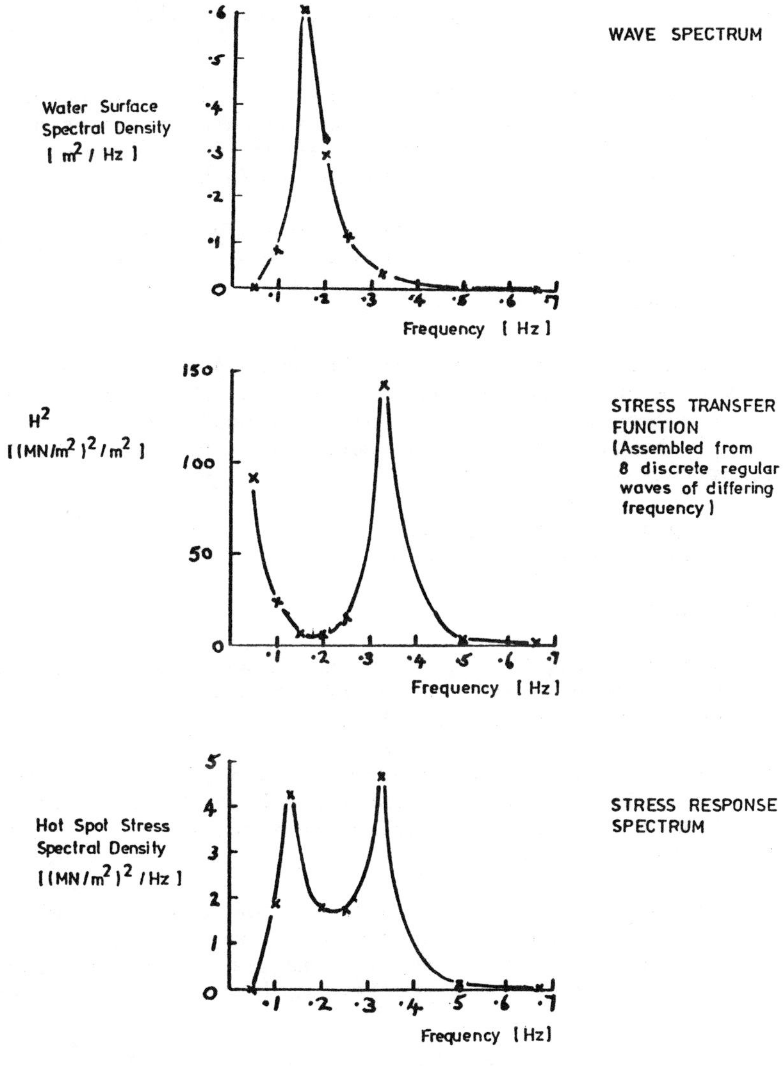

FIGURE 8 Spectral Fatigue Analysis

The damage is then summed for each wave of the applied spectrum and finally the life estimated by taking the reciprocal of the damage.

There are of course many similar formulae which are equally applicable [6].

In practice, FATJACK allows stress concentration factors to be applied separately to the bending and axial stresses at the ends of each tube. The combined stresses are then calculated at 8 points around the circumference of the tube.

For each of these points a stress transfer function may be assembled and the damage calculated and accumulated.

CONCLUSIONS

To the date of publication of the paper 4 full scale structures have been analysed on this suite of programs. Unfortunately, confidentially prohibits any publication of results. In all cases, however, local modes are found to exist at remarkably low frequencies.

The computations have proved economic and the input data reasonably simple to prepare.

REFERENCES

1. Bretscheider, C.L. "Revisions in Wave Forecasting, Deep and Shallow Water". Proc. of 6th Conf., on Coastal Engineering, ASCE, Council on Wave Research, 1958.

2. Hasselmann, M.S. "Measurements of Wind-Wave Growth and Swell Decay during the Joint North Sea Wave Project (JONSWAP)". Deutsche Hydrogrophische Zeitschrift, Reihe A (80), No. 12, 1973.

3. Hallam, M.G., N.S. Heaf and L.R. Wootton. "Dynamics of Marine Structures: Methods of Calculating the Dynamic Response of Fixed Structures subject to Wave and Current Action". Report UR8, CIRIA/UEG, 6 Storey's Gate, London SW1P 3AU.

4. Bathe K-S, E.L. Wilson. "Solution Methods for Eigenvalue Problems in Structural Mechanics", Int. J. Numerical Method in Eng., Vol.6, 1973.

5. Goudreau, G.L., and R.L. Taylor. "Evaluation of Numerical Integration Methods in Elasto Dynamics," Computer Methods in Applied Mechanics and Engineering, Vol.2, 1973.

6. Williams, A.K., and E. Rinne. "Fatigue Analysis of Steel Offshore Structures". Proc. Instn. Civ. Engrs. Part 1, 1976, 60, Nov.

RELIABILITY ANALYSIS APPLIED TO OFFSHORE STRUCTURES

R. S. Lamb

Atkins Research and Development, England.

ABSTRACT

General reliability methods are described, with the presentation of the method considered most appropriate for offshore application. The potential of reliability analysis in offshore engineering is described, and examples given of probable areas of application. Attention is drawn to the major problems currently standing in the way of its widespread application.

The current application of the method in a major offshore research project is reported.

1. INTRODUCTION

Much of the rapid expansion of offshore work has been taking place at a time when designers lacked long-term experience of the area as a basis for design judgement. The global factors of safety (or practical safety factors) in current use have been derived empirically by observing the performance of generally much smaller land-based structures, or of offshore structures under the different design conditions of the Gulf of Mexico.

It is arguable whether these factors are appropriate for structures in other areas of the world.

Ample justification for the need for care in extrapolating existing codes of practice beyond their original terms of reference can be found in the history of failures in many other branches of engineering.

While the offshore engineer has normally been fully aware of the problem, he has had no practically applicable tools available, other than 'engineering judgement', to assess where his current design practice is likely to become unrealistic.

The development of new limit state codes of practice should now provide a framework on which to develop methods which will organise engineering judgement along rational lines, and alleviate the major problems of extrapolation.

The fundamental parameters for the assessment of overall structural performance, or of an individual component is undoubtedly that of reliability, or its complement - overall risk of failure. It is important, however, to distinguish between two types of risk.

1) Structural risks which can be controlled by the adoption of suitable factors of safety, and design methods.
2) Inherent system risks, which are to a large extent independent of the structure design calculations, and more a fundamental property of the type of structure, its location, and usage. In this category must be included ship collision, terrorist attack, fire, blowout, human design error, human operating error etc.

In offshore design, category 2 risks are of major importance. Consideration must be given to the control of these potential hazards, although the dangers cannot be removed completely. In particular, consideration should be given to ensuring that should such events occur, the damage is not catastrophic.

This paper is concerned primarily with category 1 risks due to the inherent variability of the materials and loading assumed in the design. The reliability achieved against category 1 risks is termed total notional reliability, as it does not represent the overall reliability of the system.

The application of reliability techniques to any branch of engineering has been hindered by three main problems.

1) An accurate evaluation of the reliability of a component can only be achieved if a substantial volume of relevant data is available on all the parameters that affect the design, such that the statistical distribution can be assessed.
2) The numerical methods and computational power necessary have only recently become available, and even now rigorous methods are possible only in very limited cases.
3) Land based structures have developed comparatively slowly, and in general very few failures have occurred. Thus there has been no necessity for developing more advanced ways of quantifying and controlling the actual reliability or risk of fail-

ure achieved using existing standards or codes of practice.

This paper describes the State of the Art in Reliability Analysis, and what can be achieved relevant to offshore engineering using the method. An account of work at present being undertaken by Atkins Research and Development using the techniques is included.

2.1 Principles of Advanced First Order Second Moment Methods

Reliability analysis sets out to determine the probability of a limit state of either unserviceability or collapse being exceeded. For engineering applications the limit state must be definable, not necessarily explicitly, however, in terms of a finite number of design input parameters.

Thus, using the notation of CIRIA report No.63 [1] for the simple case of the limit state of collapse of a cylindrical steel tie in tension, the limit state function 'Z' can be expressed explicitly in terms of 'D' - the diameter of the tie, 'fy' - the yield stress of steel, and 'L' - the applied axial loading as:-

$$Z = \frac{\pi D^2}{4} \times fy - L \qquad (1)$$

where collapse is defined as $Z \leq 0$.

For more complex design problems Z cannot be explained explicitly, and must be written as:

$$Z = g(x_1, x_2, x_3 \ldots x_n) \qquad (2)$$

where x_{1-n} are all the design input parameters occurring in the engineering calculation of the limit state conditions.

Thus equation (1) expressed in the form of equation (2) is:-

$$Z = g(D, fy, L) \qquad (3)$$

Each design input parameter in the 'g' function of equation (2) or (3) has an associated mean value and statistical distribution (referred to as probability density function or p.d.f. in this paper), and the fundamental problem to be solved by any method of reliability analysis is to determine the p.d.f. of Z. With this information the probability of $Z < 0$ can be determined and hence the probability or risk of failure. The probability that $Z > 0$ is of course the reliability.

If the 'g' function is anything other than trivial, then approximate numerical methods must be applied to determine the p.d.f. of Z. The principal approximation made in first order second moment methods is as the name implies, the linearisation of the 'g' function.

If $Z = g(x)$ (4)

for the case of x representing a single variable,

$$\text{then} \qquad Z = g(x^*) + \frac{(x - x^*)}{1!} g'(x^*) + \frac{(x - x^*)^2}{2!} g''(x^*) + \ldots$$

$$\simeq g(x^*) + (x - x^*)\, g'(x^*) \qquad (5)$$

where $g'(x^*) = \frac{\partial g}{\partial x}$ evaluated at $x = x^*$.

When Z is a function of several variables $X \equiv x_1, x_2, x_3 \ldots x_n$

$$\text{then} \qquad Z \simeq g(x_1^*, x_x^*, \ldots x_n^*) + \sum_{i=1}^{i=n} (x_i - x_i^*) g_i'(\underline{x}^*) \qquad (6)$$

where $g_i'(\underline{x}^*) = \partial g / \partial x_i$ evaluated at the point $\underline{x}^* = (x_1^*, x_2^*, \ldots x_n^*)$

Equation (6) may be rewritten as:

$$Z\ Y = k_o + \sum_{i=1}^{i=n} k_i x_i \qquad (7)$$

which is clearly linear in all x. Moreover, the particular values of x_i are realisations of the basic random variables. Under these conditions and provided that all the x are independent, it can be shown (3) that the mean and standard deviation of Y are given exactly by:-

$$m_y = k_o + \sum_{i=1}^{i=n} k_i\, m_i \qquad (8)$$

$$\text{and} \qquad \sigma_y = \left[\sum_{i=1}^{i=n} k_i^2\, \sigma_i^2 \right]^{\frac{1}{2}} \qquad (9)$$

where m_i and σ_i are the mean and standard deviation of x_i, respectively.

The mean and standard deviation of Z are thus given approximately by:

$$m_z \simeq m_y \qquad (10)$$

and

$$\sigma_z \simeq \sigma_y \tag{11}$$

with $\frac{\partial g}{\partial x_i}$ evaluated at $x_j^* = m_j$ for all j.

Thus by utilising a first order approximation to the design function, approximate expressions (10) and (11) can be derive relating the statistical properties of the design input variables to the p.d.f. of the limit state function Z, by means of the partial differentials of the limit state function evaluated (usually numerically, as the 'g' function is seldom explicit) at a point X*. The term second moment is applied to the theory because of the form of equation (9).

The principal conceptual difficulty with the method is that of deciding the most suitable point on the limit state function surface, (X*) at which the linearisation should be performed, such that errors due to ignoring higher order terms are minimised. In advanced first order second moment reliability theory, the point X* is derived as the point of maximum probability of failure. The determination of this point on the surface is a problem of linear programming and several iterative techniques [2] are available to solve this problem.

3. APPLICATION OF RELIABILITY ANALYSIS TO OFFSHORE STRUCTURES

3.1 Practicability of State of the Art Techniques

Flint and Baker [4] have investigated whether the techniques or reliability analyses are sufficiently advanced to be applied to offshore structures.

Evaluation of the 'notional' failure probability of a structure (a measure of that portion of the risk within the code drafting committee's direct control) is performed in three stages:-

1) Collect data on the statistical variation of design input parameters (for example, yield stress of steel, lack of straightness of compression members).
2) Choose the limit state criteria for which the probability of exceeding the relevant limit is to be assessed (for example, for a simple tie member, the limit state of collapse is that the applied stress shall not exceed the yield stress).
3) Evaluate the probability that the limit state will be exceeded, by considering the statistical variation of the design input variables in the limit state function.

Flint and Baker made several simplifying assumptions during the project, but the major problem area for offshore structures, that of undertaking stage 3 on a dynamically sensitive structure, was investigated thoroughly. It was concluded that the method was practical, and their work is being utilised and extended by Atkins Research and Development as described in section (4).

Thus it is hoped that in the future a new generation of limit state codes applicable to offshore structures will be drafted based on reliability criteria. The following problems, however, should be noted:-

3.1.1 Judging socially and economically acceptable levels of reliability Partial safety factors in the limit state design function can be modified to ensure that the correct level of reliability is obtained. By repeating the process over a range of structures, a series of partial safety factors, linked to socially - and economically - acceptable levels of reliability, could be derived for a code. Adequacy of a structure has to be judged against socially and economically acceptable levels of reliability, and these acceptable levels are very difficult to judge. There may be a tendency to increase levels of reliability over those implied today causing increased costs of construction.

3.1.2 Tendency to over-design due to shortage of information The shortage of statistical input data in certain areas will require the use of organised engineering judgement in applying reliability techniques. If information is crude, this could be reflected as a high partial safety factor, to cover ignorance as well as the inherent variability of the phenomenon. Unless care is taken, this can lead rapidly to uneconomic jacket design. This is not a criticism of the method, but merely demonstrates the 'acts of faith' involved in the existing regulations.

3.2 Possible Applications

The method is complex to manipulate, and initially, its use is likely to be limited to deriving rational safety factors in codes. Updating these factors would take place when more advanced analytical techniques become available, more efficient quality control is achieved, or structures outside the range originally envisaged by the code are conceived. This represents the true strength of the method.

It is unlikely that in the short term, reliability analysis will be used to change the overall levels of reliability that are inherent in the codes of practice used today. What is more likely is that the technique will show means by which partial safety factors can be efficiently grouped, thus reducing design effort, and the values that should be applied in order to achieve a risk of failure commensurate with the consequences of failure in all components of the structure. (i.e. say between the ultimate axial capacity of the piles, and the failure of a node).

Many applications of the method would prove useful. In the case of a partially damaged structure, for example, a series of decisions must be made about closing down and evacuation. At present, they have to be made on an 'engineering judgement' basis. Evaluation of the reliability of the damaged structure would form a rational basis on which these decisions could be made.

4. AN APPLICATION OF THE METHOD

Atkins Research and Development are currently engaged on a design sensitivity study in which advanced first order second moment reliability techniques are being applied to a jacket designed for the northern North Sea. The basic aim of the study is to investigate the importance of sensitivity of a steel jacket design to uncertainties in the design process.

These uncertainties may be due to: no full scale corroboration of laboratory tests, inadequate formulation of the problem or inherent variability of the phneomenon. Such uncertainties are conventionally covered by adopting a conservative estimate of the behaviour of the structure.

Considerable research effort is being currently devoted to tackle these areas of uncertainty, and it is suggested that the two criteria which should be applied in order to assess the possible benefit of such work are:

1. Greater consistency in safety (consistent reliability)
2. Direct cost benefit in terms of reduced cost of weight of structure or piling.

Criteria (1) is a goal which is fundamental to all rationally based design procedures. The way in which more consistent reliability can be achieved will generally be by improving the method of analysis of the phenomenon, improved quality control or more relevant testing, such as full-scale studies. Thus by any one of these methods, improvements can be made such that a greater uniformity of performance of structural components is achieved.

Current design practice utilises partial safety factors, or implied partial safety factors, which reflect a component due to lack of understanding of the phenomenon or lack of relevant performance data.

Considering a number of similar components designed using present-day knowledge and understanding, normally embodied in a code of practice, some members will be more likely to fail than others. This may be due to an inherently random phenomenon such as steel strength, or could be due to the inadequacy of the design formulation to predict the performance of the component under a particular range of conditions. Improved design methods, or better quality control can be envisaged such that the variation in performance of components is reduced. Also, with the assumption that the reliability achieved using the existing design method is adequate, partial safety factors can be reduced. Thus, whilst the average performance of components is reduced and the structure is reduced in weight or cost the probability of failure of an individual component is maintained constant.

Criteria (1) suggests that any structural performance phenomena which is known to have a significant uncertainty due to lack of knowledge and understanding should be investigated in order to obtain greater consistency in safety through all structural components. Whilst this is a worthwhile aim, criteria (2) is the means by which priorities should be decided.

The purpose of this study is to investigate the areas where research is most needed in order to improve the estimation of the safety of steel jackets. The basis for judgment will be that structures redesigned in the light of improved knowledge or understanding, shall have the same total notional reliability (or notional risk of failure) as the original structure. The improvement in the structure will be noted as a reduction in the structural weight.

4.1 Method

A flow chart of the basic method used in this study is shown in figure 1.

One reference structure typical of current practice in the northern North Sea has been chosen as the base of this study. This is a limitation of the study but it was considered more important to carry out a thorough investigation of all the implications on the design of the structure of changed design parameters, or partial safety factors due to improved knowledge or understanding.

Thus the implications of changed input parameters will be carried through all major aspects of design including static, dynamic, fatigue, foundation, loadout, installation and hydrostatic analysis. Only after an investigation of all these areas of design can the full implications or true sensitivity to a change in a property be derived.

5. CONCLUSIONS

1. Reliability techniques will find increasing application in engineering practice, and the mathematical tools are available to apply approximate methods to complex offshore structures.

Figure 1 Flow Chart for Sensitivity Study

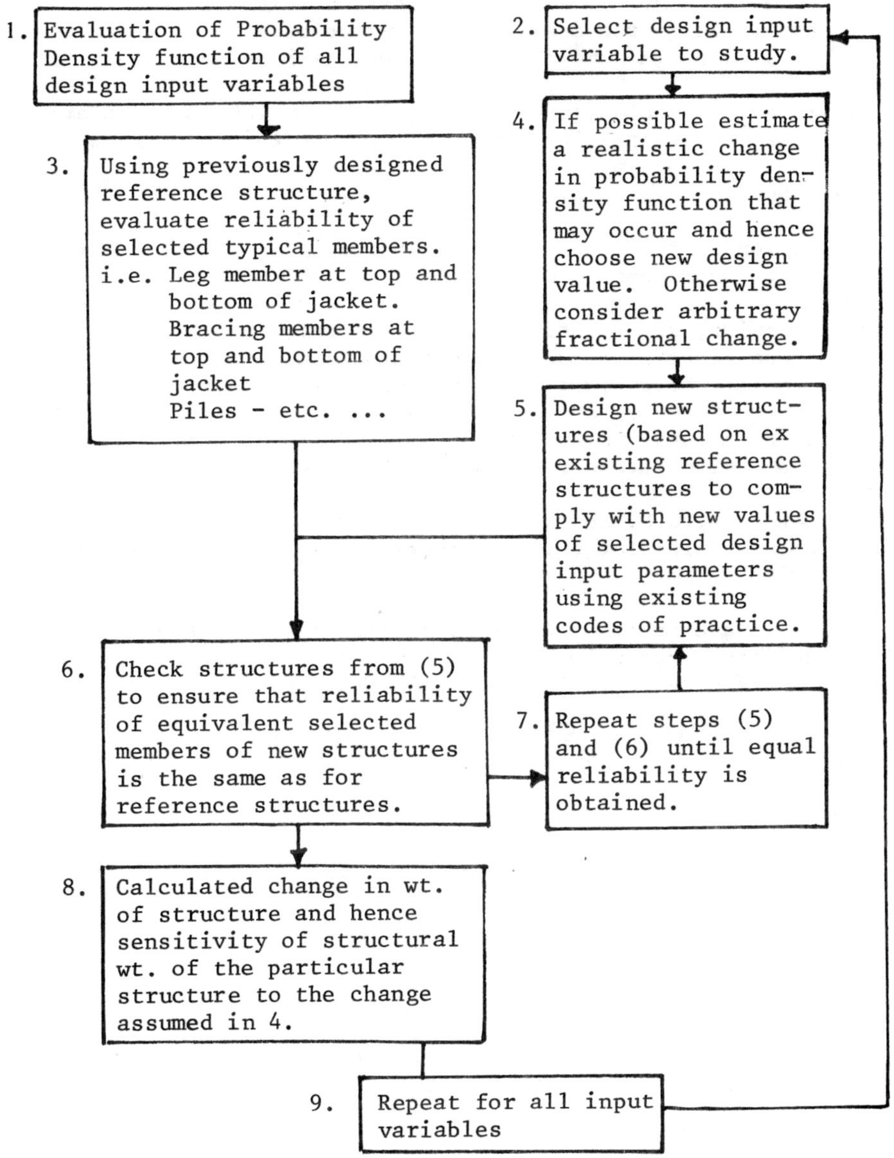

2. Realistic values of notional reliability will only be produced provided adequate statistical information is available on the design input variables. Such information is lacking in some major areas of offshore design.

3. As a matter of urgency, data collection should be undertaken on an organised basis to improve this situation.

REFERENCES

1. Rationalisation of Safety and Serviceability Factors in Structural Codes. CIRIA Report No. 63.

2. Rackwitz, R. Principles and methods for a practical probabilistic approach to structural safety. Sub-Committee for First Order Reliability Concepts for Design Codes of the Joint CEB-CECM-CIB-FIP-IABSE Committee on Structural Safety, December 1975 (to be published by the Comite European due Beton 1975).

3. Benjamin, J.R. and C.A. Cornell Probability statistics and decision for civil engineers. McGraw Hill (New York), 1970.

4. Flint and Baker Rationalisation of Safety and Serviceability Factors in Structural Codes. Supplementary Report on Offshore Installations CIRIA Oct. 1976.

DEEP WATER RESONANCE PROBLEMS IN THE MOORING SYSTEM OF THE TETHERED PLATFORM

E.M.Q. Røren and B. Steinsvik

Aker Engineering A/S, Oslo, Norway.

INTRODUCTION

The trend in offshore exploration drilling is to move into deeper and deeper water. As a result of this achievement several deep water production concepts have been put forward lately. One of these new concepts is the tethered or vertically moored platform. At the moment several oil companies are seriously considering the tethered platform for the development of offshore oil and gas fields in deep water.

However, when moving into very deep water the tethered platform will be faced with some of the same problems which any fixed platform will be faced with in deep water.

Apart from technical challenges related to the marine production riser, these deep water problems are associated with resonance at the lower natural periods of the platform system.

These lower natural periods, i.e. natural periods in heave, roll and pitch, have a tendency, in deep water,of moving towards periods at which the wave energy spectra contain a considerable amount of energy.

THE TETHERED PLATFORM

Fig. 1 shows a tethered platform for which the consequences of resonance at periods of 5-6 seconds were studied closely by Aker Engineering.

The platform shown in Fig. 1 is to some degree similar to a semisubmersible with four large vertical columns and two rather small pontoons. In addition to the four large diameter columns, one slender column extends from each pontoon to the deck. Horizontal and vertical trusses give the platform the

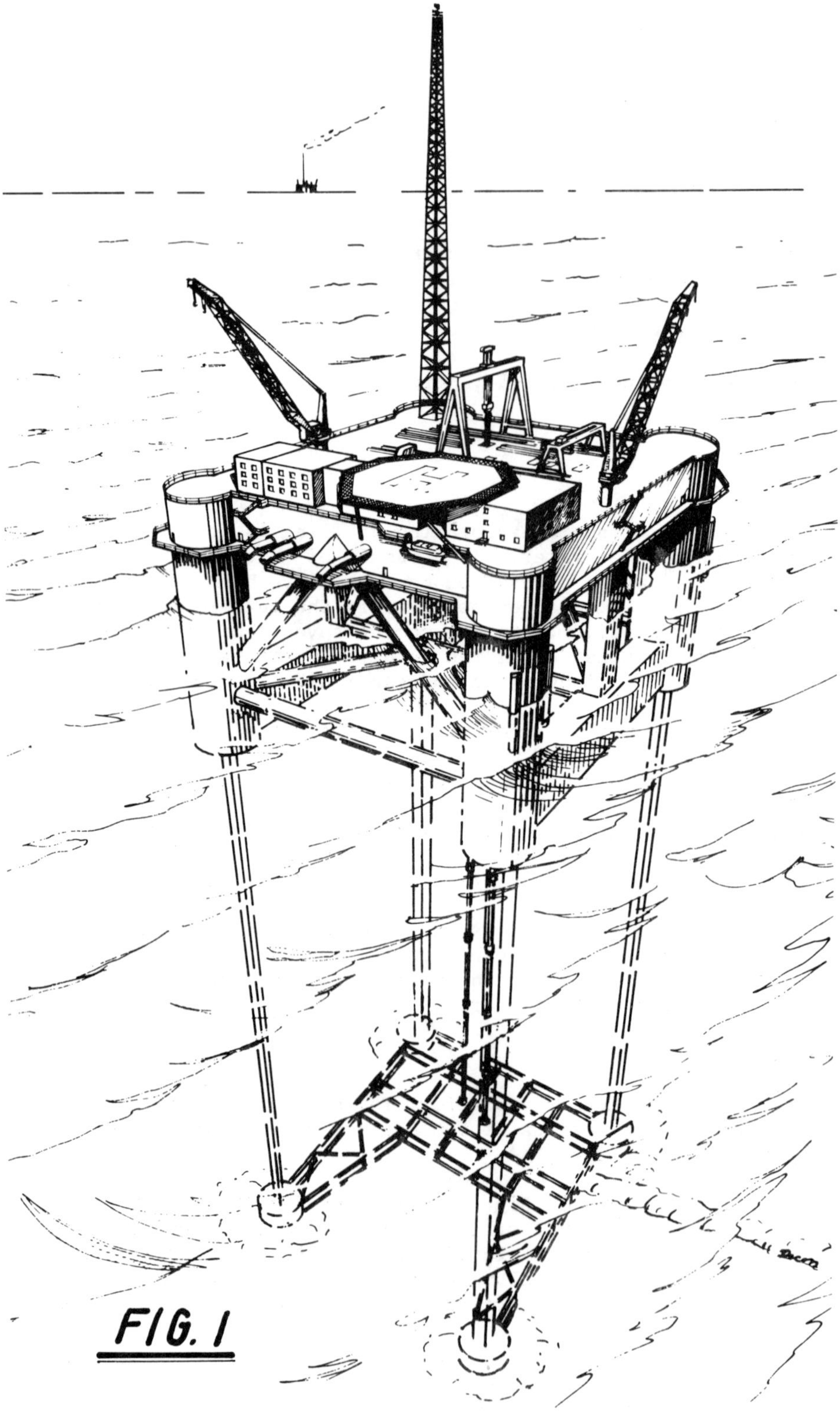

FIG. 1

required structural integrity.

The shown platform has a displacement of 31000 tonnes and a net pay-load capacity of 10000 tonnes and a steel weight of 11000 tonnes. This platform will be moored against vertical motions. This is done by connecting three or more mooring cables to each of the four corner columns as well as to a seafloor anchor. These lines will be given a predetermined tension by deballasting the platform. This mooring system will reduce the heave, roll and pitch motions to virtually zero but will allow for horizontal motion. Any mooring line can be pulled for inspection or replacement as long as the weather condition and service facilities allow for such an operation. The limiting wave height for this operation is 5 metres or Beaufort Condition 6.

As seen from Fig. 1 an integrated deck structure is proposed. This is done in order to obtain a high pay-load capacity. All the heavy processing equipment i.e. separators, turbo-generators, injection pumps, etc. are placed on the lower deck together with some of the permanent drilling equipment. Fuel-oil and potable water will be placed in the four corner columns.

Model tests have been carried out at NSMB in Wageningen for this type of platform. These tests proved that the platform can survive the 100 year design condition for a typical northern North Sea location, both in moored and free-floating mode.

For further reference see Ref. 1 given at the end of the Paper.

The platform used in the analysis later described has the following main particulars:

Operating draft:	31.0 m
Large column diameter:	14.0 m
Pontoon height:	11.0 m
Pontoon breadth:	8.5 m
Lower deck elevation above baseline:	55.8 m
Upper deck elevation above baseline:	63.5 m
Large column centreline distance, longitudinal:	67.2 m
Large column centreline distance, transverse:	64.0
Displacement:	31000 tonnes
Platform mass:	20000 tonnes
Mooring pretension:	8000 tonnes

The extreme wave design condition is as follows:

Significant wave height:	16.0 m
Maximum wave height:	30.0 m

ELEMENTS IN THE MOORING SYSTEM

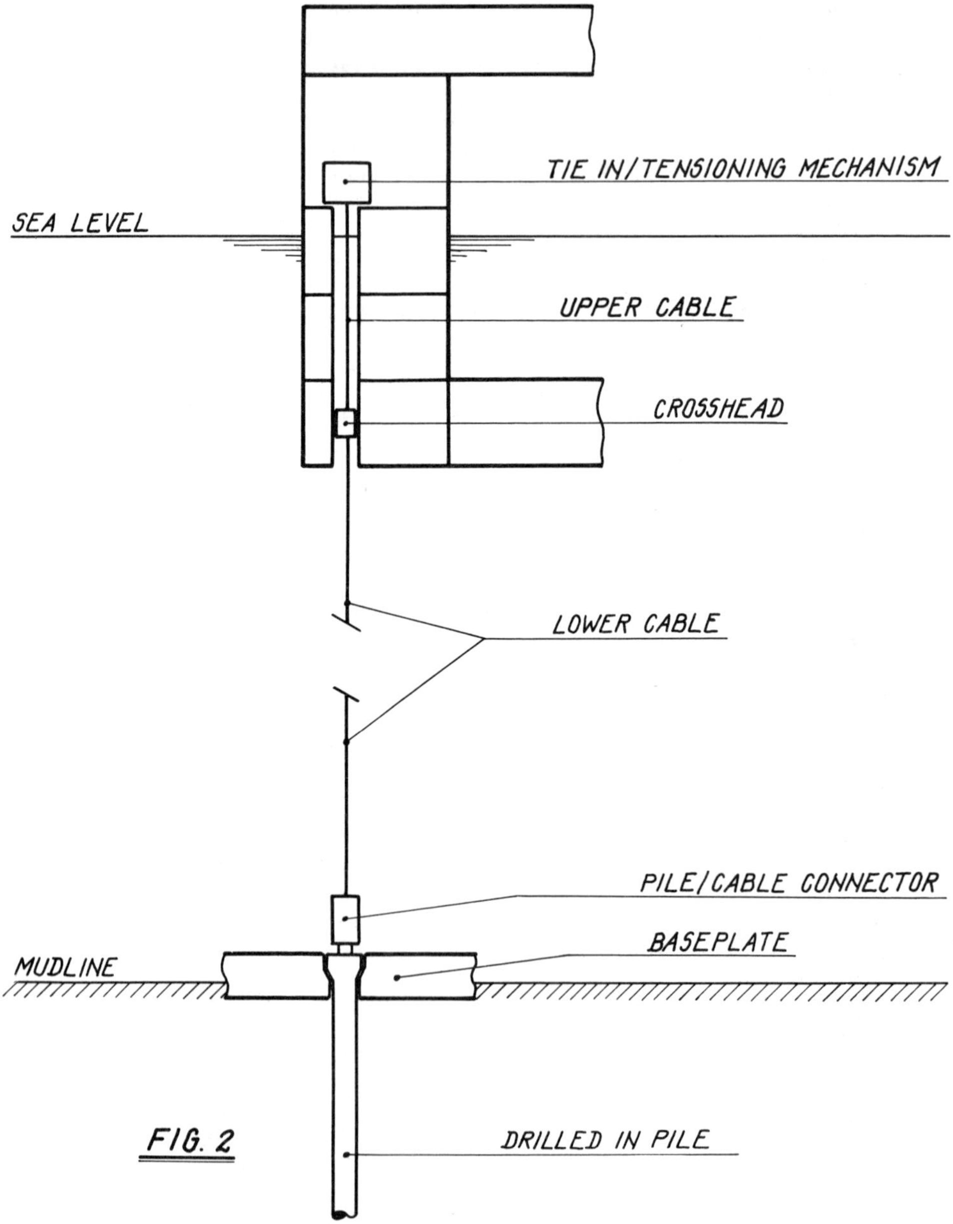

FIG. 2

Zero crossing period: 14.7 sec.

In addition to this design spectrum, wave scatter diagram for the northern North Sea was used in the fatigue analysis.

MOORING SYSTEM

Fig. 2 shows the general lay-out of one mooring line in one platform corner. All other mooring lines have equivalent lay out.

The mooring cables are of parallel wire construction and are fully protected against sea water. The diameter of the wires is 7 mm and the number of wires in each cable can be as high as 400. The wires are protected by a polyethylene duct which is filled by an anticorrosive inhibitor. The reason for selecting this type of cable is its high and constant modulus of elasticity.

In the analysis described later the following mooring system was used:

Water depth:	1000 m
No. of cables per leg:	5
No. of wires per cable:	360
Modulus of elasticity of cables:	0.201 MN/mm
Breaking strength:	1600 N/mm
Steel area per leg:	69272 mm
Longitudinal stiffness per leg:	14 MN/m
Weight of one cable:	110 tonnes
Total weight of cables (20):	2200 tonnes

NATURAL PERIODS OF MOTIONS

The natural periods of the described platform system, neglecting the coupling terms in the mass, added-mass and restoring matrices which have marginal influence on natural periods, are given by:

$$T_n = 2\pi \sqrt{\frac{M_n + A_n}{K_n}}$$

where:

T_n = natural period of the nth degree of freedom
M_n = mass or inertia in the nth degree of freedom
A_n = added mass or inertia in the nth degree of freedom
K_n = restoring stiffness for the nth degree of freedom.

The calculated natural periods of the platform and mooring system described earlier together with an example in 300 m of water are as follows:

	Water Depths 1000 m	300 M
T_1 (Surge)	141.0 cec	76.5 sec
T_2 (Sway)	162.0 sec	86.5 sec
T_3 (Heave)	5.0 sec	2.7 sec
T_4 (Roll)	5.6 sec	3.0 sec
T_5 (Pitch)	4.8 sec	2.6 sec
T_6 (Yaw)	113.0 sec	68.0 sec

The calculated natural periods in surge, sway and yaw are comfortably high. They are far too high to cause any resonance problems due to the lack of energy in the wave energy spectra at these periods.

The natural periods in heave, roll and pitch are in a period band which for most geographical locations have a substantial amount of energy in the wave energy spectra. This could be a very serious problem, and in this situation there are two obvious alternatives to reduce the problem:

1. Increase the mooring stiffness so that the natural periods in heave, roll and pitch are satisfactorily low.
2. Determine the consequence of resonance in heave, roll and pitch with the proper damping.

Regarding alternative 1 it has been argued for a long time what the lower cut-off period should be, i.e. the lowest wave period which is to be considered. Assuming this period to be 3 seconds and that the natural periods in heave, roll and pitch should be below this period, this would require a 250% increase mooring line longitudinal stiffness. In other words, the weight of the mooring cables would increase from 2200 tonnes to 7700 tonnes. This seems to be a very expensive way to solve the problem since this solution reduces the pay-load by about 50%. Therefore, one should take a close look on alternative 2.

THE ANALYSIS

The Equations of Motion

The simplified equations of motion result from equating the external forces on the structure (assumed to be prevented from moving) to the reactive motion forces acting on the structure (which is assumed to be moving in a calm fluid). This results in a second order differential equation in matrix form:

$$(M_1 + A_1)\ddot{n} + (B_1 \frac{8}{3\pi} \dot{n}_a + B_2)\dot{n} + (C_1 + C_2)n = F.\exp(-i\omega t)$$

where:

M_1 - is the structure mass matrix
A_1 - is the added mass matrix
B_1 - is the viscous damping matrix
B_2 - is the potential damping matrix
C_1 - is the hydrostatic restoring matrix
C_2 - is the anchor restoring matrix
$\ddot{n}, \dot{n}, n$ - are the complex vectors of structure acceleration, velocity and displacement
$\dot{n}_a$ - is the real velocity amplitude vector
F - is the complex forcing vector
ω - circular frequency

All the matrices are symmetrical except the viscous damping matrix which is unsymmetrical due to the way the velocity amplitude must be included in the damping matrix.

A special anchor restoring matrix for taut line has been developed. This restoring matrix assumes straight anchor lines and includes restoring due to the change in direction of the pretension force during motion and restoring due to the material stiffness of the mooring lines.

For further reference see [2].

Hydrodynamic Model

The platform structure was divided into two different hydrodynamic models.

The large volume of the structure, i.e. the columns and pontoons, was modelled into a three-dimensional sink-source model consisting of 322 elements. This model was used for all calculations by the computer program NV 459 developed by Det norske Veritas. The computer program NV 459 is based upon diffraction theory, [3]. The remaining part of the structure, i.e. the trusses, was modelled into cylindrical elements ranging in length from 1.6 m to 3.2 m. This model was used for the calculations by the Morison theory computer programs SEARESPONS and NV 407 B. The computer program SEARESPONS was developed within Aker Engineering and solves the equations of motion both in a linear and a nonlinear manner, [2]. NV 407 B was developed by the Det norske Veritas, [4].

Calculation of Transfer Functions

NV 459 was used to compute the added mass and potential damping coefficients for the columns and pontoons. These coefficients were computed at five wave periods. Some of these coefficients were found to be constant while other varied

HEAVE RESPONSE IN REGULAR WAVES

45° HEADING

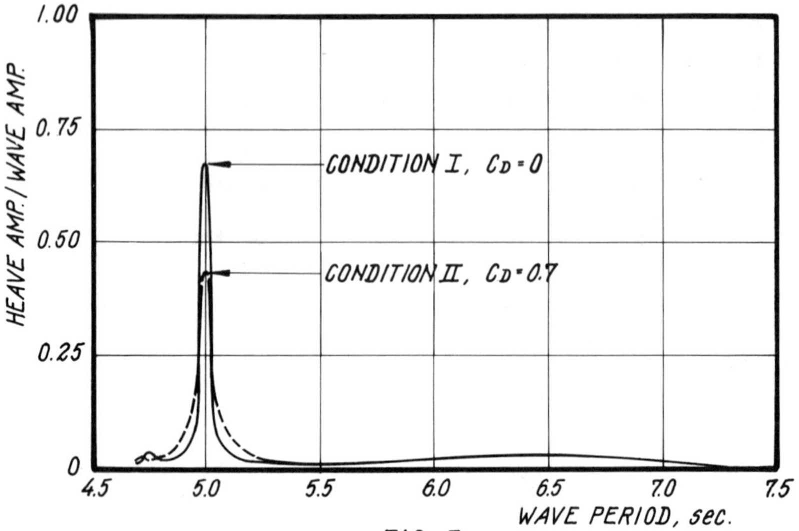

FIG. 3

ROLL RESPONSE IN REGULAR WAVES

45° HEADING

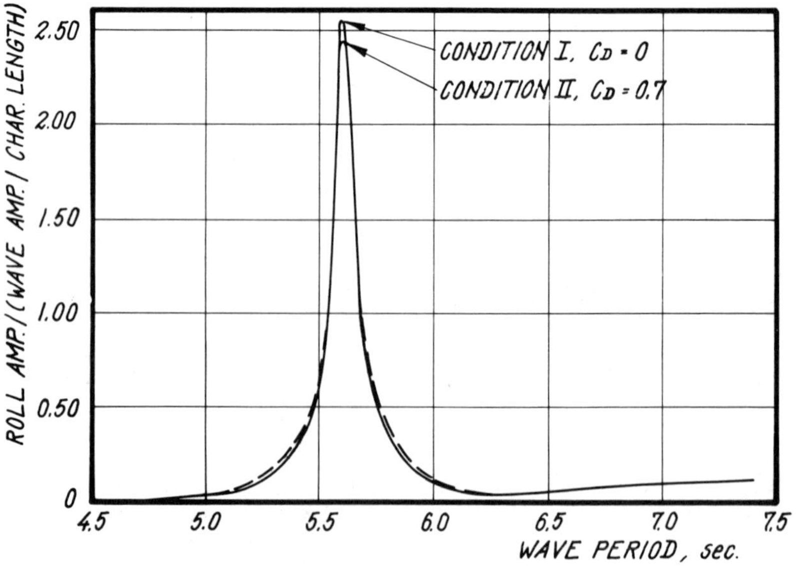

CHARACTERISTIC LENGTH = 67.2 m

FIG. 4

greatly with the wave period. In addition to the added mass and potential damping coefficients, NV 459 computed the exciting forces on the columns and pontoons at different wave headings.

The viscous damping force matrix for the trusses was computed by SEARESPONS. Also the hydrostatic restoring matrix and the anchor restoring matrix was computed by SEARESPONS. NV 407 B was used to compute the added mass matrix and exciting forces for the trusses at 15 wave periods. The exciting forces were computed in two steps:

1: Two-dimensional $C_m = 1$ and $C_d = 0$

2: Two-dimensional $C_m = 0$ and $C_d = 0.7$ *

These computed coefficients/matrices and exciting forces were put together in two conditions, and the equations of motion were solved by the linearized method described in [2]. These two conditions were:

1. Including added mass, potential damping and exciting forces as computed by NV 459 for the columns and pontoons. Added mass and mass exciting forces on the trusses are also included. Viscous damping and exciting forces on the trusses were not included.
2. As condition 1 but including viscous damping and exciting forces.

The resulting equations of motion from these two conditions were solved for 50 wave periods from 4.7 to 7.5 seconds with a wave amplitude of 2.0 m. In this manner two sets of transfer functions for the six degrees of freedom were developed for different wave headings. Figs. 3, 4 and 5 show the computed transfer functions for heave, roll and pitch, respectively for the two conditions. As it can be seen, the introduction of viscous effects on the trusses has reduced the height of the resonance peaks.

These transfer functions were used for computing the dynamic mooring leg force. Fig. 7 shows an example of the leg force transfer function.

<u>Spectral Analysis and Long-Term Distribution of Mooring Forces</u>

An inhouse computer program, SEASPECTRA, was used to calculate the mooring leg force in irregular waves. In these calculations the two-parameter Pierson-Moskowitz wave energy spectrum was used. Fig. 6 shows the mooring leg force in irregular waves for one of the four mooring legs in 45 degrees wave heading. Similar transfer functions were calculated for all

* A C_d-value of 0.7 is selected according to Ref. 5.

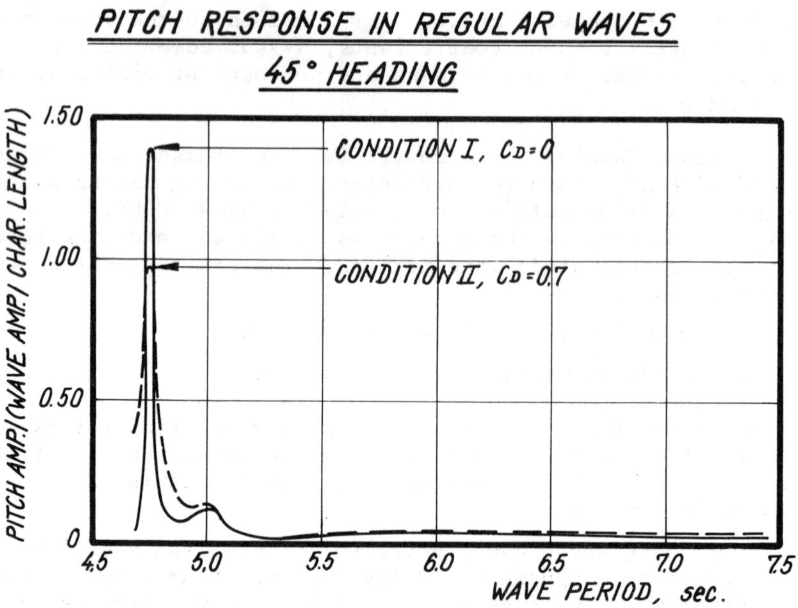

CHARACTERISTIC LENGTH = 67.2 m

FIG. 5

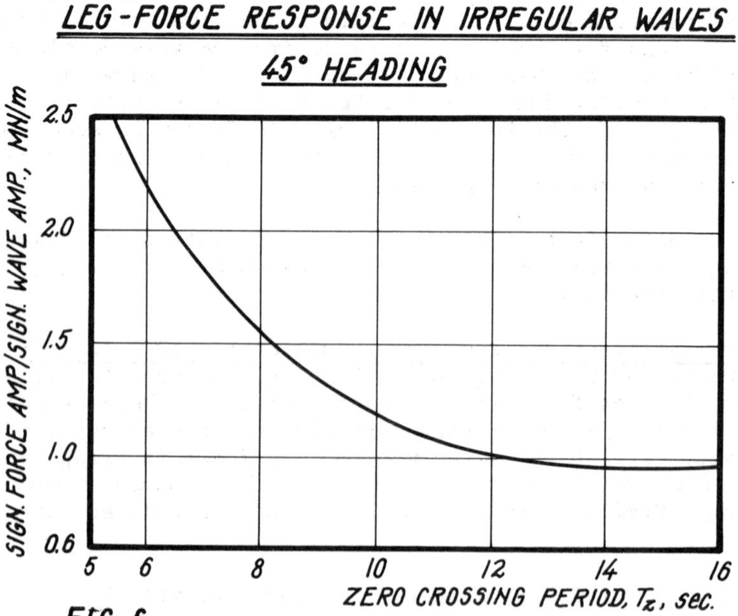

FIG. 6

four legs for eight wave headings. The term "Significant Force Amplitude" used in Fig. 6, is analogous to the significant wave amplitude and is equal to two times the root mean square value of the response spectrum.

These transfer functions in irregular seas together with a wave scatter diagram were used to calculate the long-term distribution of individual dynamic mooring forces. The computer program SEASPECTRA was used for these calculations also. Based upon a 20 year stress history the long-term analysis resulted in the following stress blocks:

i	sigma	n_i
1	575	9×10^0
2	496	9×10^1
3	416	9×10^2
4	340	9×10^3
5	262	9×10^4
6	186	9×10^5
7	110	9×10^6
8	37	9×10^7

where:

sigma = double amplitude stress (N/mm^2) in stress block i

n_i = number of cycles in stress block i

Fatigue Analysis

Figs. 8 and 9 show four different S-N curves. Curve Ia and IIa are according to DIN 1073 (Deutsche Industri Normen) and AWS (American Welding Society), respectively, and are valid for cables in air. Curves Ib and IIb are the same curves but modified to be valid for cables in sea water. In our opinion curve Ia is the most correct to use since it is supported by a large number of fatigue tests for this type of cable and since the cable is fully protected against sea water. These S-N curves together with the calculated stress blocks were used to calculate Miner's cumulative usage factors. The result of these calculations was as follows:

S-N Curve	Usage factor	Fatigue life (years)
Ia	1.252	16.0
Ib	2.746	7.3
IIa	1.853	10.8
IIb	2.272	8.8

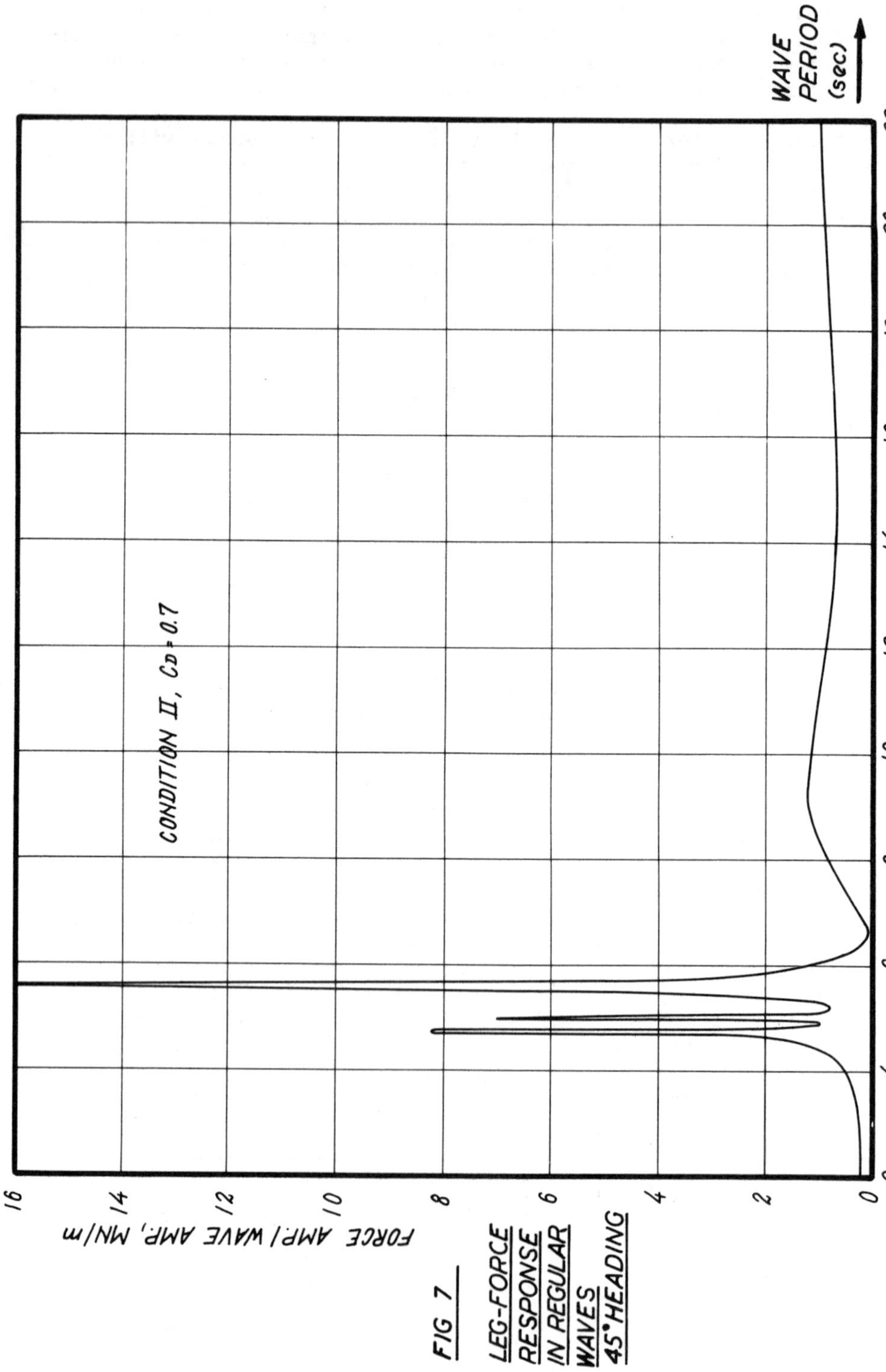

FIG 7

LEG-FORCE RESPONSE IN REGULAR WAVES 45° HEADING

S-N - CURVES

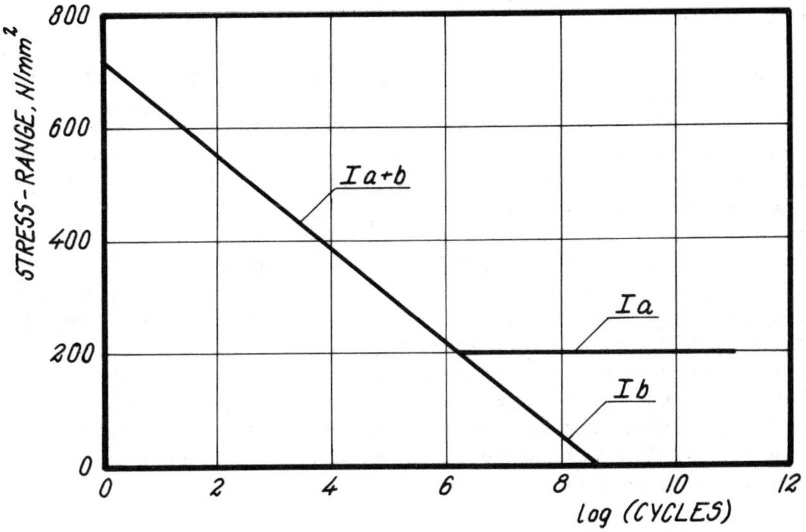

FIG. 8

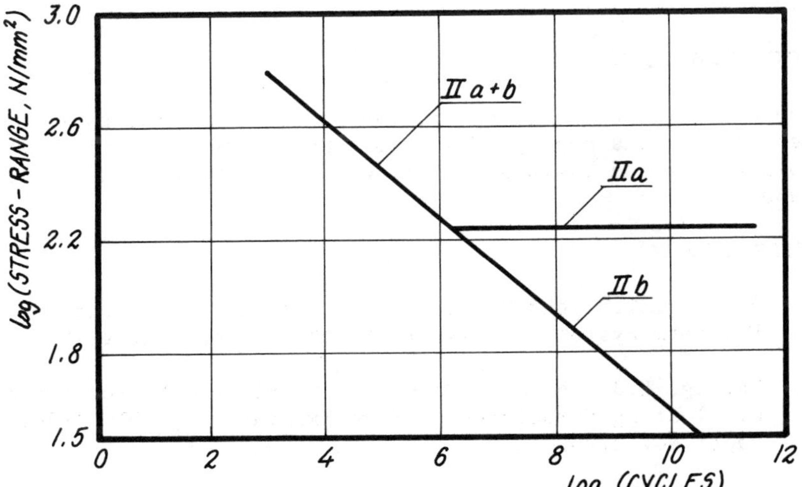

FIG. 9

Earlier fatigue calculations, where resonance has not been a problem, have resulted in fatigue lives of about 100 years. The presence of the resonance peaks have reduced the fatigue life drastically.

CONCLUSIONS

This paper has outlined a linearized method of determining the mooring force resonance peaks in the mooring system of a tethered platform by combining diffraction and Morison theory. Transfer functions for the dynamic mooring force have been developed and through spectral analysis, long-term statistics and fatigue analysis the fatigue life has finally been established. The calculated resonance peaks have a great impact on the fatigue life of the mooring cables, however, by increasing the steel area of the mooring cables slightly the fatigue life will be increased considerably. The introduction of viscous effects from the trusses has reduced the resonance peaks and by introducing other sources of damping like viscous effects from the columns and pontoons, internal damping within the structure and damping in the seafloor foundation the resonance peaks would be further decreased and hence fatigue life improved.

The deep water resonance problem in the mooring system can be solved in a satisfactory manner to give an adequate fatigue life, but one has to be very careful throughout all evaluations.

ACKNOWLEDGEMENT

The authors wish to thank Mr. Jan Vugts, Shell International Petroleum Maatschappij, The Hague, Netherlands for his valuable advice and assistance during the work which led to this paper.

REFERENCES

1. Addison, G.D. and B. Steinsvik, "Tethered Production Platform System", Offshore North Sea Conference 1976.

2. Natvig, B.J. and J.W. Pendred, "Nonlinear Motion Response of Floating Structures to Wave Excitation", OTC 1976, Paper No. 2796.

3. Løken, A.E. and O.A. Olsen, "Diffraction Theory and Statistical Methods to Predict Wave Induced Motions and Loads for Large Structures" OTC 1976, Paper No. 2502.

4. Langfeldt, J.H., V. Egeland and S. Gran, "NV 407 B Motion and Loads for Drilling Platform User's Manual", Det Norske Veritas Report No. 76-63-S.

5. Olsen, O.A. "Investigation of Drag Coefficient Based on Published Literature", Det Norske Veritas Report No. 74-2-5

APPLICATIONS OF BOUNDARY ELEMENTS IN THE ANALYSIS OF OFFSHORE STRUCTURES

C. Brebbia and R. Nakaguma

Southampton University England and IPT, Sao Paulo - Brazil

INTRODUCTION

In recent years analysts have become aware of the limitation of finite elements, specially their inability to model infinite domains and produce accurate results for stress concentration problems. In addition the method is expensive for three dimensional problems. These shortcomings have prompted some researchers [1,2] to investigate the applications of boundary solutions for elastostatic problems. Considerable work in this method has been carried out at Southampton University where the technique was developed into what is now called the boundary element method [3,4,5,6]. The method is very well suited to solve three dimensional problems as it reduces the problem to a boundary solution, i.e. only elements on the boundary need to be defined. By contrast, finite elements require many interior nodes, and approximations involving discontinuity of stresses are generally made. The method is also of great interest for soil-structure interaction problems, because it permits the analysis of infinite and semi-infinite domains, such as the soil.

Further still, boundary element techniques can be shown to be a special case of weighted residuals [7] and as such it is not difficult to relate to finite elements as other 'domain' methods. Hence boundary elements can be easily combined with finite elements in general purpose programs. In this paper the boundary element technique is described in terms of weighted residuals and matrix algebra for the three dimensional elastostatic case. The development of different types of boundary elements is described but the applications are restricted to constant elements, i.e. elements on which the displacements and surface tractions are assumed to be constant.

Boundary elements can be successfully applied to analyse some of the problems present in offshore structures. The first two applications presented in this paper are meant to show the accuracy of the technique by comparing numerical results with known solutions. The last two examples illustrate how the technique can be used for solving some complex offshore structural problems. The method is applied to study stress concentrations at the intersection of cylindrical shells of the type used for oil tanks and to determine the state of stress at the joint of tubular members.

The simplicity of the theoretical development is emphasized throughout together with the accuracy of the results. Because of this and the possibility of solving a large range of problems which can not be accurately analysed using finite elements, the boundary element technique offers an interesting alternative. In addition it is possible to combine both methods.

BASIC RELATIONSHIPS

The principle of virtual displacements for linear elastic problems can be written as,

$$\int_\Omega (\sigma_{jk,j} + b_k) u_k^* \, d\Omega = \int_{\Gamma_2} (p_k - \bar{p}_k) u_k^* \, d\Gamma \qquad (1)$$

$$(\ )_{,j} = \frac{\partial (\ )}{\partial x_j}$$

where u_k^* are the virtual displacements identically satisfying the homogeneous boundary conditions $\bar{u}_k^* \equiv 0$ on Γ_1. If we now interpret u_k^* as a weighting function which does not satisfy these conditions on Γ_1 the expression can be written as,

$$\int_\Omega (\sigma_{jk,j} + b_k) u_k^* \, d\Omega = \int_{\Gamma_2} (p_k - \bar{p}_k) u_k^* \, d\Gamma + \int_{\Gamma_1} (\bar{u}_k - u_k) p_k^* \, d\Gamma \qquad (2)$$

where

$p_k^* = \alpha_{nj} \sigma_{jk}^*$ are the surface forces or tractions corresponding to the u_k^* system.

We will assume that the strain-displacement relationships are linear, i.e.

$$\varepsilon_{ij} = \frac{1}{2}\left(\frac{\partial u_i}{\partial x_j} + \frac{\partial u_j}{\partial x_i}\right) = \frac{1}{2}(u_{i,j} + u_{j,i})$$

$$\text{and} \quad \varepsilon_{ij}^* = \frac{1}{2}\left(\frac{\partial u_i^*}{\partial x_j} + \frac{\partial u_j^*}{\partial x_i}\right) = \frac{1}{2}(u_{i,j}^* + u_{j,i}^*) \qquad (3)$$

and that the material properties are also linear, i.e. in tensor form,

$$\sigma_{ij} = d_{ij}^{k\ell}\, \varepsilon_{k\ell} \quad , \quad \sigma^*_{ij} = d_{ij}^{k\ell}\, \varepsilon^*_{k\ell} \tag{4}$$

Hence we can now integrate (2) by parts which gives,

$$\int_\Omega b_k\, u^*_k\, d\Omega - \int_\Omega \sigma_{jk}\, \varepsilon^*_{jk}\, d\Omega = - \int_{\Gamma_2} \bar{p}_k\, u^*_k\, d\Gamma - \int_{\Gamma_1} p_k\, u^*_k\, d\Gamma + \int_{\Gamma_1} (\bar{u}_k - u_k)\, p^*_k\, d\Gamma$$

Integrating by parts once more one obtains,

$$\int_\Omega b_k\, u^*_k\, d\Omega + \int_\Omega \sigma^*_{jk,j}\, u_k\, d\Omega = - \int_{\Gamma_2} \bar{p}_k\, u^*_k\, d\Gamma - \int_{\Gamma_1} p_k\, u^*_k\, d\Gamma + \int_{\Gamma_1} \bar{u}_k\, p^*_k\, d\Gamma + \int_{\Gamma_2} u_k\, p^*_k\, d\Gamma \tag{5}$$

One now looks for fundamental solutions satisfying the equilibrium equations, usually of the type,

$$\sigma^*_{jk,j} + \Delta^i = 0 \tag{6}$$

where Δ^i is the Dirac delta function and represents a unit load at the internal point i in the ℓ direction. This type of solution will produce for each direction 'ℓ' the following equation,

$$u^i_\ell + \int_{\Gamma_1} \bar{u}_k\, p^*_k\, d\Gamma + \int_{\Gamma_2} u_k\, p^*_k\, d\Gamma = \int_\Omega b_k\, u^*_k\, d\Omega + \int_{\Gamma_1} p_k\, u^*_k\, d\Gamma + \int_{\Gamma_2} \bar{p}_k\, u^*_k\, d\Gamma \tag{7}$$

u_ℓ : represents the displacement at i in the 'ℓ' direction.

In general we can write for the point 'i'

$$u_\ell^i + \int_\Gamma u_k \, p_k^* \, d\Gamma = \int_\Gamma p_k \, u_k^* \, d\Gamma + \int_\Omega b_k \, u_k^* \, d\Omega \qquad (8)$$

where $\Gamma = \Gamma_1 + \Gamma_2$.

Note that u_k^* and p_k^* are the fundamental solutions, i.e. the displacements and tractions due to a unit concentrated load at the point 'i' in the 'ℓ' direction. If we consider unit forces acting in the three directions, expression (8) can be written as,

$$u_\ell^i + \int_\Gamma u_k \, p_{\ell k}^* \, d\Gamma = \int_\Gamma p_k \, u_{\ell k}^* + \int_\Omega b_k \, u_{\ell k}^* \, d\Omega \qquad (9)$$

where $p_{\ell k}^*$ and $u_{\ell k}^*$ represent the tractions and displacements in the k direction due to unit forces acting in the direction ℓ. Equation (9) is valid for the particular point 'i' where these forces are applied.

FUNDAMENTAL SOLUTION

The fundamental solution for a three dimensional isotropic body is

$$u_{\ell k}^* = \frac{1}{16\pi G(1-\nu) r}\left[(3-4\nu)\Delta_{\ell k} + r_{,\ell}\, r_{,k}\right] \qquad (10)$$

$$p_{\ell k}^* = -\frac{1}{8\pi(1-\nu) r^2}\left[\frac{\partial r}{\partial n}\{(1-2\nu)\Delta_{\ell k} + 3r_{,\ell}\, r_{,k}\} - (1-2\nu)\{r_{,\ell}\,\alpha_{nk} - r_{,k}\,\alpha_{n\ell}\}\right]$$

$$\frac{\partial r}{\partial n} = \sum r_{,i}\, n_i$$

where n is the normal to the surface of the body, $\Delta_{\ell k}$ is the Kronecker delta, r is the distance from the point of application of the load to the point under consideration and α_n are the direction cosines. (see figure 1). Note that

$$r_{,\ell} = \frac{\partial r}{\partial x_\ell} = \frac{x_\ell^{(k)} - x_\ell^{(i)}}{r}$$

(k) refers to the coordinate of point k.

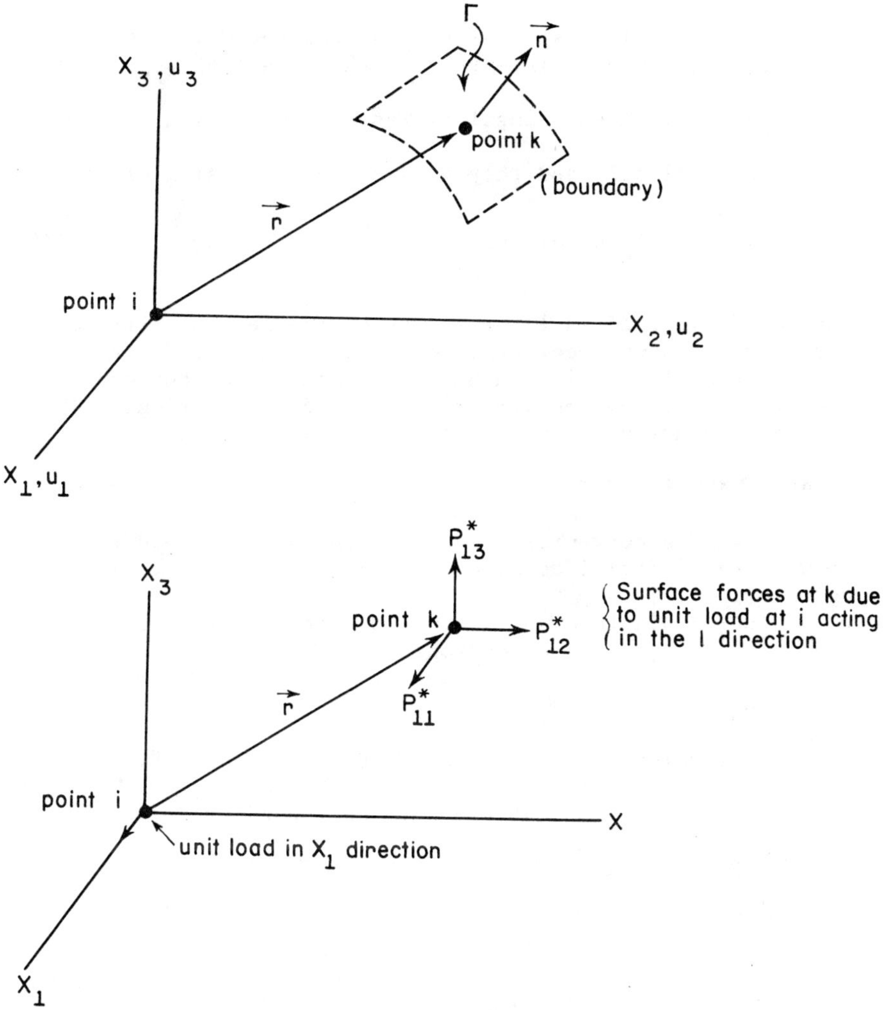

Three dimensional case

Fig. 1

BOUNDARY POINT

When equation (9) is specialised for the boundary a singularity will occur in the integrals of the type $\int_{\Gamma_2} u_k\, p^*_{\ell k}\, d\Gamma$ or $\int_{\Gamma_1} u_k\, p^*_{\ell k}\, d\Gamma$. These singularities can be studied and integrated analytically and they produce an expression of the type,

$$c^i u^i_\ell + \int_\Gamma u_k\, p^*_{\ell k}\, d\Gamma = \int_\Gamma p_k\, u^*_{\ell k}\, d \;+ \int_\Omega b_k\, u^*_{\ell k}\, d\Omega \qquad (11)$$

The coefficient c^i is ½ for smooth boundaries but different for boundaries with corners. It is difficult to determine it analytically but it can be easily found once the equations are discretized by applying rigid body considerations. This will be discussed in what follows.

MATRIX FORMULATION

It is now more convenient to work with matrices than to carry on with the indicial notation. We can define the following

$\underset{\sim}{u}^i = \begin{Bmatrix} u_1 \\ u_2 \\ u_3 \end{Bmatrix}$: displacement vector at point 'i' with components $x_1\ x_2\ x_3$ directions

$\underset{\sim}{u}$ = displacement vector at any point on boundary Γ.

$\underset{\sim}{p} = \begin{Bmatrix} p_1 \\ p_2 \\ p_3 \end{Bmatrix}$: tractions at any point on boundary Γ.

$\underset{\sim}{b} = \begin{Bmatrix} b_1 \\ b_2 \\ b_3 \end{Bmatrix}$ = body forces at any point on domain Ω

$\underset{\sim}{p}^* = \begin{bmatrix} p^*_{11} & p^*_{12} & p^*_{13} \\ p^*_{21} & p^*_{22} & p^*_{23} \\ p^*_{31} & p^*_{32} & p^*_{33} \end{bmatrix}$: matrix whose coefficients, $p^*_{\ell k}$, are the forces in k direction due to unit force at 'i' acting in the 'ℓ' direction

$\underset{\sim}{u}^* = \begin{bmatrix} u^*_{11} & u^*_{12} & u^*_{13} \\ u^*_{21} & u^*_{22} & u^*_{23} \\ u^*_{31} & u^*_{32} & u^*_{33} \end{bmatrix}$: matrix whose coefficients $u^*_{\ell k}$ are the displacements in the 'k' direction due to a unit force at 'i' acting in the 'ℓ' direction.

Equation (11) can now be expressed in matrix form as follows,

$$c^i \, \underset{\sim}{u}^i + \int_\Gamma \underset{\sim}{p}^* \, \underset{\sim}{u} \, d\Gamma = \int_\Gamma \underset{\sim}{u}^* \, \underset{\sim}{p} \, d\Gamma + \int_\Omega \underset{\sim}{u}^* \, \underset{\sim}{b} \, d\Omega \qquad (12)$$

We can assume that the boundary is divided into elements (figure 2) and that the u and p functions can be approximated on each element using the following interpolation functions,

$$\underset{\sim}{u} = \underset{\sim}{\Phi}^T \underset{\sim}{u}^n = \begin{bmatrix} \underset{\sim}{\Phi}^T & \cdot & \cdot \\ \cdot & \underset{\sim}{\Phi}^T & \cdot \\ \cdot & \cdot & \underset{\sim}{\Phi}^T \end{bmatrix} \underset{\sim}{u}^n$$

$$\underset{\sim}{p} = \underset{\sim}{\Phi}^T \underset{\sim}{p}^n = \begin{bmatrix} \underset{\sim}{\Phi}^T & \cdot & \cdot \\ \cdot & \underset{\sim}{\Phi}^T & \cdot \\ \cdot & \cdot & \underset{\sim}{\Phi}^T \end{bmatrix} \qquad (13)$$

Note that we have assumed the same functions for $\underset{\sim}{u}$ and $\underset{\sim}{p}$. In general they may not be the same and it may be more consistent to take the functions for $\underset{\sim}{p}$ of one order less than those for $\underset{\sim}{u}$.

The Φ functions can be considered as the standard two dimensional finite element type functions. (see reference [8]). $\underset{\sim}{u}^n$ and $\underset{\sim}{p}^n$ are the nodal displacements and tractions.

We can now substitute those functions into (12) to obtain for a particular nodal point,

$$\underset{\sim}{c}^i \, \underset{\sim}{u}^i + \sum_{j=1}^{n} \left\{ \int_{\Gamma_j} \underset{\sim}{p}^* \, \underset{\sim}{\Phi}^T \, d\Gamma \right\} \underset{\sim}{u}^n = \sum_{j=1}^{n} \left\{ \int_{\Gamma_j} \underset{\sim}{u}^* \, \underset{\sim}{\Phi}^T \, d\Gamma \right\} \underset{\sim}{p}^n$$

$$+ \sum_{s=1}^{m} \left\{ \int_{\Omega_s} \underset{\sim}{u}^* \, \underset{\sim}{b} \, d\Omega \right\} \qquad (14)$$

$\sum$ j=1 to n indicates summation over the n elements of the surface

Γ_j is the surface of 'j' element.

Note that we have also considered that the value was divided into m internal cells or elements over which integrals corresponding to the body forces have to be computed. These are not

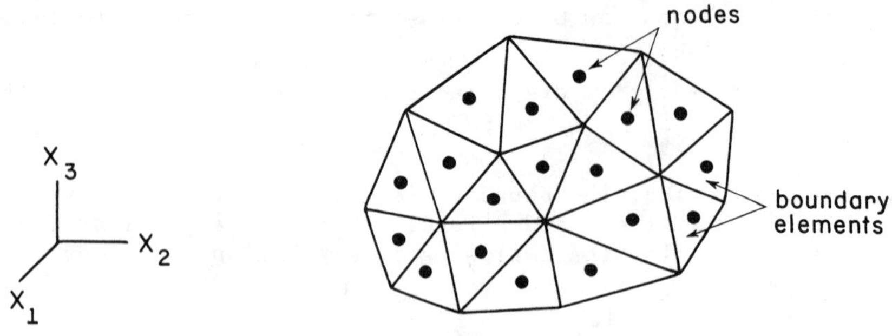

3 dimensional body divided into constant boundary elements

Fig.2a

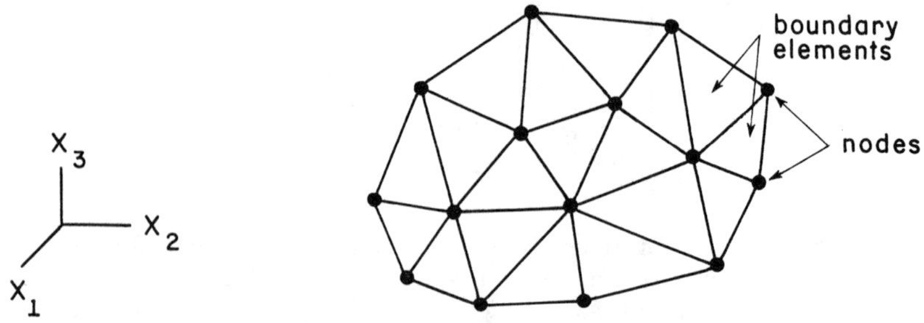

3 dimensional body divided into linear boundary elements

Fig.2b

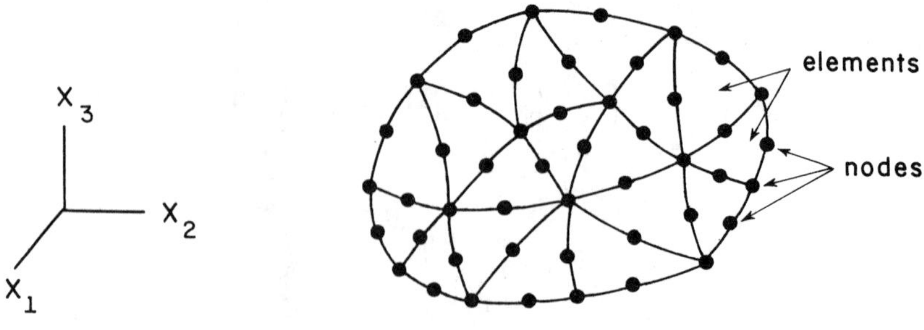

3 dimensional body divided into quadratic boundary elements

Fig.2c

finite elements but simply regions where the integration (usually numerical) is carried out. Once this is done the problem is reduced to a boundary problem.

The integrals are generally solved numerically and the functions Φ expressed in some homogeneous system of coordinates such as the ξ_i system of figure 3. The coordinates need to be transformed from the ξ_i system to the global x_i system.

It is usual now to apply numerical integration which implies that the integrals in equation (14) are replaced by summations, such that

$$\underset{\sim}{c}^i \underset{\sim}{u}^i + \sum_{j=1}^{n} \left\{ \sum_{k=1}^{\ell} |G| \, w_k \, \underset{\sim}{p}^* \, \underset{\sim}{\phi}^T \right\} \underset{\sim}{u}_j = \sum_{j=1}^{n} \left\{ \sum_{k=1}^{\ell} |G| \, w_k \, \underset{\sim}{u}^* \, \underset{\sim}{\phi}^T \right\} \underset{\sim}{p}_j$$

$$+ \int_{\Omega} \underset{\sim}{u}^* \, \underset{\sim}{b} \, d\Omega \qquad (15)$$

ℓ : number of integration points
w_k : weight at the integration points, p_i^* is p* function at the integration points
u_i : is u* function at the integration points

Note that the body force term needs also to be integrated numerically but the integral is carried out in the domain and not on the boundary. Note that in addition to boundary elements it is convenient to define internal cells (figure 4) for integration of the body forces. These cells are used only for the numerical integration of the body force terms and should not be confused with finite elements. If there are m of these cells we can write

$$\int_{\Omega} \underset{\sim}{u}^* \, \underset{\sim}{b} \, d\Omega = \sum_{s=1}^{m} \left\{ \sum_{p=1}^{P} \underset{\sim}{u}^* \, \underset{\sim}{b})_p \, w_p \right\} V_s = \underset{\sim}{b}^i \qquad (16)$$

where w_p are the weighting coefficients for the numerical

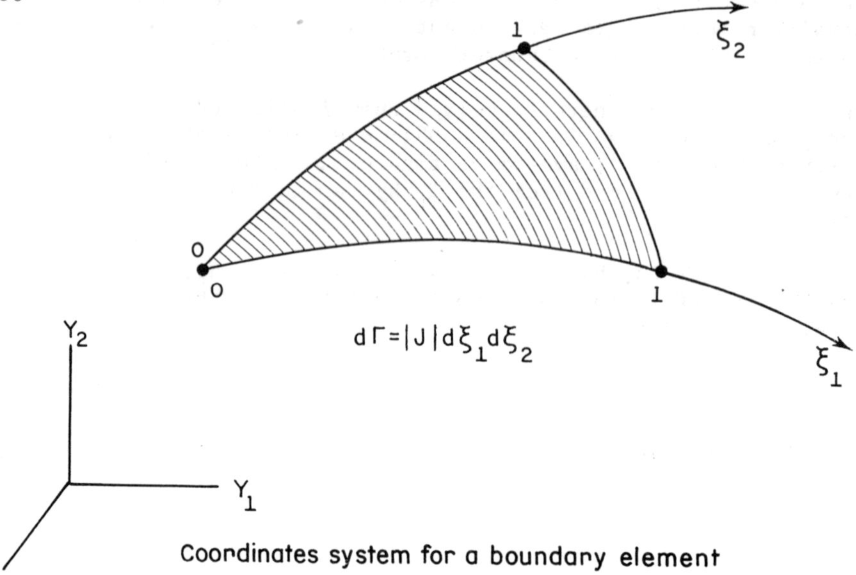

Coordinates system for a boundary element
Fig.3

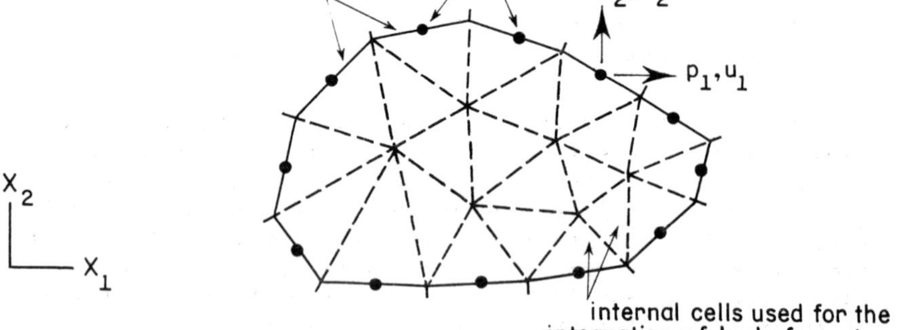

Two dimensional body divided into boundary elements and internal cells
Fig.4

integration and V_s is the volume of the cell under consideration. Evaluation of the body force terms produces vector b^j corresponding to the 'j' element. A quintic triangular integration formula (see |8|) was used throughout.

Equation (15) corresponds to a particular node 'i'. The terms in p* and u* relate the 'i' node with the element 'j' over which the integral is carried out. We call these integrals $\hat{\underset{\sim}{H}}_{ij}$ and $\underset{\sim}{G}_{ij}$ and they are now 3 × 3 matrices. Hence one has,

$$\underset{\sim}{c}^i \underset{\sim}{u}^i + \sum_{j=1}^{n} \hat{\underset{\sim}{H}}_{ij} \underset{\sim}{u}_j = \sum_{j=1}^{n} \underset{\sim}{G}_{ij} \underset{\sim}{p}_j + \underset{\sim}{b}^i \tag{17}$$

This equation relates the value of $\underset{\sim}{u}$ at node 'i' with the values of $\underset{\sim}{u}$'s and $\underset{\sim}{p}$'s at all the nodes on the boundary, including 'i'.

One can write equation (17) for each 'i' node obtaining 2 × n equations where n is the total number of nodes. Let us now call

$$\underset{\sim}{H}_{ij} = \hat{\underset{\sim}{H}}_{ij} \qquad \text{when } i \neq j \tag{18}$$

$$\underset{\sim}{H}_{ij} = \hat{\underset{\sim}{H}}_{ij} + \underset{\sim}{c}^i \qquad \text{when } i = j \tag{19}$$

In general we can write for the node 'i'

$$\sum_{j=1}^{n} \underset{\sim}{H}_{ij} \underset{\sim}{u}_j = \sum_{j=1}^{n} \underset{\sim}{G}_{ij} \underset{\sim}{p}_j + \underset{\sim}{b}^i \tag{20}$$

SYSTEM OF EQUATIONS

Numerical evaluation of equations (15) will produce a system of equations for the node under consideration. Repeating for all the nodes gives a final system of equations that can be written as,

$$\underset{\sim}{H}\,\underset{\sim}{U} = \underset{\sim}{G}\,\underset{\sim}{P} + \underset{\sim}{B} \tag{21}$$

where $\underset{\sim}{U}$ are the displacements and $\underset{\sim}{P}$ the values that the distributed tractions take at all the boundary nodes. $\underset{\sim}{B}$ contains the body forces.

It is important to point out that the diagonal coefficients in the H matrix can be obtained by applying rigid body conditions. If we assume to have unit rigid body displacements in all directions equation (21) becomes,

$$\underset{\sim}{H}\,\underset{\sim}{I} = \underset{\sim}{0} \tag{22}$$

where $\underset{\sim}{I}$ is a unit vector. Hence the diagonal terms of $\underset{\sim}{H}$ are simply,

$$h_{ii} = - \sum_{i \neq j} h_{ij} \tag{23}$$

Note that n_1 values of displacements and n_2 values of tractions are known, hence one has a set of n unknowns in the above equation. Reordering the equations i.e. with the unknowns on the left hand side vector $\underset{\sim}{X}$, we obtain,

$$\underset{\sim}{A}\,\underset{\sim}{X} = \underset{\sim}{F} + \underset{\sim}{B} \tag{24}$$

where $\underset{\sim}{X}$ contains the unknown displacements and tractions.

INTERNAL POINTS

Once the boundary values are known we can compute the internal values of displacements and stresses. The displacements at a point are given by

$$u^i = \int_\Gamma \underset{\sim}{u}^* \underset{\sim}{p}\, d\Gamma - \int_\Gamma \underset{\sim}{p}^* \underset{\sim}{u}\, d\Gamma \tag{25}$$

or for 'ℓ' component

$$u_\ell = \int_\Gamma u^*_{\ell k}\, p_k\, d\Gamma - \int_\Gamma p^*_{\ell k}\, u_k\, d\Gamma \tag{26}$$

For an isotropic medium the stresses can now be calculated by differentiating u at the internal point, i.e.

$$\sigma_{ij} = \frac{2G\nu}{1-2\nu}\, \Delta_{ij} \frac{\partial u_\ell}{\partial x_\ell} + G \left(\frac{\partial u_i}{\partial x_j} + \frac{\partial u_j}{\partial x_i} \right) \tag{27}$$

After carrying out the differentiation we can obtain,

$$\sigma_{ij} = \int_{\Gamma} \left\{ \frac{2G\nu}{1-2\nu} \Delta_{ij} \frac{\partial u^*_{\ell k}}{\partial x_\ell} + G \left(\frac{\partial u^*_{ik}}{\partial x_j} \quad \frac{\partial u^*_{jk}}{\partial x_i} \right) \right\} p_k \, d\Gamma$$

$$- \int_{\Gamma} \left\{ \frac{2G\nu}{1-2\nu} \Delta_{ij} \frac{\partial p^*_{\ell k}}{\partial x} + G \left(\frac{\partial p^*_{ik}}{\partial x_j} + \frac{\partial p^*_{jk}}{\partial x_i} \right) \right\} u_k \, d\Gamma \tag{28}$$

All derivatives are taken at the internal points.

We can reduce this expression to

$$\sigma_{ij} = \int_{\Gamma} D_{kij} \, p_k \, d\Gamma - \int_{\Gamma} S_{kij} \, u_k \, d\Gamma \tag{29}$$

where the third order tensor components D_{kij} and S_{kij} are

$$D_{kij} = \frac{1}{r^\alpha} \left\{ (1-2\nu)\{\Delta_{ki}\, r_{,j} + \Delta_{kj}\, r_{,i} - \Delta_{ij}\, r_{,k}\} + \right.$$

$$\left. + \beta \, r_{,i} \, r_{,j} \, r_{,k} \right\} \frac{1}{4\alpha\pi(1-\nu)} \tag{30}$$

$$S_{kij} = \frac{2\mu}{r^\beta} \left\{ \beta \frac{\partial r}{\partial n} \left[(1-2\nu)\Delta_{ij}\, r_{,k} + \nu(\Delta_{ij}\, r_{,j} + \Delta_{jk}\, r_{,i}) \right.\right.$$

$$\left. - \gamma \, r_{,i} \, r_{,j} \, r_{,k} \right] + \beta\nu(\alpha_{ni}\, r_{,j}\, r_{,k} + \alpha_{nj}\, r_{,i}\, r_{,k})$$

$$+ (1-2\nu)(\beta\alpha_{nk}\, r_{,i} + \alpha_{nj}\, \Delta_{ij} + \alpha_{ni}\, \Delta_{jk})$$

$$\left. - (1-4\nu)\alpha_{nk}\, \Delta_{ij} \right\} \frac{1}{4\alpha\pi(1-\nu)} \tag{31}$$

The above formulae apply for two and three dimensional cases, i.e.

i) For two dimensions $\alpha = 1$; $\beta = 2$; $\gamma = 4$

ii) For three dimensions $\alpha = 2$; $\beta = 3$; $\gamma = 5$.

APPLICATIONS

In what follows some applications of the boundary elements technique for three dimensional elasticity are presented. The surfaces in all cases were discretized using constant triangular elements, i.e. elements for which the displacements and tractions are assumed to be constant.

The examples show the accuracy and versatility of boundary elements as well as their advantages over more classical methods, such as finite elements, for solving some off-shore structural problems.

The applications to be presented are

i) Thick cylinder under internal pressure, including comparison with exact solution
ii) Cylinder under two diametrically opposed loads.
iii) Cylinder intersection problem as found in certain oil storage tanks.
iv) Study of a joint similar to those found in jacketed structures.

i) Thick Cylinder under Internal Pressure

In order to demonstrate the accuracy of the method a thick cylinder under internal pressure was analysed first. The cylinder has the physical characteristics shown in figure 5. The discretization was done using constant triangular elements. Due to symmetry it is necessary to consider only a length of 10 cm (total length is 20 cm) and one quarter of the cross sectional area.

Results for radial displacements are compared in table Ia, where good agreement is shown between the boundary elements and exact solutions [9]. Radial and circumferential stresses are compared in table Ib and Ic.

ii) Cylinder under Two Diametrically Opposed Loads

A way of determining the failure tensional stress for concrete is the test devised by Lobo Carneiro [10]. It consists in compressing a cylindrical specimen with two diametrically opposed distributed loads (for more details see reference [10]). In this analysis the total length of the cylinder was considered to be 5 cm, but only 2.5 cm was analysed as a plane stress problem due to symmetry. The section was discretized as shown in figure 6, using 76 constant boundary elements and the appropriate symmetry conditions.

Results for the normal stresses along the vertical diameter are plotted in figure 7a, where they are compared against the solution applied in reference [10], and although good agreement is shown the boundary element mesh is insufficient to represent

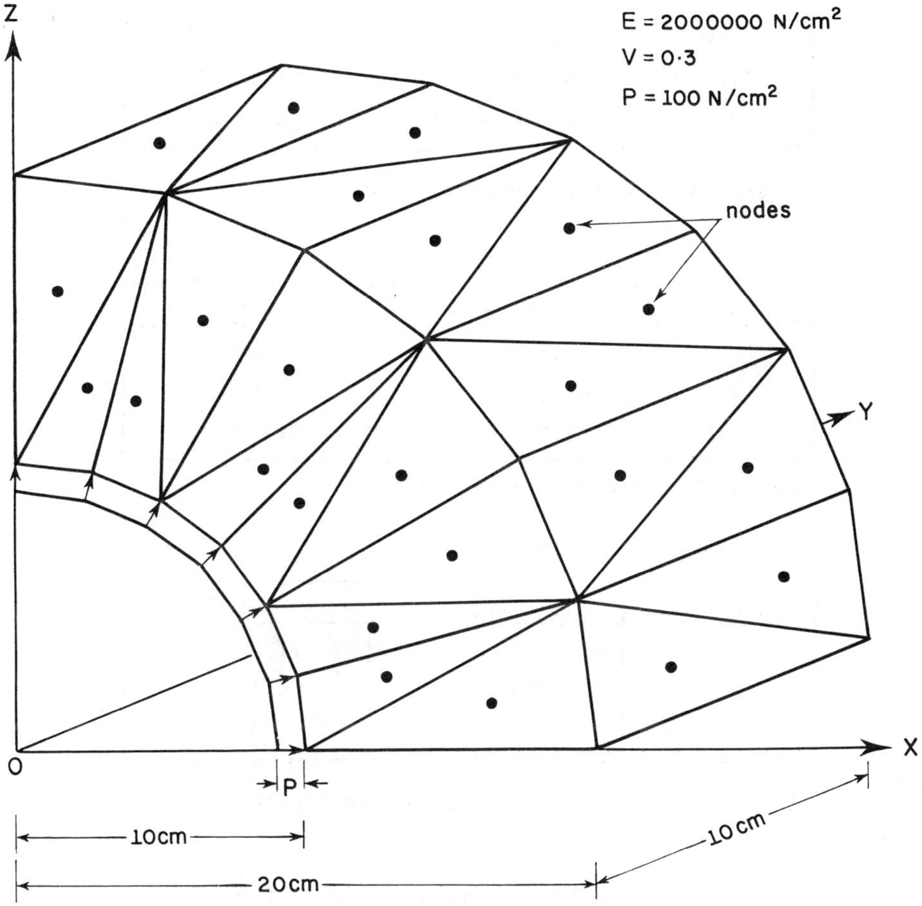

Cylinder under internal pressure
Fig.5

radius (cm)	exact solution	BEM
10.0	9.833×10^{-4}	9.459×10^{-4}
12.5	8.392×10^{-4}	8.563×10^{-4}
15.0	7.528×10^{-4}	7.669×10^{-4}
17.5	6.994×10^{-4}	7.131×10^{-4}
20.0	6.666×10^{-4}	6.456×10^{-4}

Table Ia - Radial Displacement (cm)

radius (cm)	exact solution	BEM
12.5	-5.200	-5.300
15.0	-2.593	-2.559
17.5	-1.020	-0.432

Table Ib - Radial Stress (Kg/cm^2)

radius (cm)	exact solution	BEM
12.5	11.866	11.475
15.0	9.259	9.428
17.5	7.687	7.926

Table Ic - Circumferential Stress (Kg/cm^2)

TABLE I - Thick Cylinder under Internal Pressure. Comparison between exact solutions and Boundary Element Method (BEM), constant element.

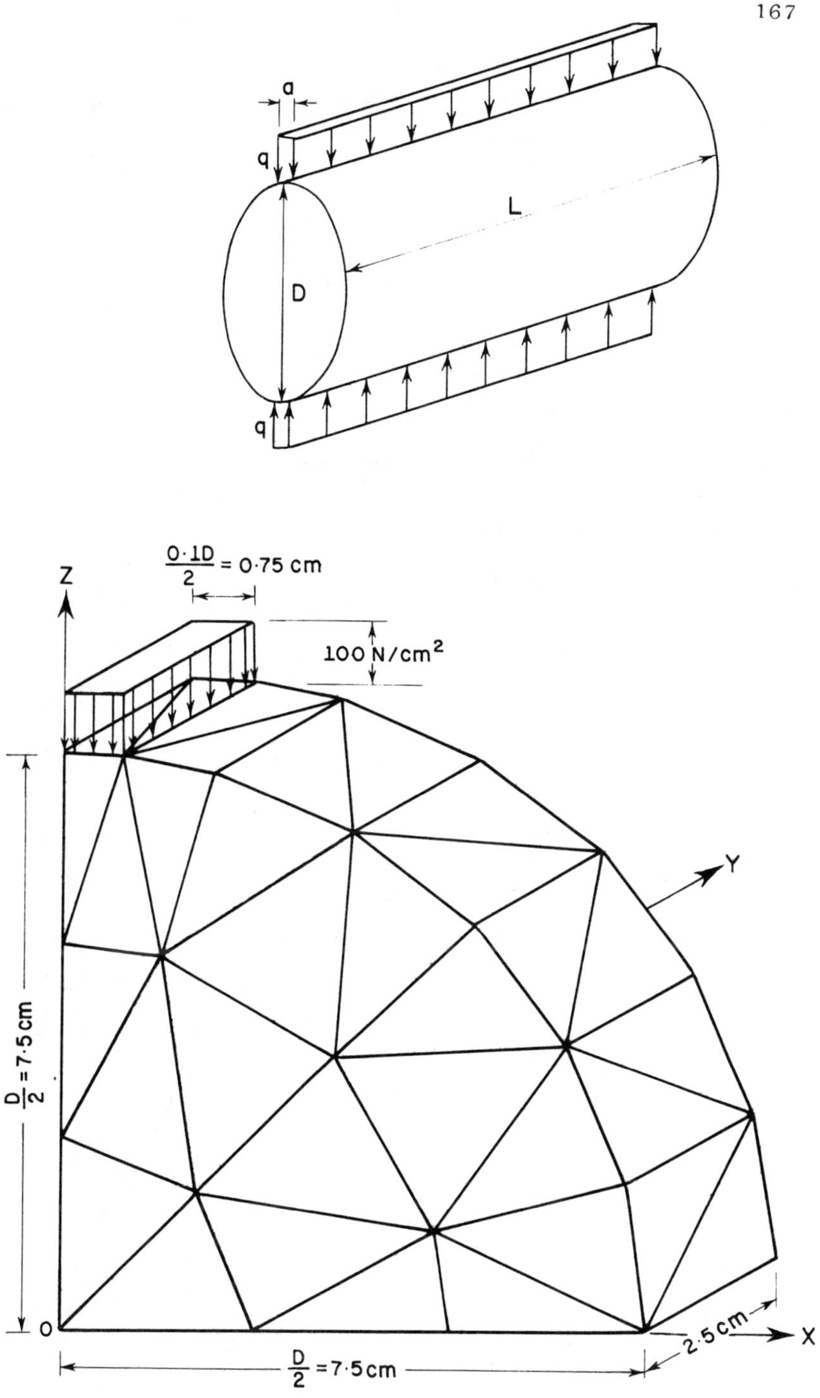

Cylinder under two diametrically opposed loads

Fig. 6

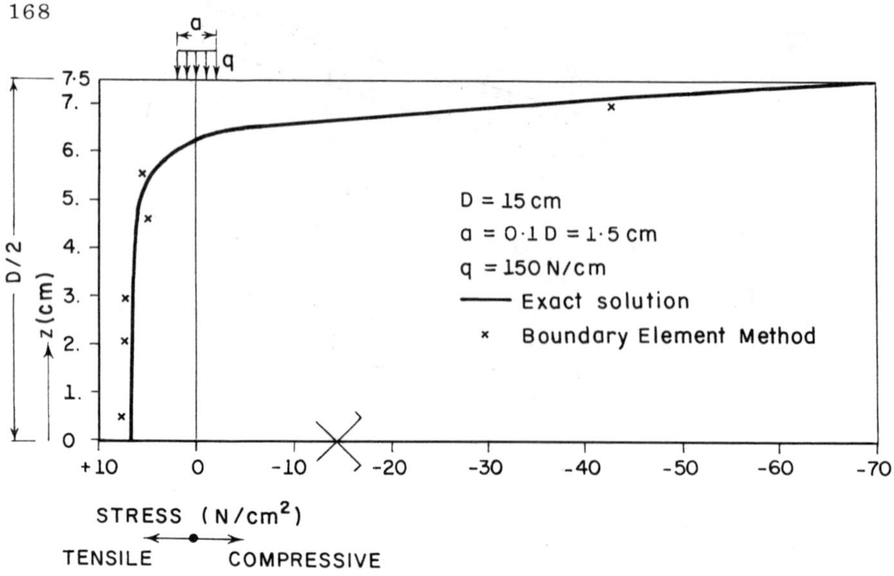

(a) Stresses along the vertical diameter

Results of the cylinder
Fig.7

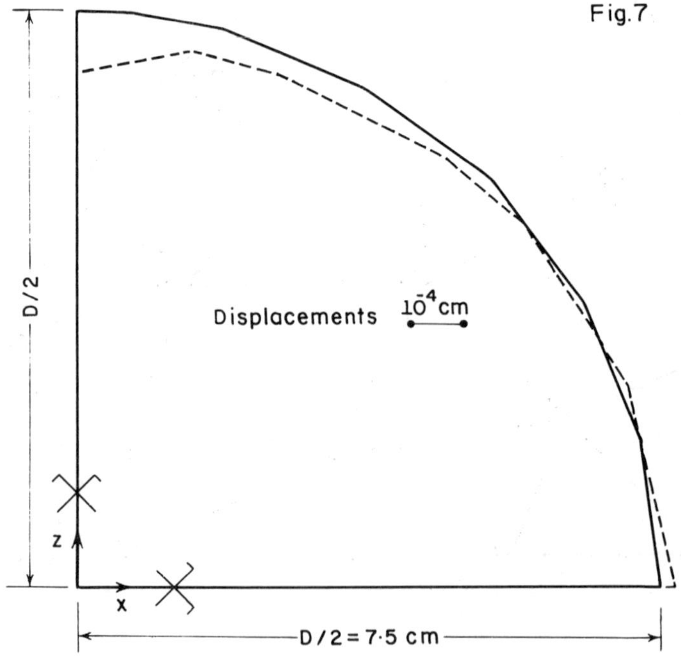

(b) Surface displacements

properly the solution and more elements are needed.

Displacements are also plotted in figure 7b and although no other solution was available, they vary as expected.

iii) Cylinder Intersection Problem

The next example illustrates how boundary elements can be applied to analyse a cylinder intersection problem of the type found in certain oil storage tanks. The cylinders under consideration intersect as shown in figure 8, and for symmetry considerations, only the region divided into elements needs to be studied. The discretization for a finite length segment is shown in figure 9. The only loads acting on the system are the internal pressures on the tanks. The region under consideration was divided into 60 constant elements and the concentration of stresses at the point of contact between cylinders was studied. Results for principal stresses on the line 1 to 5 and 6 to 9 (see figure 8) are plotted in figure 10. Large values for tensional stresses are developed at certain points, such as 1 and 5. Points 6 and 9 instead show a concentration of compressive stresses.

The application is of practical interest as cracking in oil storage tanks can have serious consequences. Boundary elements can give accurate results for internal stresses as well as for boundary tractions.

iv) Joint Study

Offshore structures have a large number of joints where several stress concentration problems occur. Boundary elements can be successfully applied to study some of these joints, specially those between thick cylinders. In figures 11 and 12 one particular type of joint is illustrated. Due to symmetry only one eighth of the structure needs to be studied.

The system was discretized using 76 boundary elements as shown in figure 13 and stresses were calculated in 5 points in the intersection (see figure 12). Results for normal stresses in the Z direction are shown in figure 14. They indicate the distribution of stresses over the intersection.

Although this example does not show regions of high stress concentration the same technique can easily be extended to the more complex offshore structures joints where the concentration of stresses is of primary importance.

CONCLUSIONS

The boundary element method is specially well suited to analyse three dimensional problems, as it reduces the dimensions of the problem, i.e. from three to two dimensions. It also offers other important advantages over 'domain' type methods such as

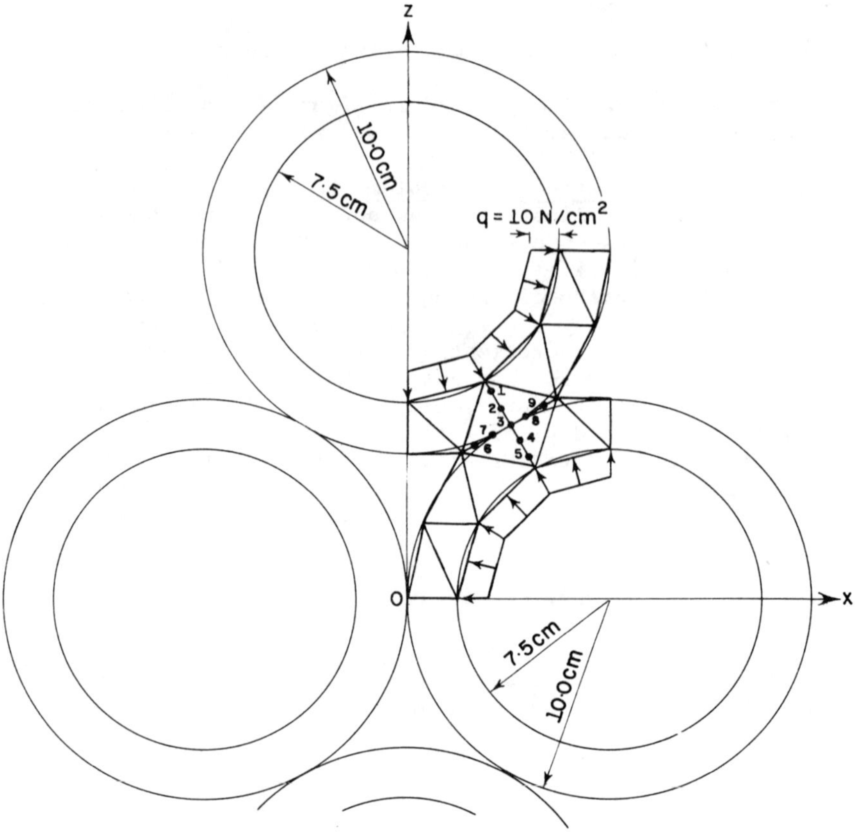

Cylinder intersection problem
Fig.8

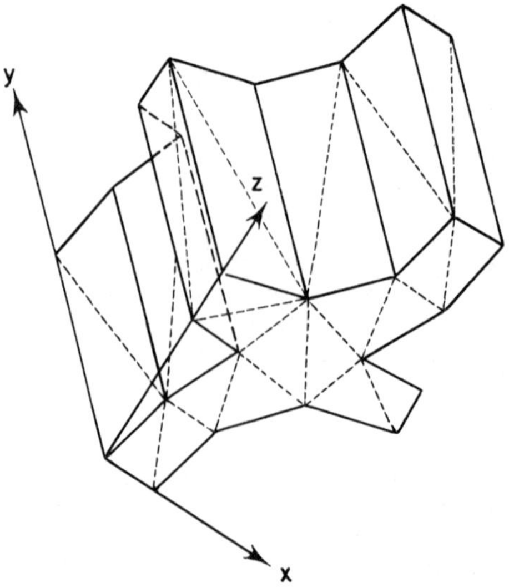

Discretisation cylinder intersection problem
Fig.9

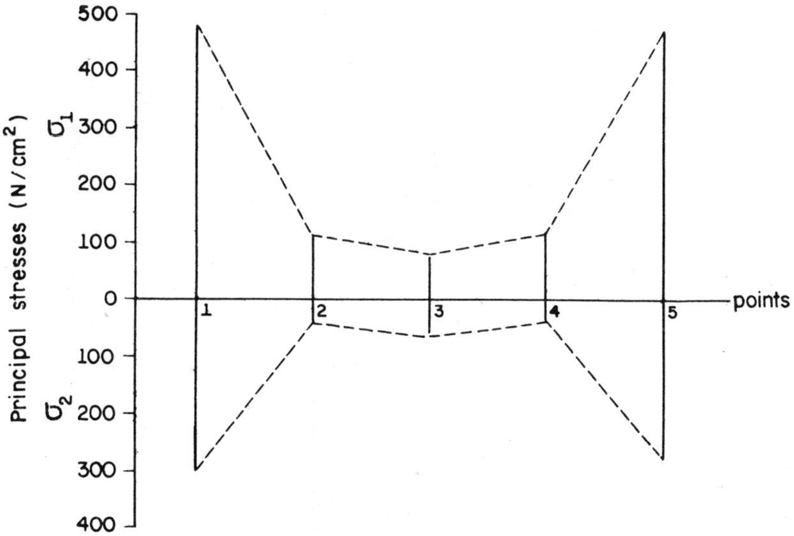

σ_1 : tensile stress

σ_2 : compressive stress

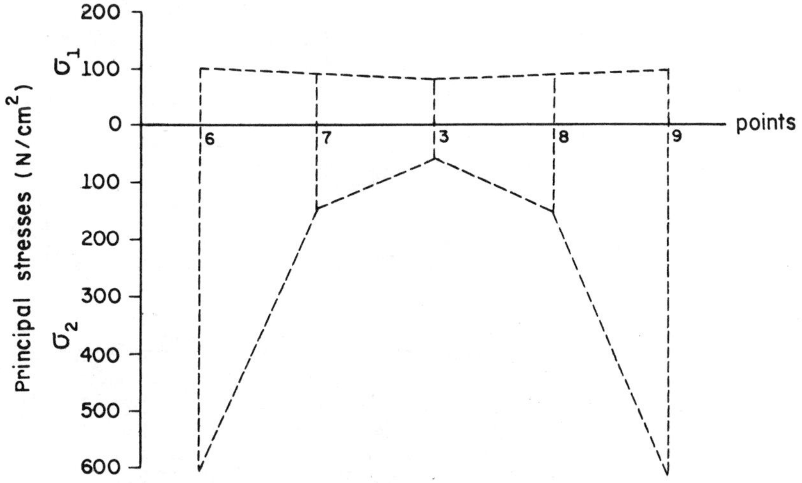

Results of the cylinder intersection problem

Fig.10

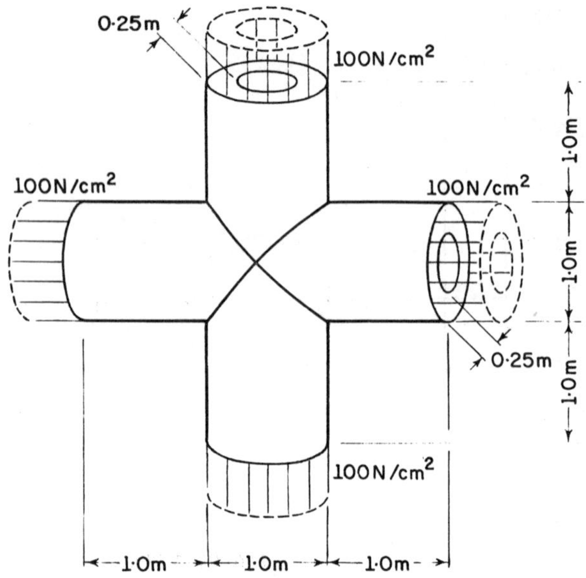

Fig.11 Joint of the type found on offshore platform

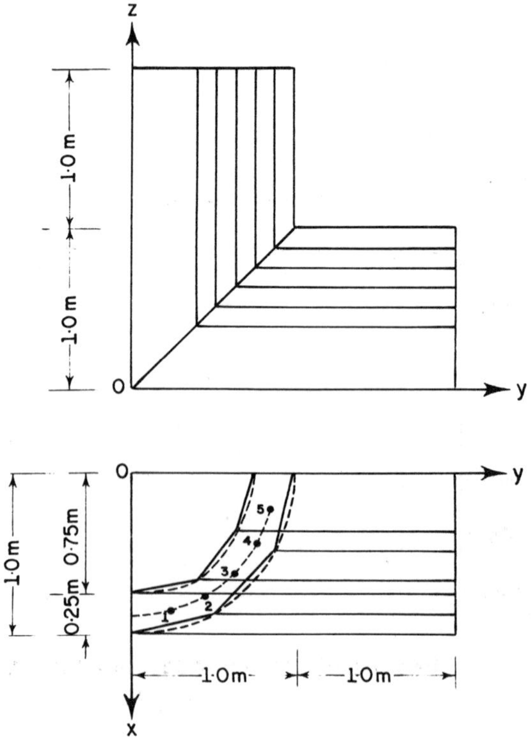

Fig.12

Projection view of the joint

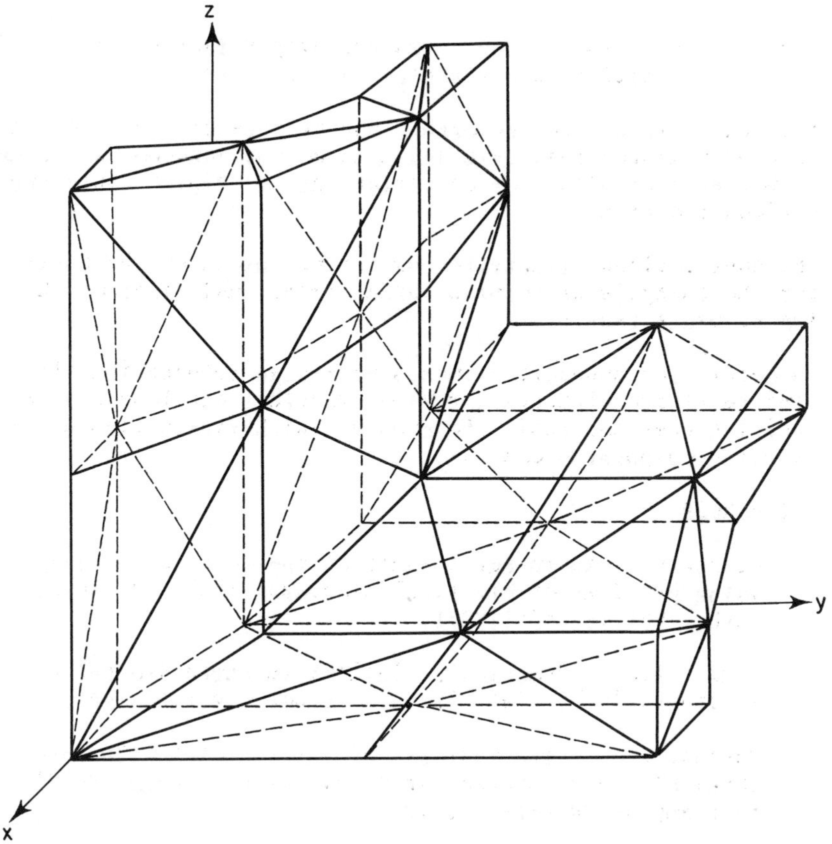

Discretisation of the joint
Fig.13

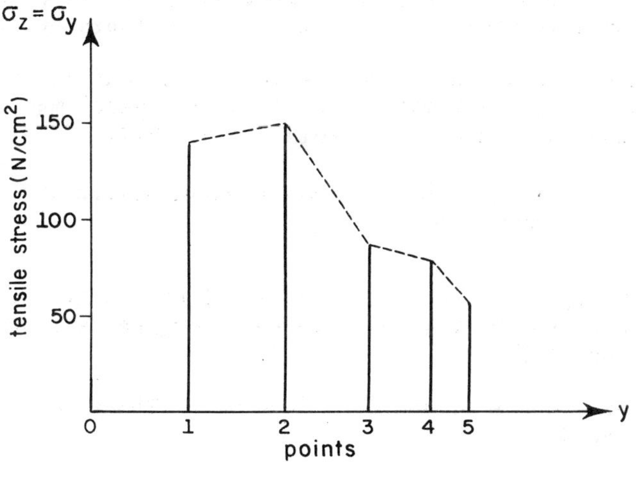

Results of the joint
Fig.14

more accurate results for stresses, proper representation of boundary conditions for infinite domains, etc.

This paper shows how the method can be interpreted as a special case of weighted residuals and extended to boundary elements with linear or higher order variations of displacements and surface tractions.

The applications illustrate how the technique can be applied for the analysis of complex three dimensional problems obtaining accurate results.

In conclusion boundary elements offers an interesting alternative to finite elements for many practical problems and in addition they can easily be combined with finite elements in existing computer packages.

REFERENCES

1. Rizzo, F. "An integral equation approach to boundary value problems of classical elastostatics", J. Appl. Math., Vol.25, 83 (1967).

2. Cruse, T.A. "Numerical solutions in three dimensional elastostatics", Int. J. Sol. Struct., Vol.5, 1969.

3. Watson, J.O. "The analysis of thick shelss with holes, by integral representation of displacement", Ph.D. Thesis, Southampton University, 1972.

4. Lachat, J.C. "A further development of the boundary integral technique for elastostatics", Ph.D. Thesis, Southampton University, 1975.

5. Butterfield, R. and G.R. Tomlin, "Integral techniques for solving zoned anisotropic continuum problems", Variational Methods in Engineering, Southampton University, 1972.

6. Brebbia, C.A. and J. Dominguez, "Application of the boundary element method for potential problems" Applied Mathematical Modelling, Vol1, No.6, 1977.

7. Brebbia, C.A. "The boundary element method", Pentech Press, London, 1978.

8. Brebbia, C.A. and J.J. Connor, "Fundamentals of finite element techniques for structural engineers", Butterworhts, London, 1973.

9. Timoshenko, S.P. and J.N. Goodier, "Theory of Elasticity", McGraw-Hill.

10. Carneiro, P.L.L.B. and A. Barcellos, "Résistance à la traction des bétons", Inst. Nacional de Tecnologia, Rio de Janeiro, Brasil, 1949.

THE INFLUENCE OF CYCLIC TEMPERATURES AND CREEP ON THE STRESS PREDICTIONS FOR CONCRETE OFFSHORE STRUCTURES

G.L. England and A. Moharram

Civil Engineering Department, Kings College, London, U.K.

INTRODUCTION

Although some aspects of creep and temperature form the subject of another paper at this conference [1], attention is focussed here on the time-variation of the stresses in oil containing structures, caused by the periodic nature of the temperatures to which they are subjected. The storage of hot oil and its subsequent replacement by sea water ballast lead to rapid changes of the temperature states within these cellular structures. In this paper, a simplified yet representative problem is considered, to examine the influence of creep and temperature on the predicted stresses in the wall sections of these containment structures. The example reflects the complete curvature restraint that results from the structural geometry. Figure 1 shows the example which has been analysed in a later section of the paper. For simplicity, the analysis is presented for the case of unidirectional stress, but may be employed readily in two- and three-dimensional stress situations when required.

An influential factor affecting the stresses and their variation in time, is the proportion of the temperature cycle for which one temperature state exists and repeats in subsequent cycles. For example, when an external wall section is subjected alternately to hot oil and cold seawater ballast on one face, while the exterior face remains constant at seawater temperature, it is the cumulative time for which the state of non-uniform temperature exists that becomes important in relation to the changing stresses. This is because stress redistribution occurs only when the temperatures vary spatially in the structure and give rise to the non-homogeneous development of creep strains. Such creep strains are by definition non-compatible and must as a result generate time-varying stresses in order that total strains - elastic, creep and thermal - may remain compatible at all times.

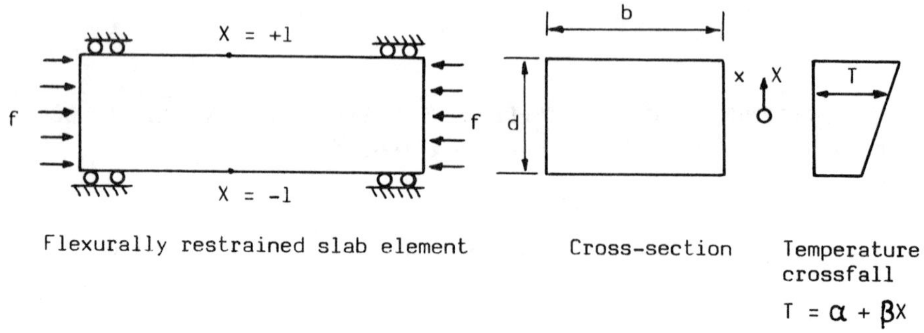

Figure 1. Simplified representation of flexurally restrained wall element of containment structure

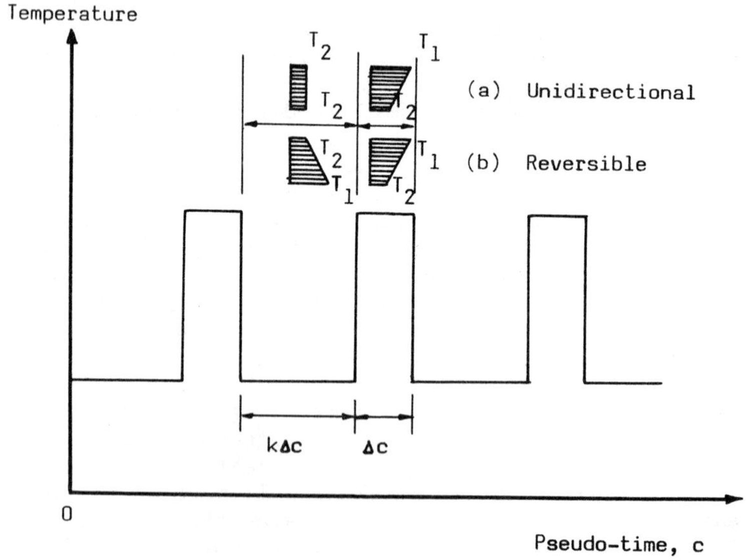

Figure 2. Nature of cyclic temperature variations and definition of cycle parameter, k.

The simple problem considered is shown in Figure 2. It is assumed that the temperature cycles repeat in a systematic manner and are of constant period with respect to pseudo-time*. during the operational lifetime of the structure. This, somewhat atypical behaviour allows a study to be made of the nature in which stresses change with time, and depend upon the cycle parameter, 'k', as shown in Figure 2. In a real situation a structure will not normally be operated in accordance with a specific value of k but will have associated with it, a spectrum of k values throughout its history of operation. From such a history it is possible to define a mean and standard deviation for all the 'k' parameters. Studies are being made currently for random distributions of k values and the results are to be presented at the forthcoming conference, Oceanology International '78, to be held in Brighton England next year [2].

Figure 3 shows diagrammatically the influence of k on the time-variation of the stresses during the raised temperature portion of the cycles for the unidirectional temperature crossfall problem. From this figure it may be observed that the stress variations in time are bounded by the initial thermo-elastic solution - curve for $k = \infty$ - and the sustained temperature crossfall solution corresponding to $k = 0$. In other words, the structure will be subjected to less severe states of stress as the result of temperature cycling than would be the case for a long period at a sustained temperature, followed by a temperature shut-down. This is true for either the unidirectional temperature crossfall or for the reversible crossfall situation associated with an internal wall between two storage chambers.

DETERMINATION OF STRESSES FOR CYCLIC VARIATIONS OF TEMPERATURE

The methods of analysis presented in a second paper [1] for cases of sustained temperatures may be used with some modification for the solution of problems in which temperature states vary in a cyclic manner in time. An independent direct approach is also available, and this depends upon theknowledge that the average power dissipated in creep over any complete temperature cycle, is a minimum with respect to variations of the stresses from the true state. In the examples presented here it is assumed that the dominant creep component is the non-recoverable or flow strain. Mention is made later of the influence of the temperature-change component of creep strain on the transient stresses.

In the numerical step-by-step procedures which proceed in a repeating manner, using the stresses from one interval to compute the creep strains over the next, for the assumption

*Pseudo-time is defined in a companion paper at this conference and represents a normalised creep parameter with respect to stress and temperature.

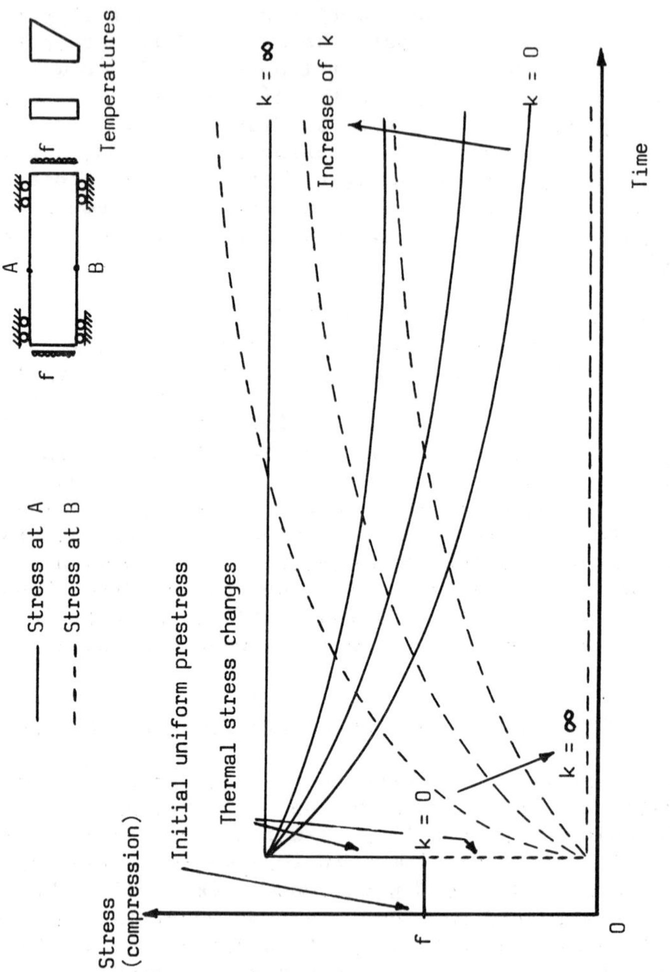

Figure 3. Diagrammatic representation of stress variation as influenced by the cycle parameter k, for unidirectional temperature crossfall.

k = 0 relates to the imposition of sustained temperature crossfall and
k = ∞ corresponds to the thermo-elastic solution.

that stresses remain approximately constant during the interval, it becomes necessary to simply bring about an additional thermo-elastic change at the end of the appropriate time interval, in order to impose or remove the effect of the temperature cross-fall. The subsequent analysis for creep then repeats as before but with modified initial stresses for the next time interval. These 'new' stresses consequently influence the creep strains in the following interval and also the rate of change of the stresses due to creep immediately following a temperature change.

A second method of analysis is based on the principle of virtual power and is outlined in a companion paper [1]. The state of stress at any time - for the sustained temperature problem - was specified as a series of terms thus,

$$\sigma = \sigma_o + \sum_{i=1}^{i=N} a_i \sigma_i \tag{1}$$

where the stresses σ, σ_o, and σ_i represent spatial variations and become simply functions of the coordinate, X, in the problem of Figure 1. σ_o is in equilibrium with the applied mechanical loading on the boundary, σ_i for i = 1 to N, represent self-equilibrating states of internal stress, and a_i represent the weighting functions of pseudo-time which proportion the stresses at all times to give the best possible approximation to the exact solution. In most problems the number of weighting parameters does not need to be large; ten may be taken as an upper limit. Frequently, acceptable solutions may be obtained by using as few as two independent stress distributions, and hence i = 2.

It is possible to utilize eq. (1),in modified form, to generate solutions to the cyclic temperature problem. For the example of Figure 1 the stress, σ, may be represented during the heated portion of the temperature cycles, as,

$$\sigma = f + a_1 X + \sum_{i=2}^{i=N} a_i \sigma_i \tag{2}$$

On removal of the temperature crossfall the state of stress would change by $E\alpha\beta X$, where E and $\bar{\alpha}$ are respectively the modulus of elasticity (assumed here to be non-temperature dependent) and the coefficient of thermal expansion of concrete. β is defined as in Figure 1. Then, for the uniform temperature portions of the cycles

$$\sigma = f + (a_1 - E\bar{\alpha}\beta)X + \sum_{i=2}^{i=N} a_i \sigma_i \tag{3}$$

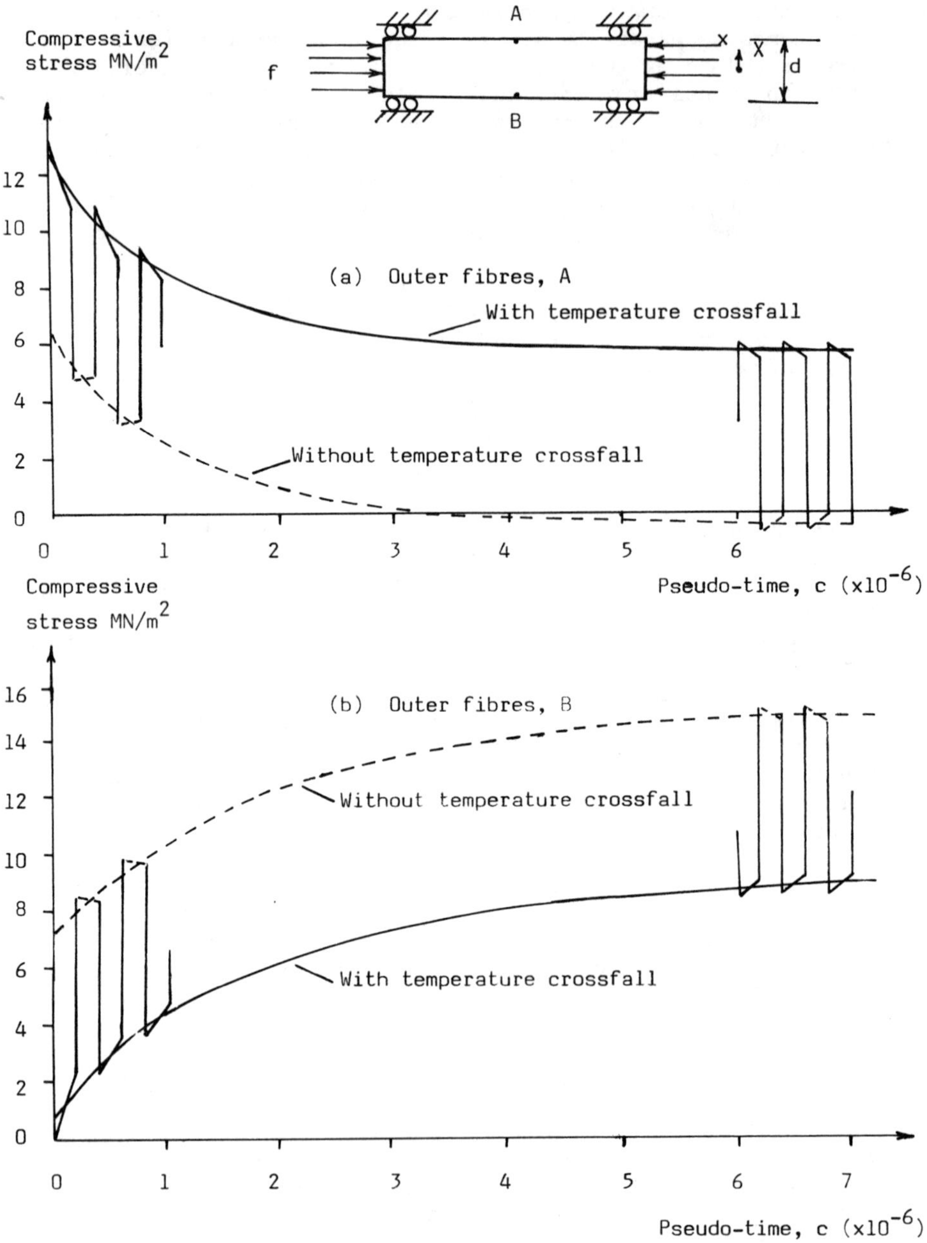

Figure 4. Time-dependent variations of stress for outer fibres of flexurally restrained wall unit subjected to cyclically varying temperatures of the unidirectional type (a) of figure 2. Data used: $T_1 = 50^oC$; $T_2 = 10^oC$; $f = 7MN/m^2$; $E = 34GN/m^2$; $= 10.10^{-6}/^oC$; $k = 1$.

Observation of eq. (3) shows that the cyclic temperature problem differs from the sustained temperature case in that at each temperature change, the weighting parameter, a_1, changes by an amount $E\bar{\alpha}\beta$, the sense depending upon whether the crossfall is removed or imposed. A similar representation is possible in the general problem whenever the self-equilibrating distribution, σ_1, represents the actual thermal stress distribution corresponding to the complete imposition of the new temperature state.

Eq. (3) has the advantage over the step-by-step form of analysis that it is simpler to incorporate into a computer analysis. The solution for the a_i functions result from a set of first order differential equations in pseudo-time and are easily solved in the numerical problem [3,1]. The matrix manipulations are relatively simple and permit solutions to the cyclic problem to be obtained at low cost.

Some results for the unidirectional temperature crossfall problem are shown in Figure 4. Comparisons between the step-by-step predictions and those from the virtual power approach using two self-equilibrating distributions only, $\sigma_1 = X$ and $\sigma_2 = (X^2 - 1/3)$, indicate errors of generally better than 5% and frequently less than 1%.

DISCUSSION

The results of Figures 3 and 4 show clearly that the differential thermal creep, caused by non-uniform temperatures, gives rise to significant changes of stress in prestressed sections restrained from thermal warping, and that cyclically varying temperatures influence both the transient and long-term stress states.

In many situations it will be important to know the maximum extent of the stress changes in time more precisely than the transient behaviour throughout the operational life of a structure. For these cases, solutions may be obtained from the analyses mentioned or from separate direct approaches to the steady-state stresses for either sustained [4], or cyclically varying [1,5] temperatures. Such solutions are not significantly affected by the differentiation of the total creep into its several components, since the long-term solutions depend upon the more slowly developing flow component.

There are two components of creep strain which can influence the state of stress in the shorter term and they are the recoverable or delayed elastic strain component and the temperature-change* component of creep strain [6,7]. The former component may be considered in conjunction with the short-term

* Temperature-change creep is defined as an irreversible component of creep strain which results from an upward change of temperature to a new temperature for the first time. Recovery of strain does not result in a fall in temperature.

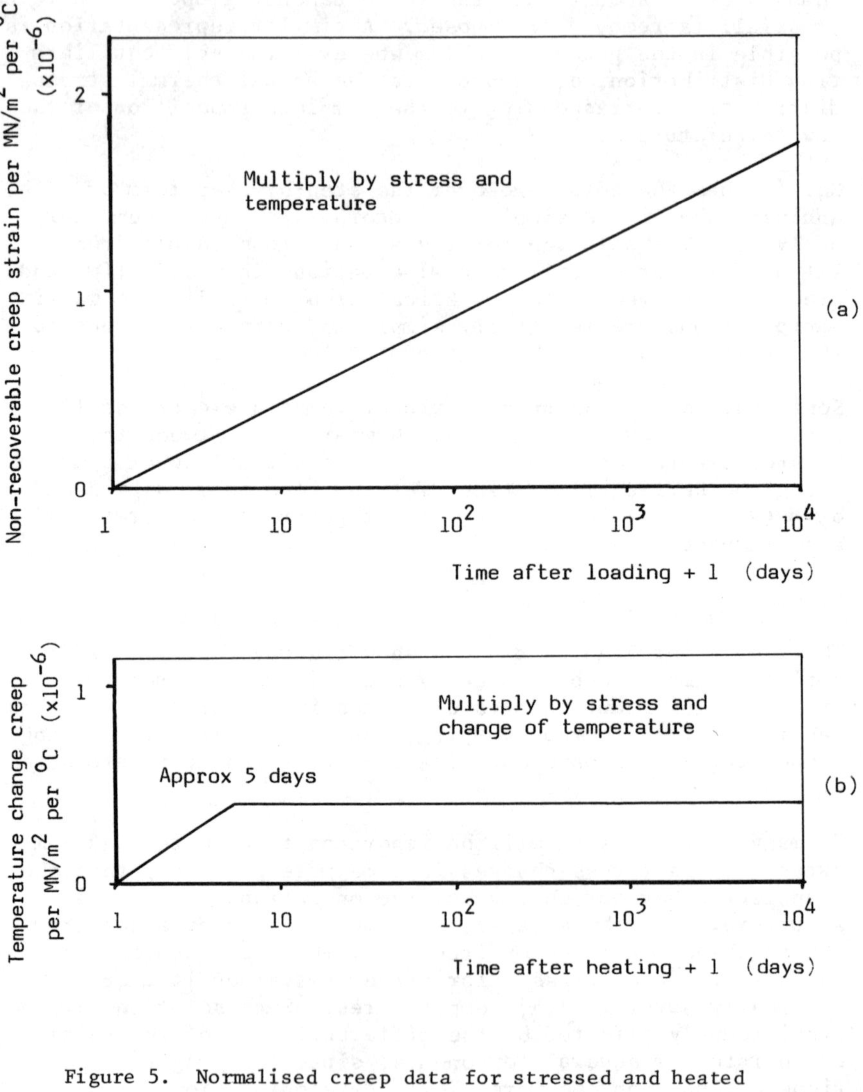

Figure 5. Normalised creep data for stressed and heated concrete.
(a) flow component
(b) temperature-change component.

elastic strain in most situations without significant loss of accuracy, and the latter component may often be included with the flow component in problems associated with raised and non-uniform temperatures. Some situations arise, however, for which it becomes more appropriate to take note of the separate effects of flow and temperature-change creep. For example, in those cases where temperature crossfalls show reversals with time the effect of the temperature-change creep during the first cycle is to bring about a shift of the stress solution for early times, when compared to the solution based on the flow creep term only. This shift, which is a change in the rate of stress redistribution, results from the non-symmetrical effects of the temperature-change creep over the stressed cross-section during the first temperature crossfall reversal. However, as time increases the influence of this initial effect becomes overshadowed by the flow component, and the long-term solution is insensitive to the early changes.

Figure 5 shows one representation of the temperature-change and flow creep components plotted to a logarithm scale of time. The results have been derived primarily from the work on mortar by Sanders [8] and indicate that the rate of development of temperature-change creep is greater than that of the flow component and is essentially complete after approximately five days from a change of temperature. For the purposes of design the actual form of the rapidly developing phase of the temperature change component is likely to be of less importance than the knowledge that it has a certain limiting value and that it develops much more rapidly than the flow component. Its development is thus shown by a straight line on the log plot. Although further research work is needed to establish more precisely the nature and development of the two creep components and their ranges of values for concretes of different mixes, the values quoted may be used for moderately high grade concretes.

CONCLUDING REMARKS

1. States of non-uniform temperature cause major redistribution of stresses in sections restrained from thermal warping. The section moments resulting from these stresses may be caused to change sense with passage of time.

2. Cyclically varying temperatures cause stresses to redistribute more slowly than when temperature crossfalls are sustained. The long-term limiting states for these stresses also show less severe redistribution than for the case of sustained heating followed by a temperature shutdown or reversal of the temperature crossfall.

3. Methods of analysis are available for computing the time-varying states of stress caused by cyclic changes of temperature. Approximate procedures based on the principle of virtual power allow solutions to be obtained at low cost and to any desired degree of accuracy.

REFERENCES

1. Richmond, B. and G.L. England. "The design of concrete structures for the containment of hot oil". Proc. of International Conference on Offshore Structures Engineering, Brazil Offshore '77, Rio de Janeiro, September 1977.

2. England, G.L., J.S. Macleod and A. Moharram. "Designing for creep and temperature in concrete offshore structures". Paper to be presented at forthcoming conference, Oceanology International '78, Brighton, England, March 1978.

3. England, G.L. "Analyses for creep in heated concrete structures". Research Report No. CE 75-3, University of Calgary, Canada, April 1975.

4. England, G.L. "Steady-state stresses in concrete structures subjected to sustained temperatures and loads". Nuclear Engineering & Design, Vol.3, Jan. 1966, pp.54-65 and Vol.3, No.2, February/March 1966, pp.246-255.

5. England, G.L. "Designing for creep in heated concrete structures". Proc. of Conference on Behaviour of Slender Structures". City University, London, September 1977. Session III, paper No.4

6. Hansen, T.C. and L. Eriksson, "Temperature change effect on behaviour of cement paste, mortar, and concrete under load". Journal ACI, Proc. Vol.38, No.63, April 1966, pp.489-502.

7. Illston, J.M. and P. D. Sanders. "Characteristics and predictions of creep of a saturated mortar under variable temperature". Mag. of Conc. Res. Vol.26, No.88, September 1974, pp.169-179.

8. Sanders, P.D. "The effect of temperature change on the creep of mortar under torsional loading". Ph.D. Thesis, University of London, 1973.

PART III WAVES AND OTHER FLUID INDUCED FORCES

SEA BEHAVIOUR CHARACTERISTICS OF FLOATING STRUCTURES

Sergio Hamilton Sphaier COPPE/UFRJ

Marcelo de Almeida Santos Neves COPPE/UFRJ

ABSTRACT

The paper presents long term prediction of wave induced motion and moment of a ship. The statistical method can be applied in the case of platforms, in which it was necessary to introduce the transfer function for the case.

Results for five areas of the Brasilian coast are presented. They show, that each area defines different design criteria for the strength calculation and operation.

ACKNOWLEDGEMENT

The authors are most grateful to Mr. J.H. Sanglard and Mr. M.S.S. de Barros for their help on the computational work and to Miss H.S. de Oliveira for typing the manuscript.

INTRODUCTION

In the last years some methods were developed to predict the sea behaviour of a floating structure. In the field of Naval Architecture Korvin-Kroukovsky [1] has applied the strip theory of the aerodynamics to the determination of motions and bending moments acting on a ship at sea.

The forces per unit length were determined for many transverse strip sections and the results integrated along the ship length.

In the study of the transverse two dimensional fluid motion the interaction among transverse strips can be neglected.

Besides this, some other theories were developed; Gerritsma e Beukelman [2], Kaplan [3]. At COPPE Rostovtsev and Sphaier [4] have considered the ship length influence on the fluid motion, through the breadth variation along the ship length. Nevertheless, none of those formulations satisfy the symmetry condition of Timman and Newman [5] on the coupled term. The theoretical results of those theories have showed, however, good agreement with experimental results.

Considering the ship as a slender body, the wave amplitudes and the ship motions amplitudes as small, and the wave frequency as high, Salvesen, Tuck and Faltinsen [6] have linearized the three dimensional boundary value problem, which describes the irrotational ideal fluid motion and reduced it to many two dimensional boundary value problems. Considering these same hypotheses and assuming that the Froude number was high, Ogilvie and Tuck [7] have applied a perturbation method to the study of the heave and pitch motions problem and derived a more consistent theory for the determination of the hydrodynamic coefficient of the equations of motion.

They have obtained some terms in the form of integrals to be evaluated over the free surface that are different from the ones derived in [6]. As was verified by Faltinsen [8] and Sphaier [9], the evaluation of the coefficients of the equations of motion with Ogilvie and Tuck integral terms takes more computer work time than the coefficients in the form presented in [6].

However for a ship the strip theory brings good results, this is not the case when we have a floating large structure for which the transverse dimensions are no longer small in comparison to the length. In those cases a three dimensional approach as proposed by Faltinsen and Michelsen [10] is necessary. They have used distribution of singularities over the wetted surface to represent the velocity potential. In the case of a platform, for which the structural topology is composed of several slender elements, different simplifications are possible, as proposed by Michelsen and Faltinsen [11] and Hooft [12].

Those linear formulations permit to determine the response amplitude operator (RAO) of several characteristics for a progressive wave. Supposing the actual seaway can be represented by a superposition of an infinite number of progressive waves, we can determine the sea spectrum and with the knowledge of the RAO the ship responses spectral, St. Denis and Pierson [13]. This approach describes the short term behaviour of the ship when she encounters a sea state and permit to predict the statistical distributions of the responses. But those distributions are associated to each particular sea state. Using the short term distribution and the sea state distribution, the paper presents results of long term prediction for a tanker operating in a region of the Brazilian Coast based on a method presented by Fukuda [14], Neves [15],

Neves and Sphaier [16]. To describe the sea state occurrence frequency some data of visual observation for wave height and period from Hogben and Lumb [17] are used. The Seaway will be represented by the modified two parameter spectrum of the ISSC (International Ship Structures Committee).

EQUATIONS OF MOTION

The Newton's law for a dynamical equilibrium of an oscillating body on the water free surface establishes:

$$\sum \underset{\sim}{F}_{ext} = \frac{d}{dt}\left[\int_v dm\, \underset{\sim}{V}\right]$$
$$\sum \underset{\sim}{M}_{ext} = \frac{d}{dt}\left[\int_v dm\, \underset{\sim}{V} \times \underset{\sim}{\Gamma}\right] \qquad (1)$$

where

$\underset{\sim}{F}_{ext}$ = external forces
$\underset{\sim}{M}_{ext}$ = external moments in relation to a point 0
dm = mass of an element of the body in the point P
$\underset{\sim}{V}$ = velocity vector of the point P
$\underset{\sim}{\Gamma}$ = radius vector of P in relation to 0
v = volume of the body

To describe the problem, we introduce a right hand coordinate system (0 x y z) so that the plane z = 0 coincides with the shell water free surface, z upwards. The coordinate system has a constant advance velocity U in the direction of positive x and the point 0 is located in the same vertical as the body centre of gravity (Fig. 1).

We assume that the fluid is ideal, incompressible and the fluid motion irrotational. Because of that the fluid motion can be described by a potential function ϕ, so that

$$\underset{\sim}{V} = \text{grad}\ \phi$$

and ϕ must satisfy the Laplace equation $\nabla^2\phi = 0$ in the fluid medium.

The potential ϕ is the sum of the potential due to the body advance effects, $-\ Ux + \phi_{ST}$, the potential due to the incoming waves, ϕ_{ω}, their diffraction, ϕ_{DIF} and the irradiation potential due to the ship motions effects, ϕ_{IRR}:

$$\phi(x, y, z, t) = \left[- Ux + \phi_{ST}\right] + \phi_T =$$

$$= \left[- Ux + \phi_{ST}\right] + \phi_\omega + \phi_{DIF} + \phi_{IRR} \qquad (2)$$

The potential function ϕ must satisfy the following conditions:

- on the body $\frac{\partial\phi}{\partial n} = V_n$ or $\frac{D}{Dt} F_c(x, y, z, t) = 0$

where:

$F_c(x,y,z,t) = 0$ - is the equation of the body surface
V_n - the component of the body velocity in the direction of the normal to the body surface.
n - is the normal to the body

- on the free surface

  a - kinematics condition $\frac{D}{Dt} F_{SL} = 0$

  b - dynamical condition $p = p_{atm} = \text{const.}$

where

$F_{SL}(x,y,z,t) = z - \zeta(x,y,t)$ is the free surface equation
p - pressure
p_{atm} - atmospheric pressure

- on fixed boundary; for instance, shallow water

$$\frac{\partial\phi}{\partial n} = 0$$

- in infinite regions, radiations conditions.

The pressure in each point of the fluid medium is determined by the Bernoulli equation

$$\frac{p}{\rho} + \frac{\partial\phi}{\partial t} + gz + \frac{1}{2} |\nabla^2\phi|^2 \frac{1}{2} U^2 = F(t) \qquad (3)$$

where

ρ - mass density
g - gravity acceleration
$F(t)$ - a time depending function

Due to the difficulties to obtain a solution of this problem

Fig. 1 Wave and ship axes convention

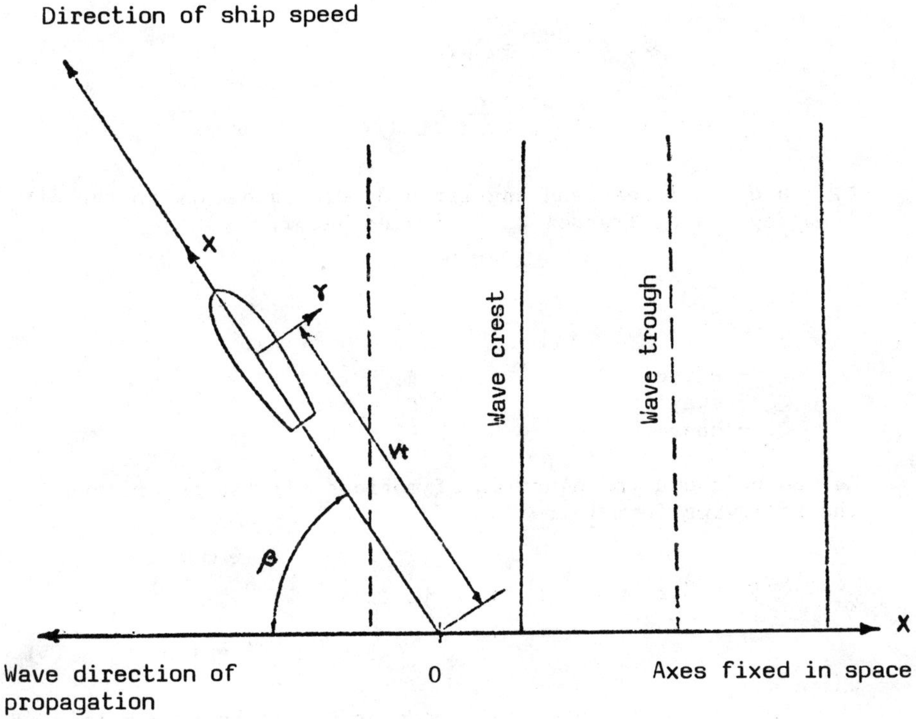

we introduce some simplifying assumptions. If we assume that the incident waves are progressive waves having small amplitudes and the ship motion responses have small amplitude too, the potential ϕ_T can be written in the following form

$$\phi_T = \phi_\omega + \varphi_7 e^{\omega_e t} + \sum_{i=1}^{6} \varphi_i \dot{\zeta}_i \tag{4}$$

where

$$\sum_{i=1}^{\cdot} \varphi_i \dot{\zeta}_i = \phi_{IRR}$$

$$\varphi_7 e^{i\omega_e t} = \phi_{DIF}$$

$$\dot{\zeta}_i = \frac{d}{dt} \zeta_i$$

ζ_i - are the linear and angular body displacements in the six degrees of freedom as indicated below.

ζ_1 - surge
ζ_2 - sway
ζ_3 - heave
ζ_4 - roll
ζ_5 - pitch
ζ_6 - yaw

As can be shown the equation of motions (1) can be written in the following form [6]:

$$\sum_{k=1}^{6} \left[(M_{jk} + A_{jk}) \ddot{\zeta}_k + B_{jk} \dot{\zeta}_x + C_{jk} \zeta_x \right] = F_j e^{i\omega_e t} ; \qquad j = 1,6 \tag{5}$$

where

M_{jk} - are the component of the inertia matrix of the body mass

A_{jk} and B_{jk} - are respectively the added mass and the damping coefficient of the body due to ϕ_{IRR}

FIG. 2

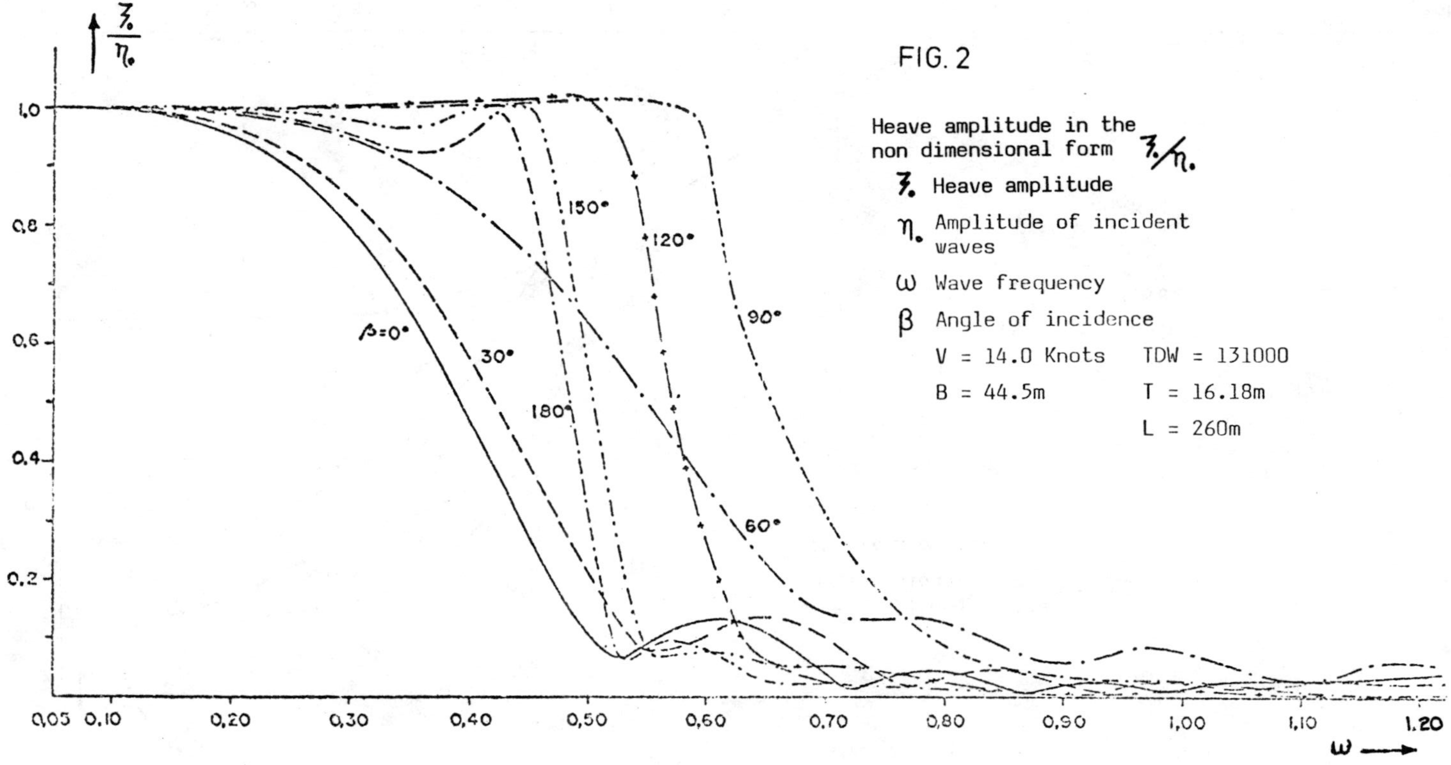

Heave amplitude in the non dimensional form ζ_0/η_0

ζ_0 Heave amplitude

η_0 Amplitude of incident waves

ω Wave frequency

β Angle of incidence

V = 14.0 Knots TDW = 131000

B = 44.5m T = 16.18m

L = 260m

FIG. 3

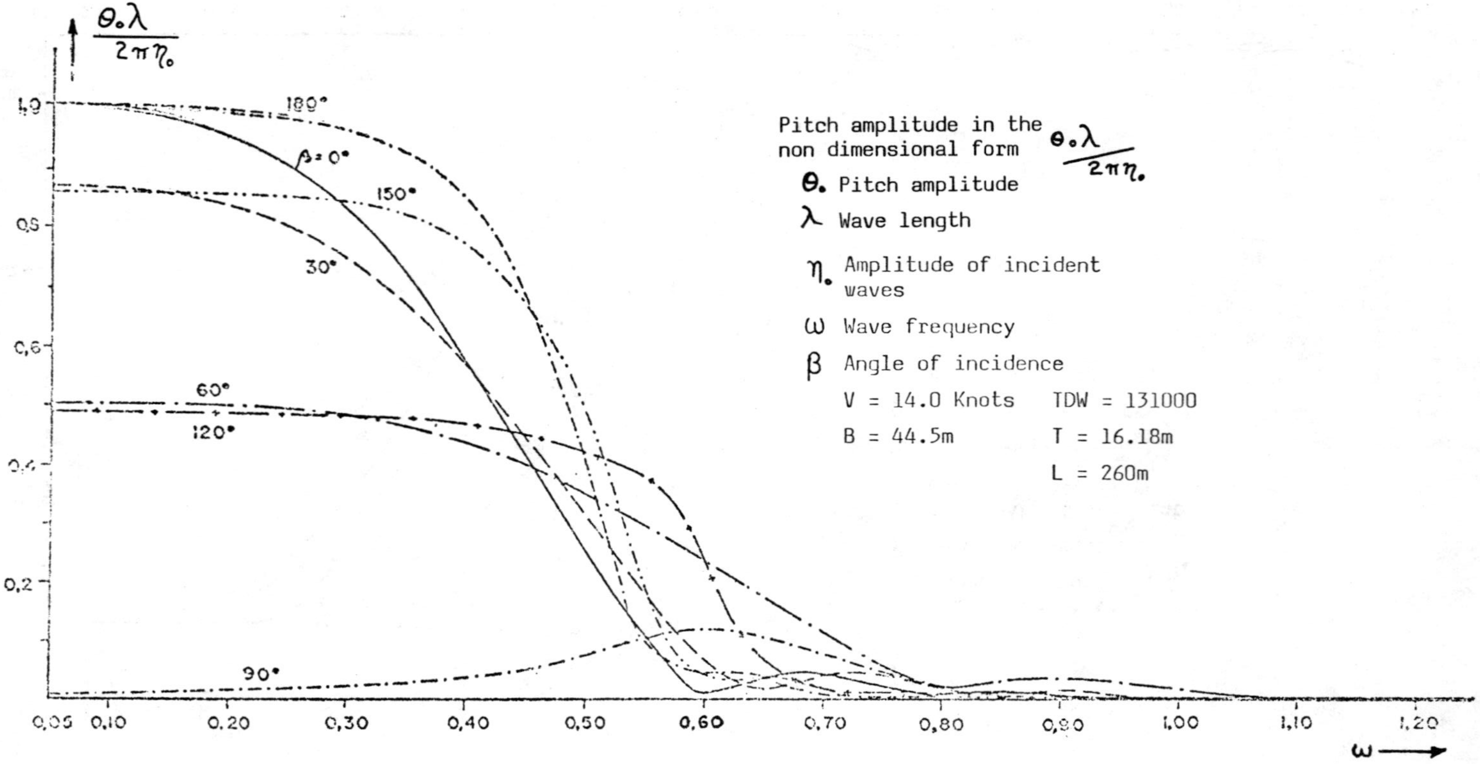

θ₀λ / 2πη₀
1,0
0,8
0,6
0,2
180°
β = 0°
150°
30°
60°
120°
90°
0,05
0,10
0,20
0,30
0,40
0,50
0,60
0,70
0,80
0,90
1,00
1,10
1,20
ω
Pitch amplitude in the non dimensional form θ₀λ / 2πη₀
θ₀ Pitch amplitude
λ Wave length
η₀ Amplitude of incident waves
ω Wave frequency
β Angle of incidence
V = 14.0 Knots
TDW = 131000
B = 44.5m
T = 16.18m
L = 260m

FIG. 4

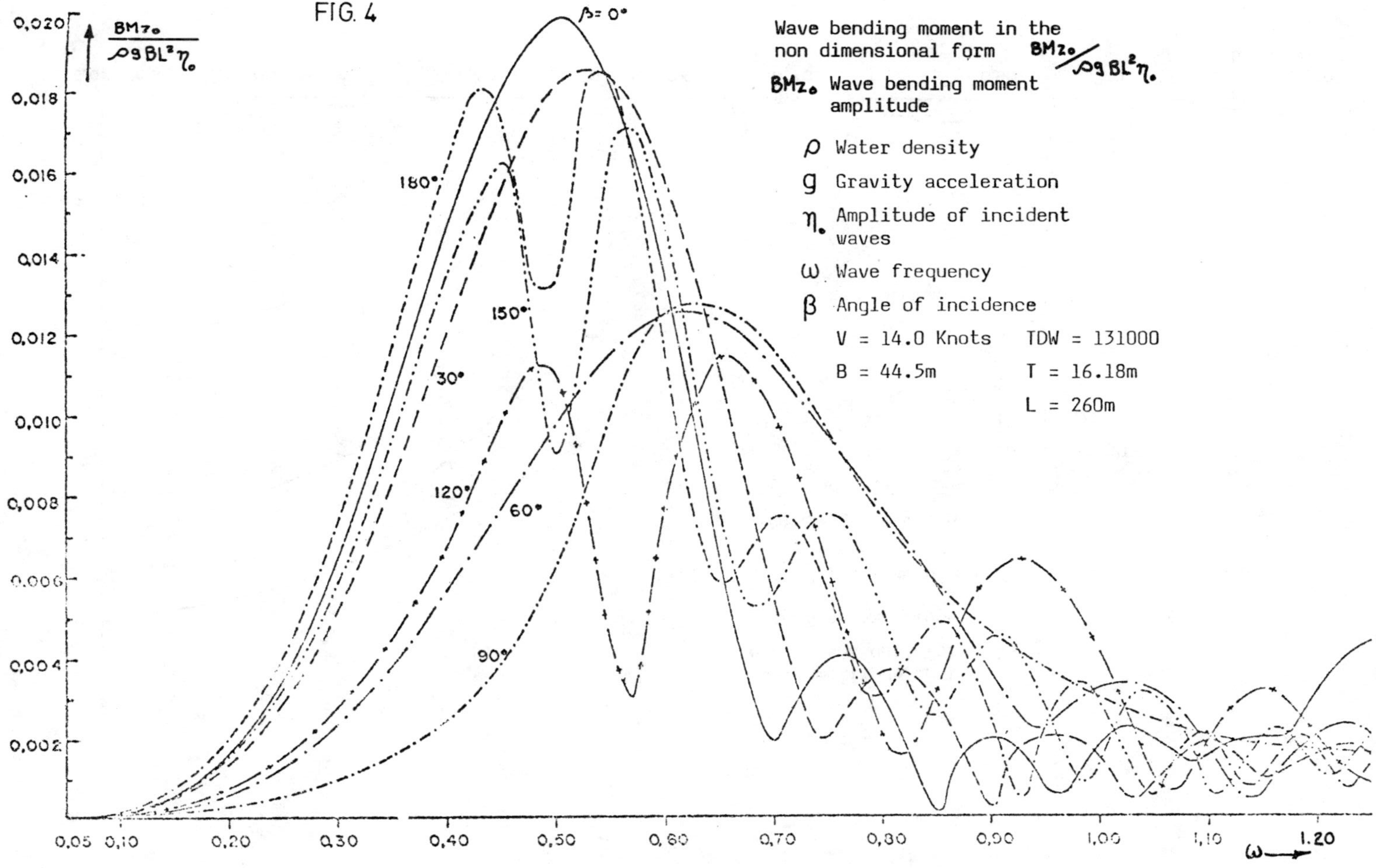
$\frac{BM_{Z_0}}{\rho g B L^2 \eta_0}$
$\beta = 0°$
180°
150°
30°
120°
60°
90°
Wave bending moment in the non dimensional form $BM_{Z_0} / \rho g B L^2 \eta_0$
BM_{Z_0} Wave bending moment amplitude
ρ Water density
g Gravity acceleration
η_0 Amplitude of incident waves
ω Wave frequency
β Angle of incidence
V = 14.0 Knots TDW = 131000
B = 44.5m T = 16.18m
L = 260m
0,020
0,018
0,016
0,014
0,012
0,010
0,008
0,006
0,004
0,002
0,05
0,10
0,20
0,30
0,40
0,50
0,60
0,70
0,80
0,90
1,00
1,10
1.20
ω

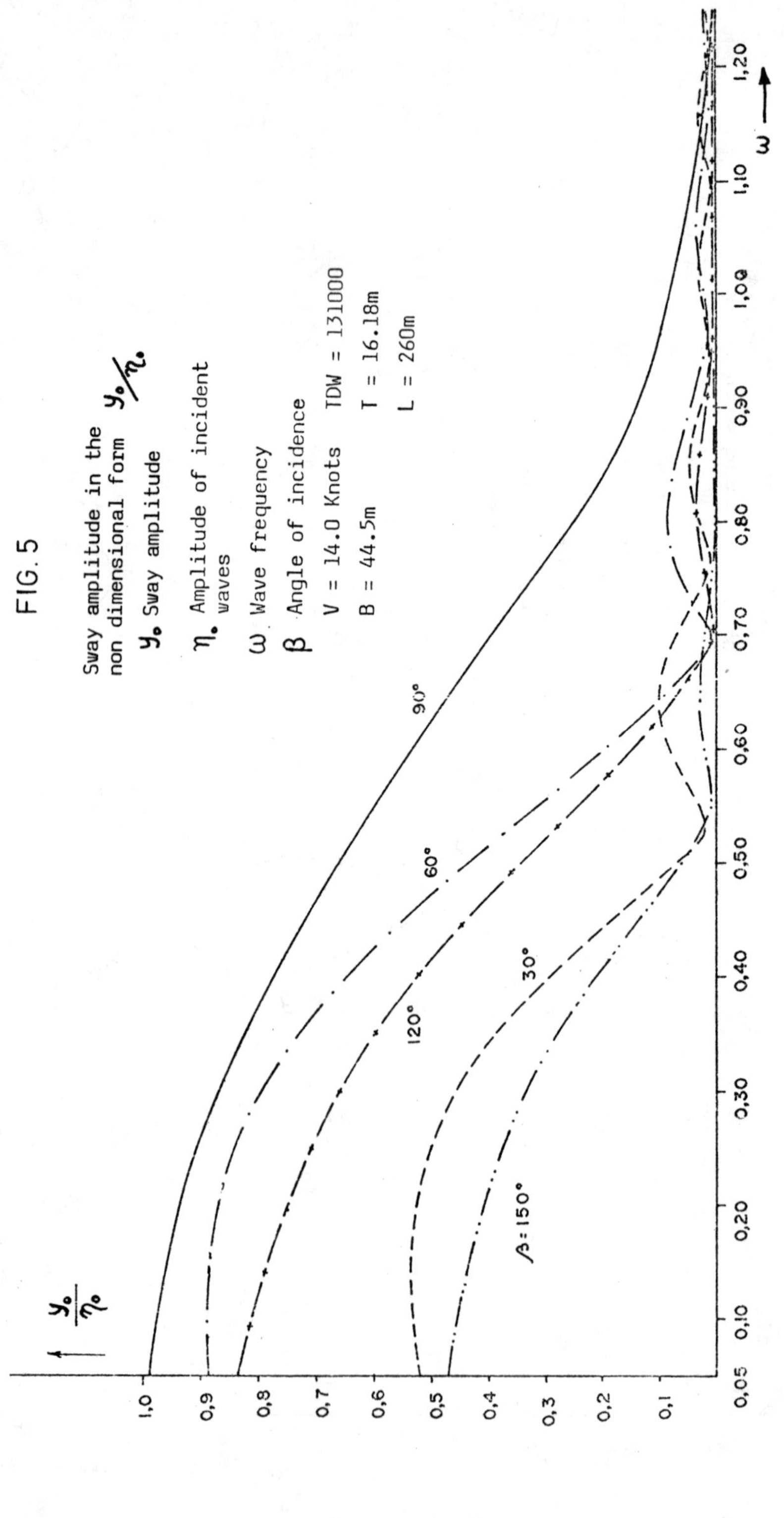
Sway amplitude in the non dimensional form y_0/η_0
y_0 Sway amplitude
η_0 Amplitude of incident waves
ω Wave frequency
β Angle of incidence
V = 14.0 Knots TDW = 131000
B = 44.5m T = 16.18m
L = 260m
90°
60°
120°
30°
β: 150°
$\frac{y_0}{\eta_0}$
1,0
0,9
0,8
0,7
0,6
0,5
0,4
0,3
0,2
0,1
0,05
0,10
0,20
0,30
0,40
0,50
0,60
0,70
0,80
0,90
1,00
1,10
1,20
ω

FIG. 5

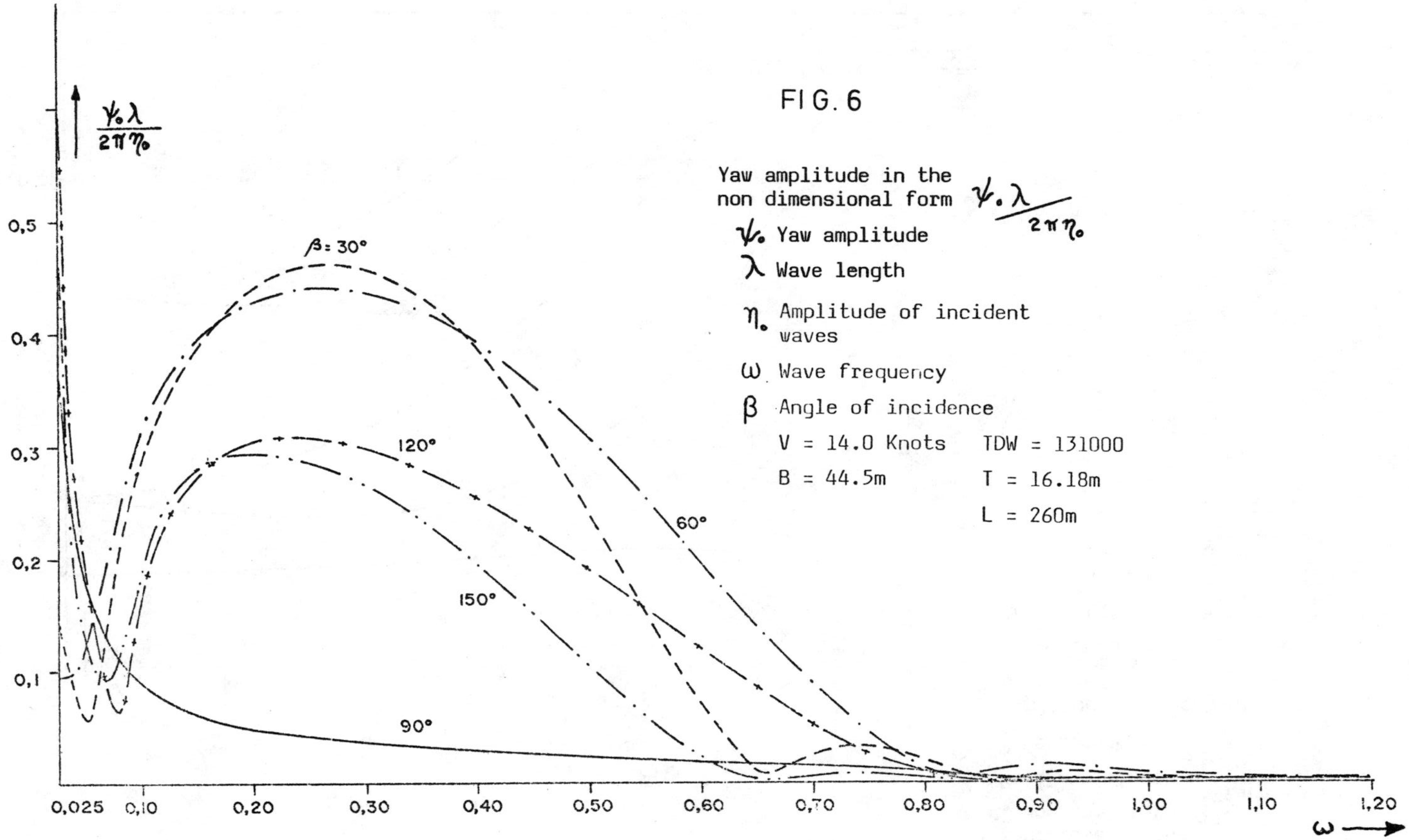
$\frac{\psi_0 \lambda}{2\pi\eta_0}$
0,5
0,4
0,3
0,2
0,1
β = 30°
120°
150°
60°
90°
0,025
0,10
0,20
0,30
0,40
0,50
0,60
0,70
0,80
0,90
1,00
1,10
1,20
ω
Yaw amplitude in the non dimensional form $\frac{\psi_0 . \lambda}{2\pi\eta_0}$
ψ_0 Yaw amplitude
λ Wave length
η_0 Amplitude of incident waves
ω Wave frequency
β Angle of incidence
V = 14.0 Knots
TDW = 131000
B = 44.5m
T = 16.18m
L = 260m

FIG. 6

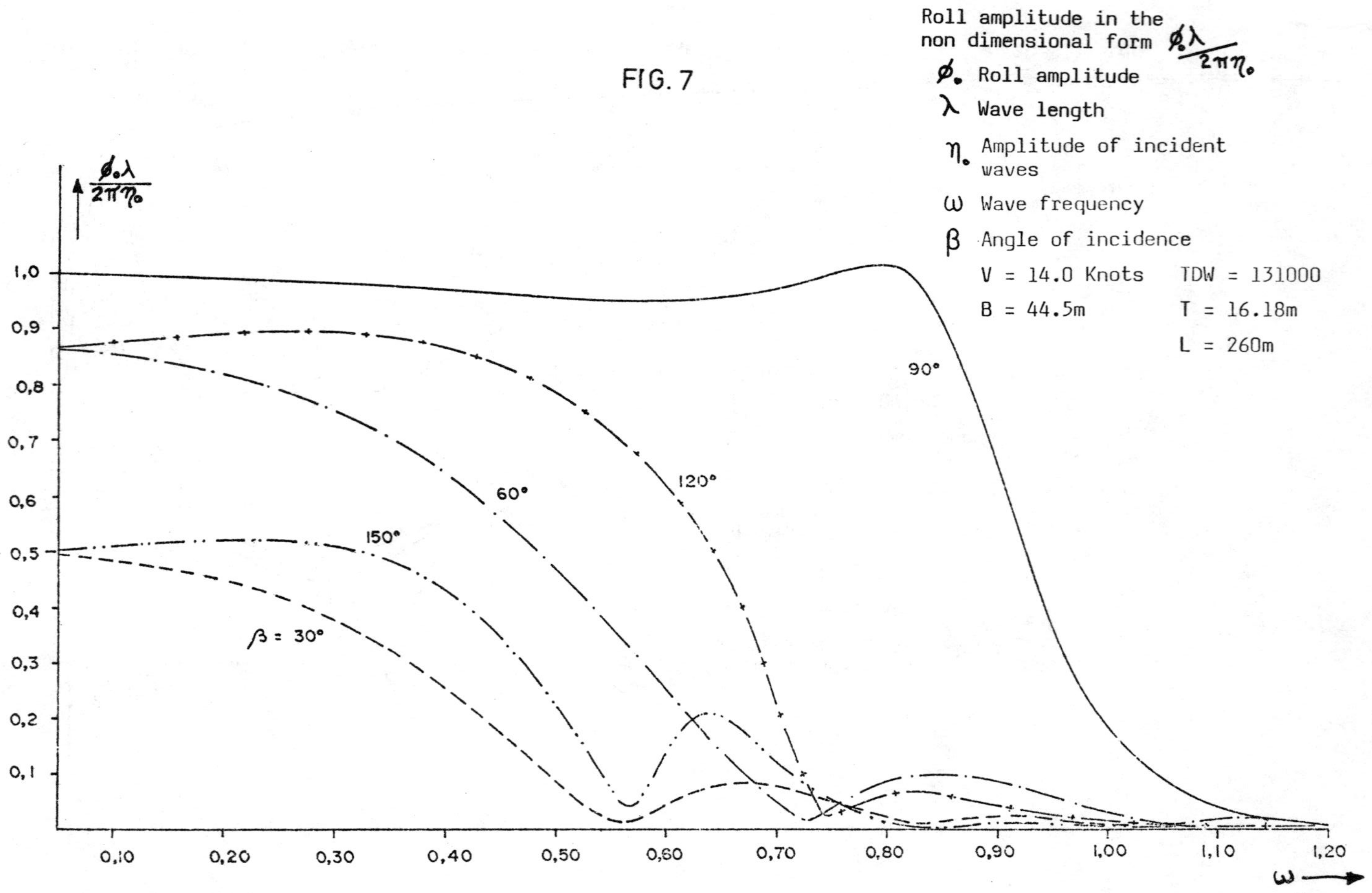
Roll amplitude in the non dimensional form $\frac{\phi_0 \lambda}{2\pi\eta_0}$
ϕ_0 Roll amplitude
λ Wave length
η_0 Amplitude of incident waves
ω Wave frequency
β Angle of incidence
V = 14.0 Knots
TDW = 131000
B = 44.5m
T = 16.18m
L = 260m
$\frac{\phi_0 \lambda}{2\pi\eta_0}$
1,0
0,9
0,8
0,7
0,6
0,5
0,4
0,3
0,2
0,1
90°
120°
60°
150°
β = 30°
0,10
0,20
0,30
0,40
0,50
0,60
0,70
0,80
0,90
1,00
1,10
1,20
ω

FIG. 7

FIG. 8

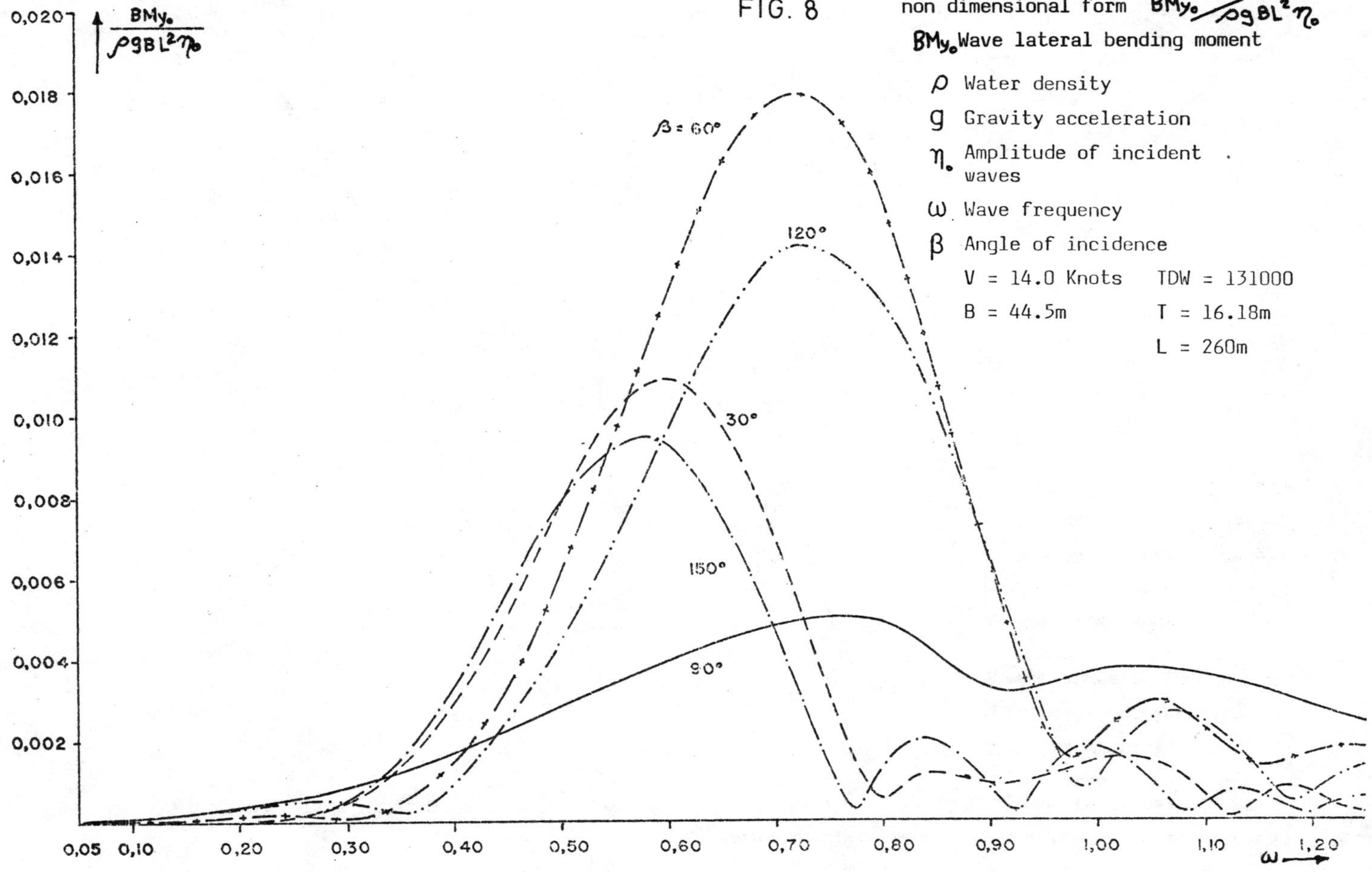

Wave lateral bending moment in the non dimensional form $BM_{y_o} / \rho g B L^2 \eta_o$

BM_{y_o} Wave lateral bending moment

ρ Water density

g Gravity acceleration

η_o Amplitude of incident waves

ω Wave frequency

β Angle of incidence

V = 14.0 Knots TDW = 131000

B = 44.5m T = 16.18m

L = 260m

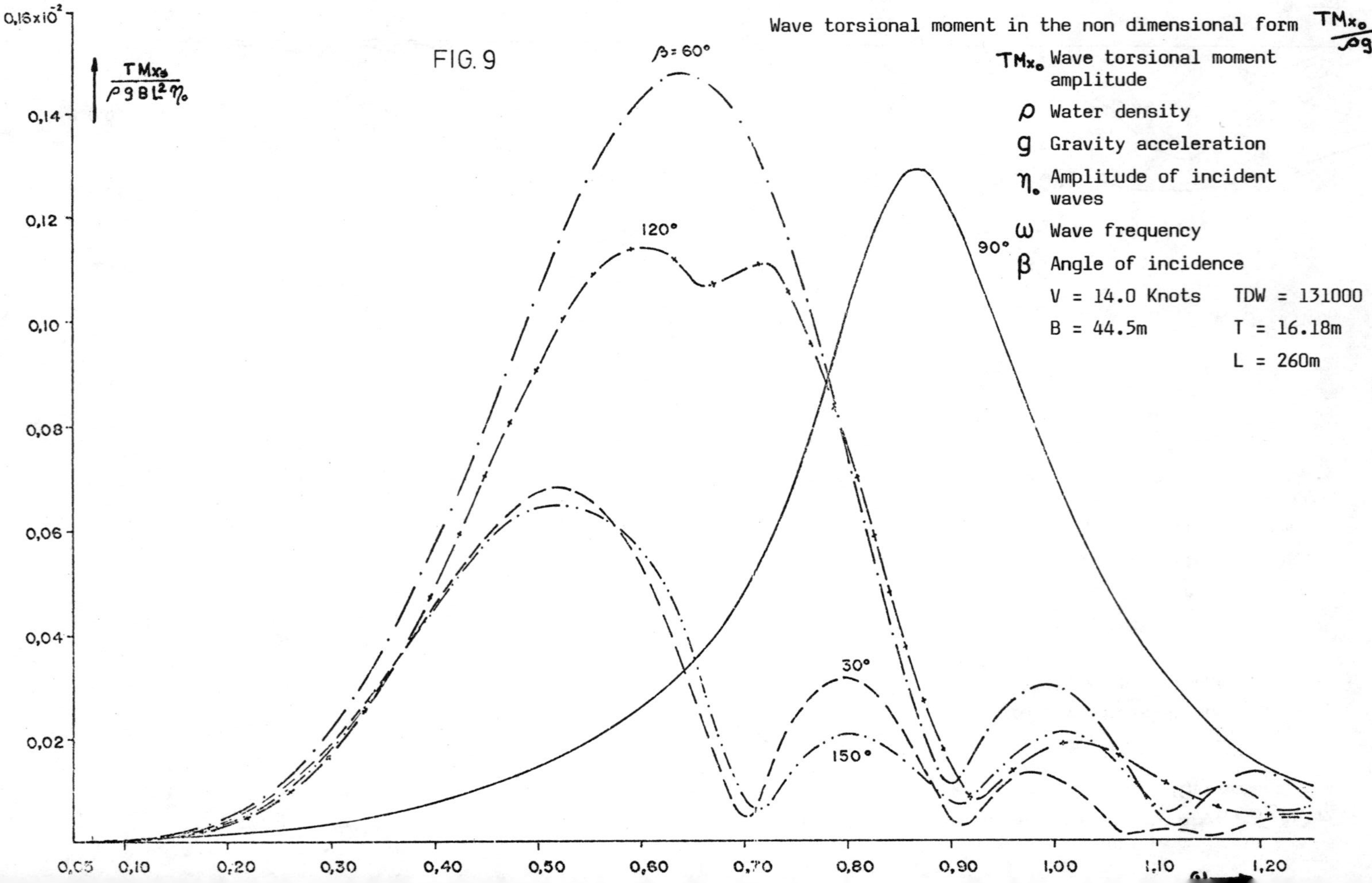

Wave torsional moment in the non dimensional form $\frac{TM_{x_0}}{\rho g B L^2 \eta_0}$
TM_{x_0} Wave torsional moment amplitude
ρ Water density
g Gravity acceleration
η_0 Amplitude of incident waves
ω Wave frequency
β Angle of incidence
V = 14.0 Knots
TDW = 131000
B = 44.5m
T = 16.18m
L = 260m
$\frac{TM_{x_0}}{\rho g B L^2 \eta_0}$
β = 60°
120°
90°
30°
150°
0,16x10⁻²
0,14
0,12
0,10
0,08
0,06
0,04
0,02
0,05
0,10
0,20
0,30
0,40
0,50
0,60
0,70
0,80
0,90
1,00
1,10
1,20
ω

FIG. 9

C_{jk} - are the hydrostatic restoration forces coefficient
F_j - are the forces and moments complex amplitudes due to ϕ_ω and ϕ_{DIF}.
ω_e - is the encounter frequency.

With the assumptions of small wave and motion amplitudes we can transfer the boundary conditions to the mean position of the body and to the plane z = 0. In the case of a ship we can consider it as a slender body. With this assumption and by considering the frequency as high the three dimensional boundary value problem can be reduced to several two-dimensional boundary value problems, the strip theory [6].

In the case of a moving platform made up of many slender elements, Hooft [2] has proposed several simplifying assumptions, for instance, the free surface effect for an immersed element can be neglected depending on the immersion height.

In the case of a body for which the dimensions do not permit a two dimensional analysis, as the case of many floating oil tanks, Faltinsen and Michelsen [10] have proposed a numerical solution utilizing a distribution of sources over the hull to represent the potential function.

$$_i(\underset{\sim}{r}) = \int_S q_i(r')\ G(r,r')\,dS \qquad i = 1,7 \tag{6}$$

where

$q_i(r')$ - the source intensity
$G(r,r')$ - the Green function; [18] and [10]
dS - an area element

RESPONSES IN REGULAR WAVES

In this paper we will use the strip theory as proposed by Kaplan [3]. Based on this formulation Raff [19] has developed later a computer program called SCORES. With the solution of the equations of motion we can calculate the forces on each ship section and then shear forces and moments.

Figures 2,3,4,5,6,7,8 and 9 present results of the ship responses in regular waves for heave pitch, vertical bending moments, sway, yaw, roll lateral bending moment and torsional moment amplitudes for a 131000 tdw tanker with the main dimensions shown in table 1. In these figures nondimensional responses are presented as function of the wave frequency ω and the angle β between the ship velocity vector and the wave celerity vector (See Fig. 1).

TABLE 1

Tanker Main Characteristics

Lpp	= 260.0 m (length between perpendiculars)
B	= 44.5 m (breadth)
T	= 16.18 m (draft)
DWT	= 131000 tonns (deadweight)
V	= 14 knots (velocity)
∇	= 155265.51 m^3 (displacement)

THE SHORT-TERM PREDICTION

As relations between the motions, bending moment and torsional moment amplitudes and the wave amplitude are linear we can apply the superposition principles to predict the ship responses in irregular seas. Hence, the response spectra can be calculated by [19]:

$$[f_i(\omega,\mu)]^2 = [T_i(\omega,\mu)]^2 \, [f(\omega,\mu)]^2 \tag{7}$$

where

$T_i(\omega,\mu)$ - the response amplitude operator (RAO)

$[f_i(\omega,\mu)]^2$ - the response spectrum

$[f(\omega,\mu)]^2$ - the sea directional spectrum

and

$$[f(\omega,\mu)]^2 = [f(\omega)]^2 \, \bar{y}(\mu) \tag{8}$$

where

$|\underline{f}(\omega)|^2$ - is the unidirectional spectrum

$|y(\mu)|$ - is the directional function

ω - wave frequency

μ - angle between the principal wave direction and the secondary waves direction

The unidirectional wave spectrum to be used here is the modified Pierson-Moskowitz Spectra:

$$\frac{[f(\omega)]^2}{H^2} = 0.11\, \omega_v^{-1} \left(\frac{\omega}{\omega_v}\right)^{-5} \exp\left[-0.44\left(\frac{\omega}{\omega_4}\right)^{-4}\right] \tag{9}$$

where

$$\omega_v = \frac{2\pi}{\tilde{T}}$$

H - significant wave height

$\tilde{T}$ - apparent mean period

The directional functions to be used are those proposed by

St. Denis and Pierson [13].

$$\bar{y}(\mu) = \frac{2}{\pi}\cos^2\mu \qquad \frac{\pi}{2} \leq \mu \leq \frac{\pi}{2}$$
$$= 0 \qquad \frac{\pi}{2} \leq \mu \leq 3\frac{\pi}{4} \tag{10}$$

If we have the RAO for a characteristic of a ship which advances with constant velocity, and she encounteres the sea wave, with an angle β between the ship velocity direction and wave celerity direction, the mean squared value $R^2_{\beta i}$ can be obtained by

$$R^2_{\beta i} = \frac{2}{\pi} \int_{-\pi/2}^{\pi/2} \int_{o}^{\infty} |f(\omega)|^2 \, |T_i(\omega,\mu)|^2 \cos^2\mu \, d\omega \, d\mu \tag{11}$$

Since the distribution of peaks or maxima of a narrow band normal distributed process follows the Rayleigh probability density function (p.d.f.), with Rayleigh constant $R_{\beta i}$ we can describe statistically the short term distribution of all the motion and wave moments acting on the ship.

LONG TERM PREDICTION

When the short term parameter "$R_{\beta i}$" is known, the probability q_β $(A_i > A_{i_L})$, that the variable A_i will be greater than a fixed A_{i_L} is given by

$$q_\beta(A_{i_L} > A_{i_L}) = \exp\left(-\frac{A^2_{i_L}}{2R^2_{\beta i}}\right) \tag{12}$$

But this is a conditional probability depending on the determined sea state with parameter $R_{\beta i}$. To obtain the probability for a long period of time, we should know the p.d.f. $p(R_{\beta i})$ of a certain sea state, i.e. of a certain combination $H_{1/3}, \tilde{T}$. If we know the p.d.f. $p(R_{\beta i})$ we have [20]:

$$Q_\beta(A_i > A_{i_L}) = \int_o^\infty \exp\left(-\frac{A_{i_L}}{2R^2_{\beta i}}\right) p(R_{\beta i}) \, dR_{\beta i} \tag{13}$$

This integral can be obtained by the sum,

$$Q_\beta(A_i > A_{i_L}) = \sum_\alpha \exp\left(-\frac{A^2_{i_L}}{2R_{\beta i\alpha}}\right) p(R_{\beta i\alpha}) \tag{14}$$

where

$p(R_{\beta i\alpha})$ is the p.d.f. that the ship in her life will be in a specific ocean area, with a short term behaviour characterized by $R_{\beta i\alpha}$.

The long term prediction will be considered for the regions printed out in Fig. 10. For these areas there is already a good number of visual observations to describe the long term wave prediction as published by Hogben and Lumb [17]. They correspond to the values of mean visual height and mean visual period. For the purpose of this paper we will use these values respectively as the significant wave height and the apparent mean period.

If we have in mind that $p(R_{\beta i\alpha})$ is the p.d.f. which describes the frequency of occurrence of the many sea states i.e. of a pair $(H_{1/3},T)$ we could say the p.d.f. $p(R_{\beta i\alpha})$ is given from the table of occurrence of the pair $(H_{1/3}, \tilde{T})$ for each region of the Brazilian shore as presented in [17].

We define p_{jk} as the long term frequency of occurrence of a sea state with mean visual height H_j and mean visual period T_k for each area, i.e. p_{jk} are the values given in the tables in [17], as shown above. We assume that β can assume seven angles β_ℓ between 0 and 2π. With this notation. We can write $R_{ijk\ell}$ as the short-term parameter of the ship characteristic i operating in a region with visual wave height H_j and apparent mean period T_k and having an incidence angle β_ℓ. (We have not considered the ship velocity variation with the roughness of the sea; although for a more detailed analysis it must be done). Then the long term probability Q_ℓ that the values A_i of i will be greater than A_{i_L}, for a constant incidence wave angle β_ℓ:

$$Q_\ell(A_i > A_{i_L}) = \sum_j \sum_k \exp\left(- \frac{A_{i_L}^2}{2R_{ijk\ell}^2} \right) p_{jk} \qquad (15)$$

Considering that the occurrence probability of each of the incidence angle β_ℓ is uniformly distributed from 0 to 2π we will have for the long term probability Q that $A_i > A_{i_L}$.

$$Q(A_i > A_{i_L}) = \frac{1}{N_\beta} \sum_\ell Q_\ell(A_i > A_{i_L}) \qquad (16)$$

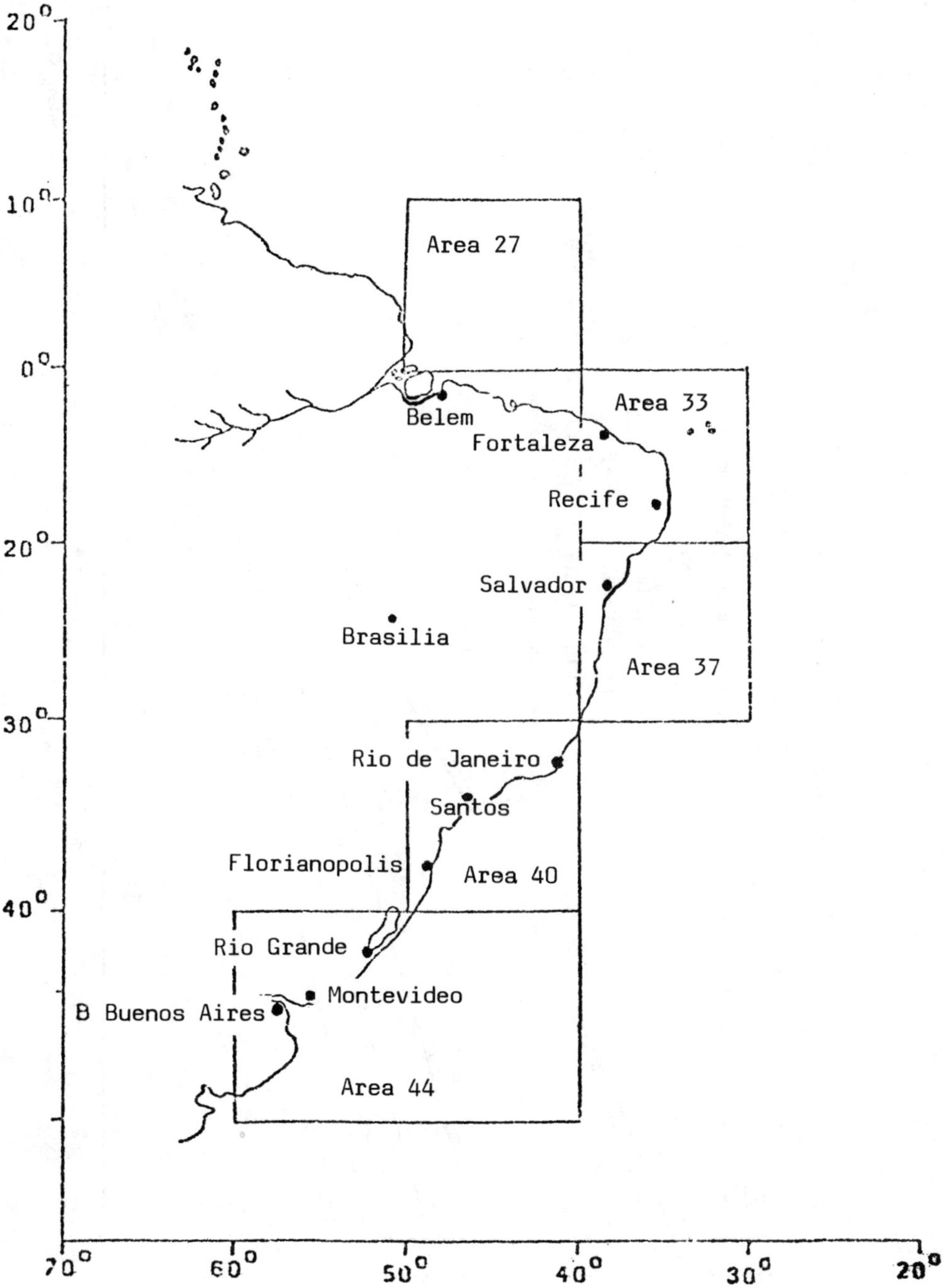
20°
10°
0°
20°
30°
40°
Area 27
Area 33
Area 37
Area 40
Area 44
Belem
Fortaleza
Recife
Salvador
Brasilia
Rio de Janeiro
Santos
Florianopolis
Rio Grande
Montevideo
B Buenos Aires
70°
60°
50°
40°
30°
20°

FIG.10

FIG. 11

Long-term distributions of heave amplitude

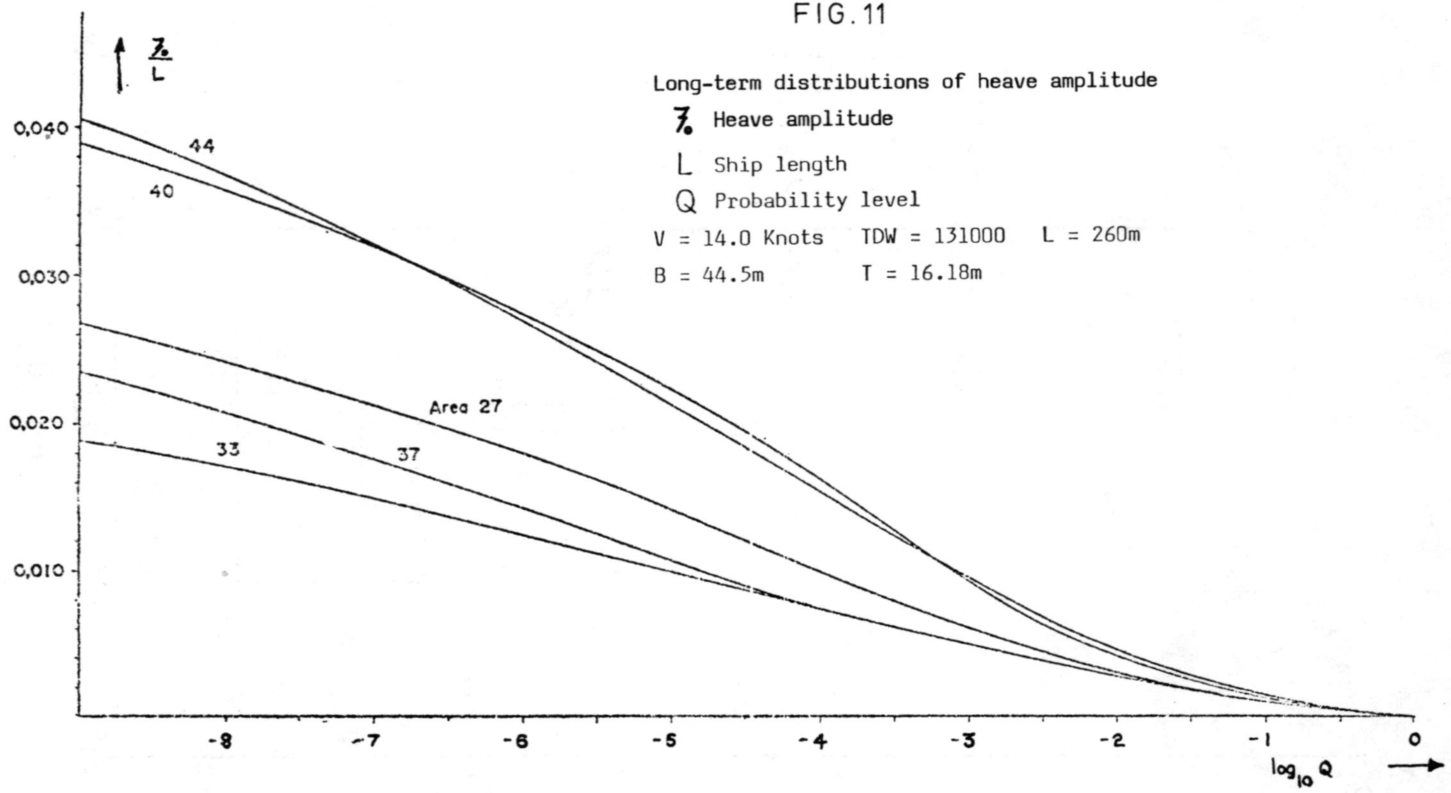

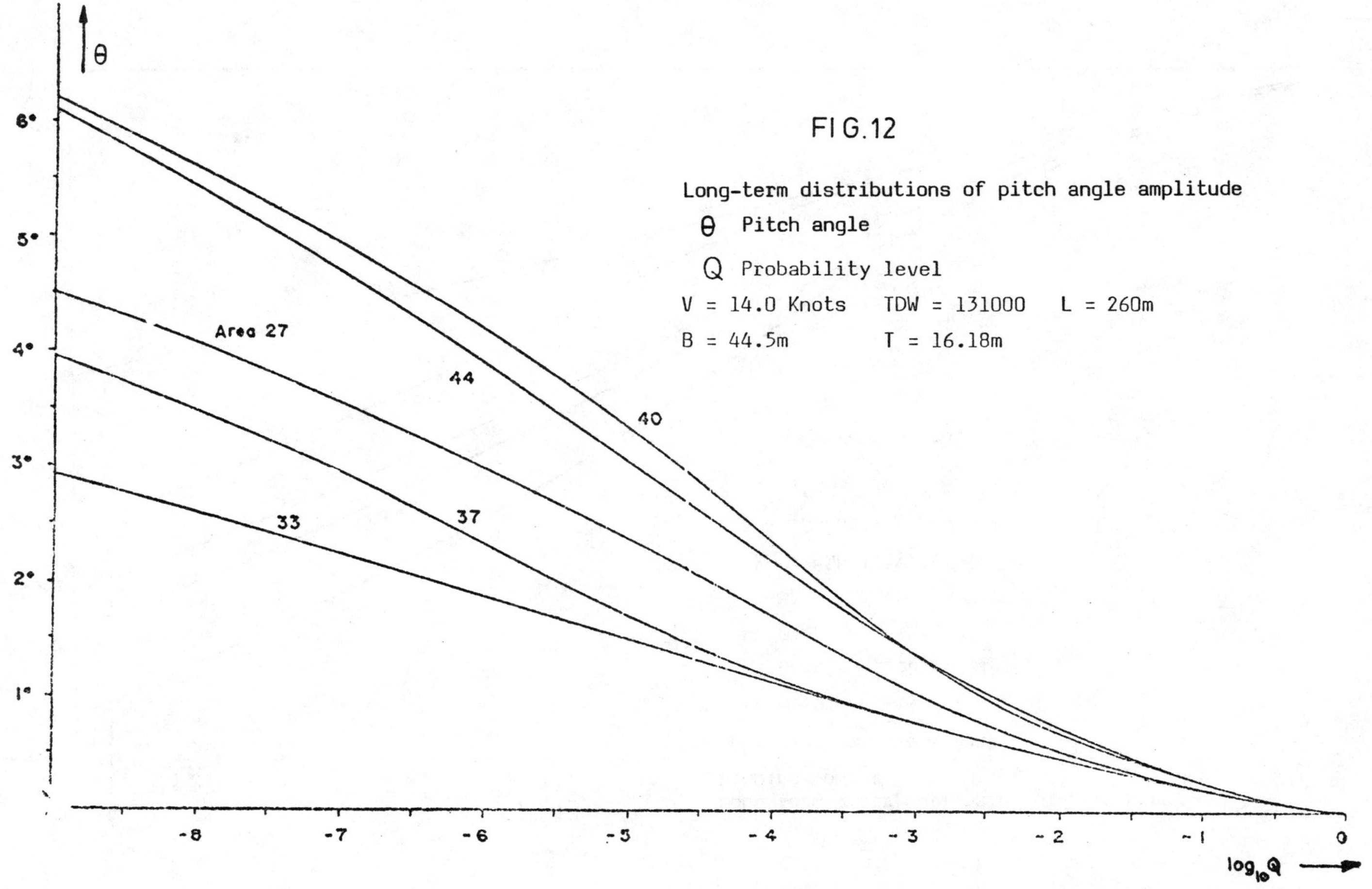

FIG.12

Long-term distributions of pitch angle amplitude

θ Pitch angle

Q Probability level

V = 14.0 Knots TDW = 131000 L = 260m

B = 44.5m T = 16.18m

FIG.13

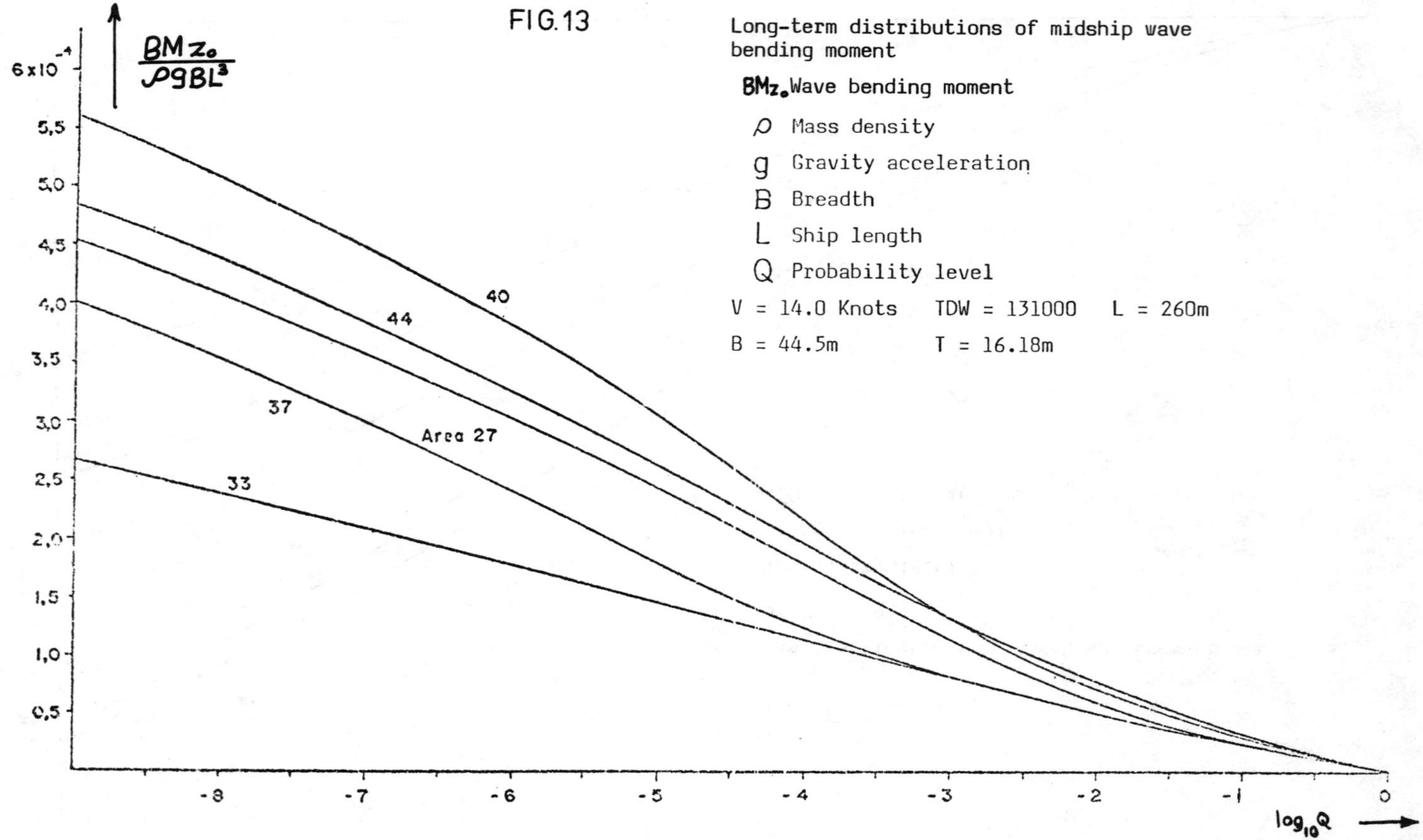
$\frac{BM_{Z_0}}{\rho g B L^3}$
6 x 10^{-4}
5,5
5,0
4,5
4,0
3,5
3,0
2,5
2,0
1,5
1,0
0,5
40
44
37
Area 27
33
-8
-7
-6
-5
-4
-3
-2
-1
0
$\log_{10} Q$
Long-term distributions of midship wave bending moment
BM_{Z_0} Wave bending moment
ρ Mass density
g Gravity acceleration
B Breadth
L Ship length
Q Probability level
V = 14.0 Knots TDW = 131000 L = 260m
B = 44.5m T = 16.18m

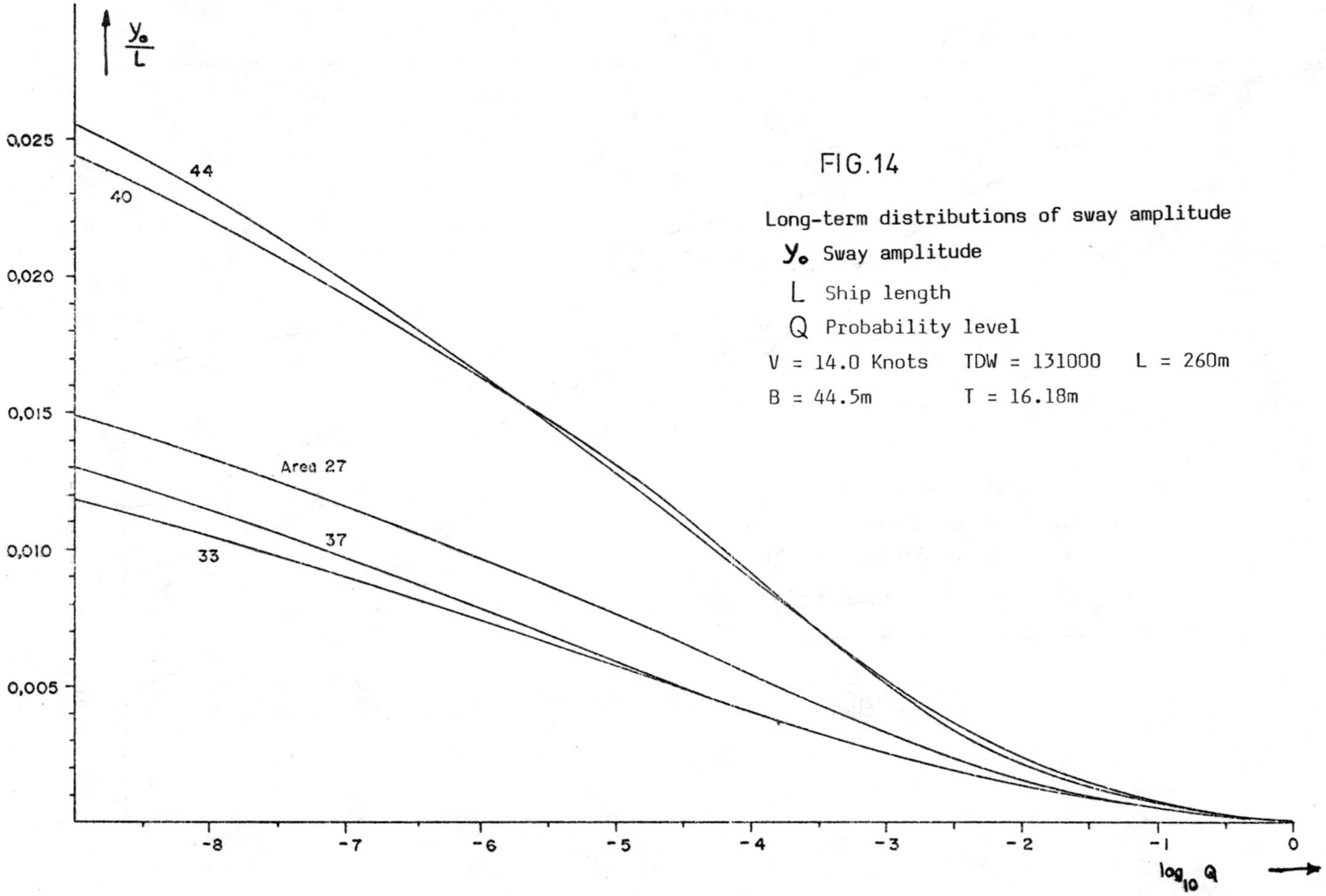

FIG.14

Long-term distributions of sway amplitude

y_o Sway amplitude

L Ship length

Q Probability level

V = 14.0 Knots TDW = 131000 L = 260m

B = 44.5m T = 16.18m

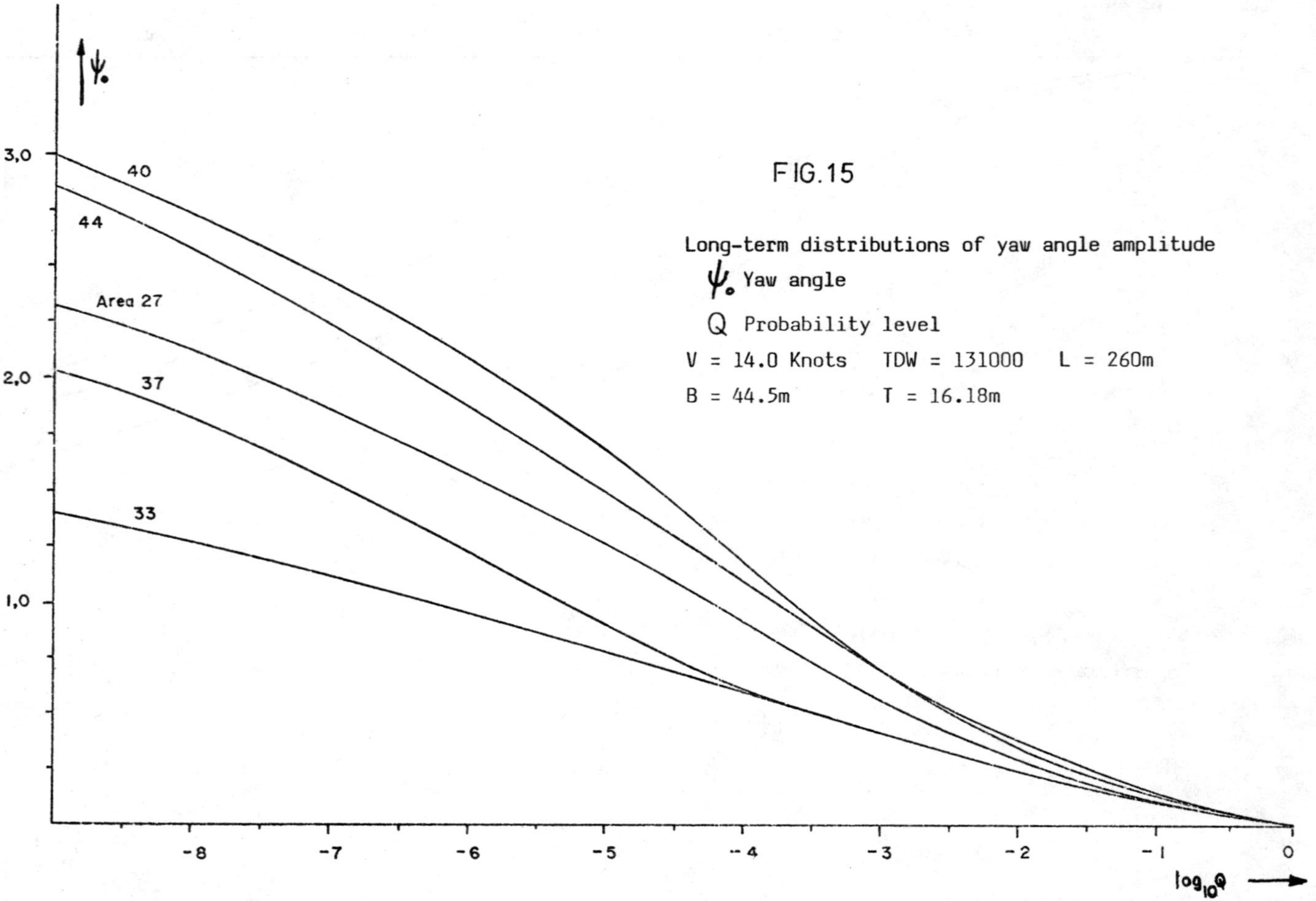

FIG.15

Long-term distributions of yaw angle amplitude

ψ_0 Yaw angle

Q Probability level

V = 14.0 Knots TDW = 131000 L = 260m

B = 44.5m T = 16.18m

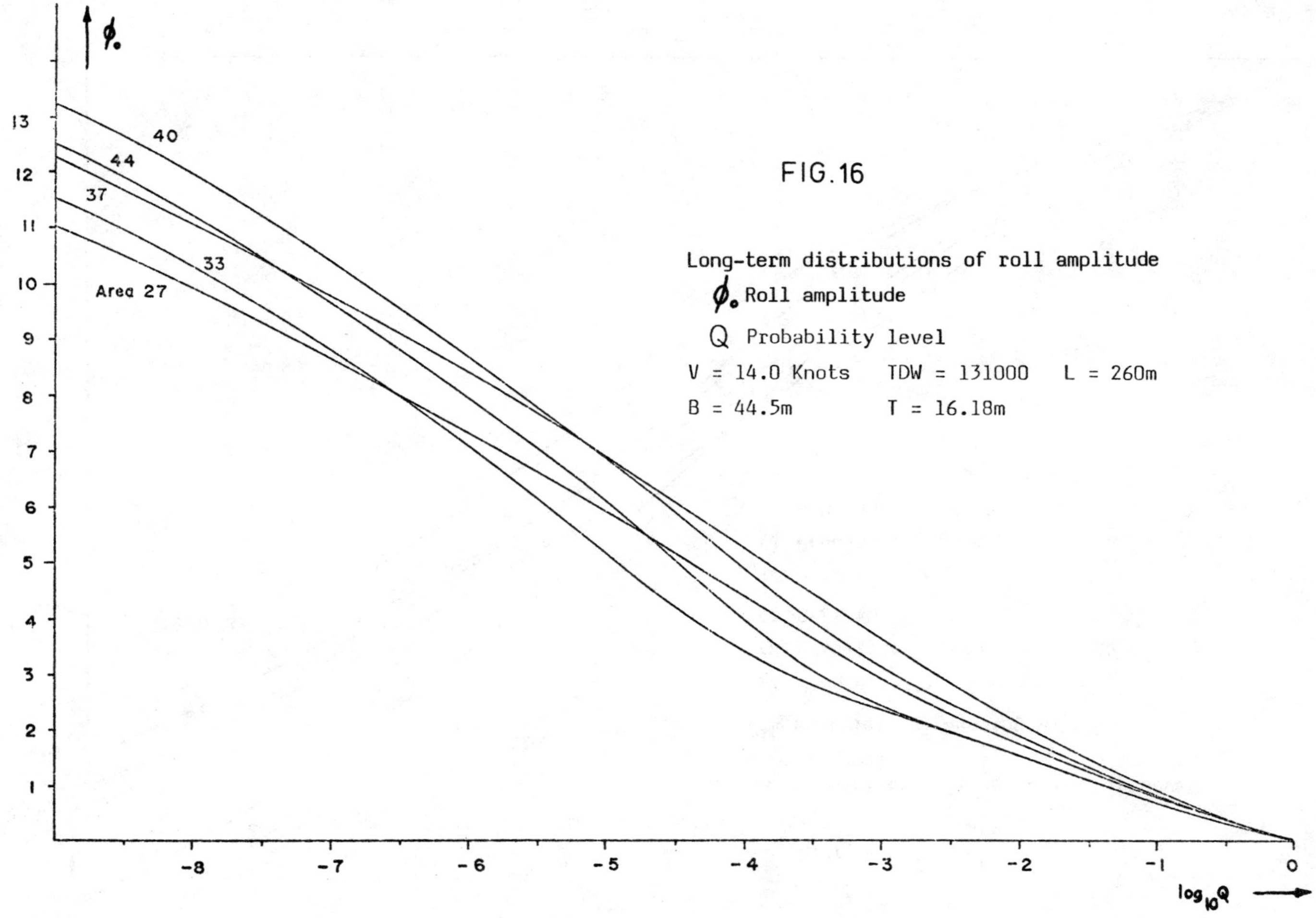

FIG.16

Long-term distributions of roll amplitude

ϕ_0 Roll amplitude

Q Probability level

V = 14.0 Knots TDW = 131000 L = 260m

B = 44.5m T = 16.18m

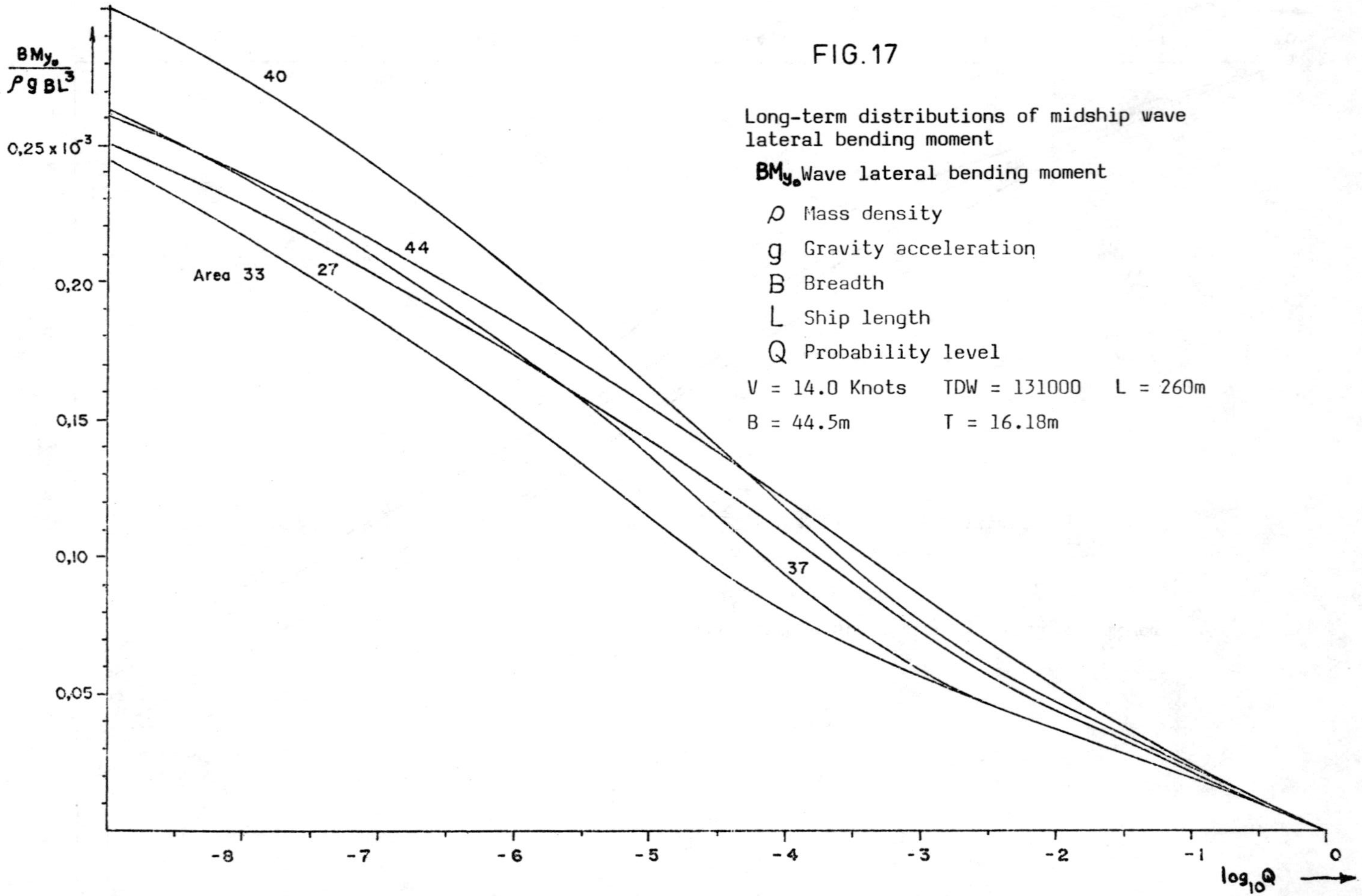

FIG.17

Long-term distributions of midship wave lateral bending moment

BM_{y_e} Wave lateral bending moment

ρ Mass density

g Gravity acceleration

B Breadth

L Ship length

Q Probability level

V = 14.0 Knots TDW = 131000 L = 260m

B = 44.5m T = 16.18m

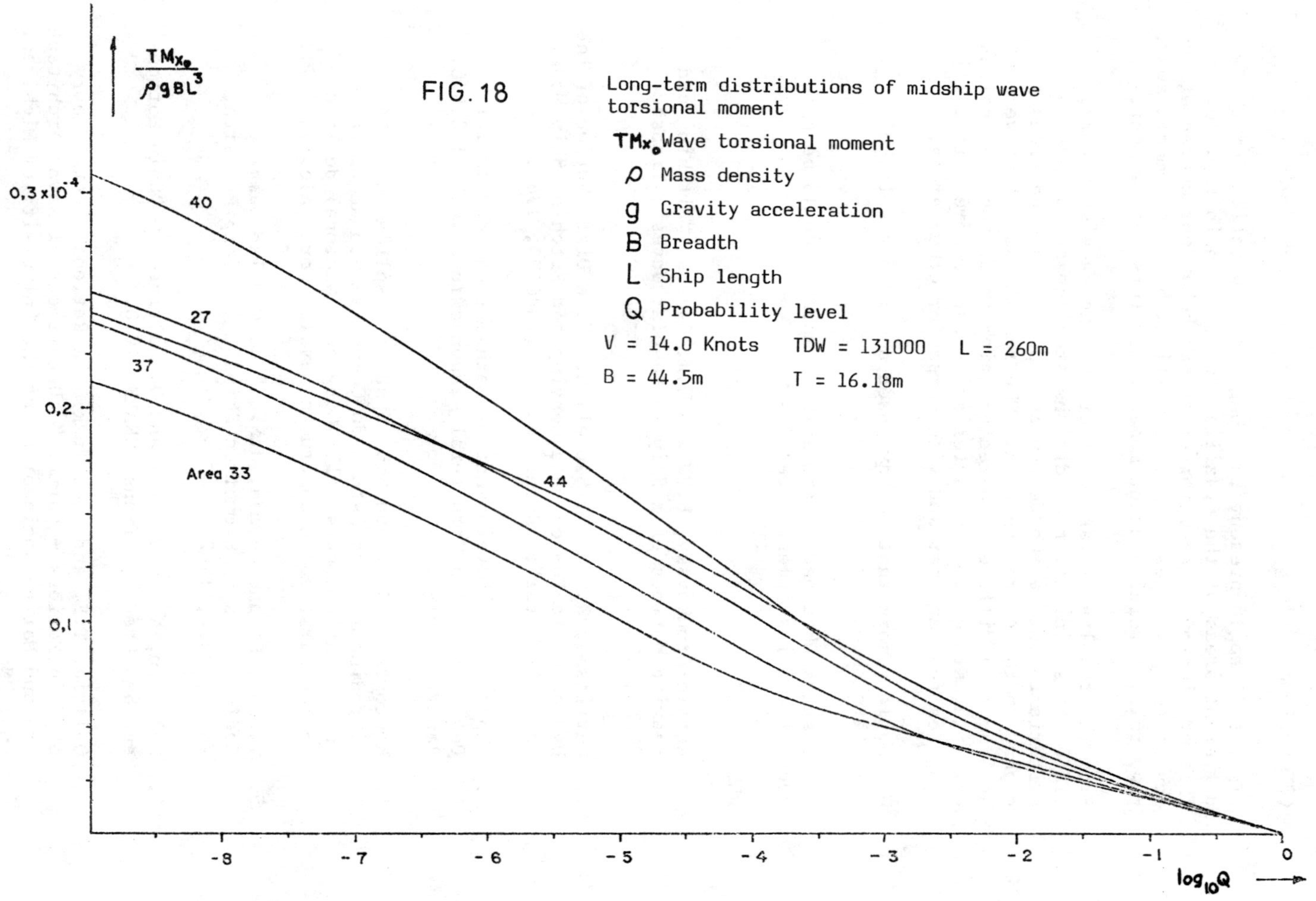

FIG. 18

Long-term distributions of midship wave torsional moment

TM_{x_0} Wave torsional moment

ρ Mass density

g Gravity acceleration

B Breadth

L Ship length

Q Probability level

V = 14.0 Knots TDW = 131000 L = 260m

B = 44.5m T = 16.18m

CONCLUSIONS

Figures 11 to 18 present the long term prediction for the different areas of the Brazilian coast shown in Fig. 10. This prediction is done for eight sea behaviour characteristics of the ship: heave, pitch, vertical bending moment, sway, yaw, roll, lateral bending moment, and torsional moment.

From the results we can see that for the same structure, different solicitation conditions are imposed so that the criteria to define the design and operations conditions depend on the operating area. The results show that we must analyse many different characteristics to define the design criteria, since some characteristics can be more critical in one area than another and vice-versa for other areas.

We should remark that we have used observed results of voluntary observing ships [17].

The results obtained by the method employed could be more accurate if instrumental results were known.

REFERENCES

1. Korvin-Kroukovsky, B.V. and W.R. Jacobs "Pitching and Heaving Motions of a Ship in Regular Waves", Transaction SNAME, 1957.

2. Gerritsma, J. and W. Beukelman. "The Distribution of the Hydrodynamic Forces on a Heaving and Pitching Ship Model in Still Water", Report No. 615, NRCTNO, 1964.

3. Kaplan, P. "Development of Mathematical Models for Describing Ship Structural Response in Waves" I.S.S.C. Report SSC-193, 1969.

4. Rostovtsev, D.M. and S.H. Sphaier "Aplicacão de Métodos Hidrodinâmicos ao Estudo do Comportamento de um navio em Ondas Longitudinais", III Congresso Nacional de Transportes Marítimos e Costrucão Naval - Porto Alegre, 1970.

5. Timman, R. and Newman, J.N. "The Coupled Damping Coefficients of Symmetric Ships", J. Ship Research, Vol.5, No.4, 1962.

6. Salvesen, N., E.O. Tuck and O. Faltinsen "Ship Motions and Sea Loads", Trans. SNAME, 1970.

7. Ogilvie, T.F. and E.O. Tuck "A Rational Strip Theory of Ship Motions - Part I " Department of Naval Architecture and Marine Engineering of the University of Michigan, Publ. No. 013, 1969.

8. Faltinsen, O.M. "A Numerical Investigation of the Ogilvie Tuck Formulas for Added-Mass and Damping Coefficients", J. Ship Research, Vol.18, No.2, 1974.

9. Sphaier, S.H. "Näherungsweise Bestimmung von Hydrodynamischer Masse und Dämpfung in Verschiedenen Koordinatensystemen", Technische Universität Berlin, Dissertation, 1976.

10. Faltinsen, O.M. and F.C. Michelsen, "Motions of Large Structures in Waves at Zero Froude Number" Symposium on Marine Vehicles, London, February, 1974.

11. Faltinsen, O. and F.C. Michelsen, "Hydrodynamic Forces Acting on Drilling Platforms" Publicacao Didática nº4.74 COPPE/UFRJ, 1974.

12. Hooft, J.P. "Hydrodynamic Aspects of Semi-Submersible Platforms" Publication No. 400 N.S.M.B., Wageningen, Holland.

13. St. Denis, M. and W. Pierson, "On the Motion of Ships in Confused Seas", Transaction SNAME, 1953.

14. Fukuda, J. "Theoretical Determination of Design Wave Bending Moments" Japan Shipbuilding and Marine Engineering, Vol.2, No.3, 1967.

15. Neves, M.A.S. "Previsão de Longo Prazo do Momento Fletor em Ondas, Incidentes em Estruturas Flutuantes - Simposium sobre Tendências Atuais no Projeto e Execucão de Estruturas Marítimas - Rio de Janeiro, UFRJ, April, 1977.

16. Neves, M.A.S. and Sphaier, S.H. "Comportamento de Navios em Ondas: Previsão de Longo Prazo para Áreas da Costa Brasileira" 5º Congresso Panamericano de Ingenieria Naval, Transporte Marítimo e Ingenieria Portuária - Caracas, Venezuela, 1977.

17. Hogben, N. and F.E. Lumb, "Ocean Wave Statistics" Ministry of Technology, NPL, London, 1967.

18. Wehausen, J.V. and E.V. Laitone, "Surface Wave" Handbuch der Physik, Vol. IX, 1960.

19. RAFF, A.I. "Programa SCORES - Ship Structural Response in Waves" I.S.S.C. - Report SSC-230, 1972.

20. Jasper, H. "Statistical Distribution Patterns of Ocean Waves and of Wave-Induced Ship Stresses and Motions with Engineering Applications", Transaction SNAME, 1956.

WAVE REGIMEN STUDY ON BRAZILIAN COAST - TENDENCIES FOR SHALLOW AND DEEP WATER

Alberto Homsi

Portobrás - INPH

ABSTRACT

In this paper the characteristics of wave agitation along the Brazilian coast for shallow and deep water, are being presented. They can be used to estimate wave motions in coastal works and offshore projects.

In the case of shallow water, the wave characteristics of agitation were determined based on wave measurements totalling more than 9000 registers of 20 minutes,each obtained from 2 to 5 places along each one of the coastal regions - South, Southeast, Northeast, the largest part being within the years 72/74. The measurements were usually taken at depths between 10 to 22 m. These data have been used to find the H_s and H_{max} distribution curves for NE, E, SE and S wind directions.

From these H_s and H_{max} distribution curves, values for 30 and 50 years of recurrence period were taken and used to construct shallow water tendency curves. By means of refraction diagrams, the shallow water wave characteristics have been transposed to deep water, and tendencies of agitation in deep water along the Brazilian coast were obtained. These results were compared with deep water wave height tendencies curves, similarly obtained from visual observation.

INTRODUCTION

This paper describes a part of the studies which are being developed in the "Instituto de Pesquisas Hidroviárias" (INPH)-PORTOBRÁS, with views to attain a better knowledge of the waves regimen along the Brazilian coast in shallow and deep water. The detailed knowledge of the wave regimen along the Brazilian coast may be only reached with the help of a complex system of wave observation stations, separated by 200 to 300 kms to gather information continuously for periods of at least

Table 1

	MEASURE-MENT PERIOD	TZ (sec)	NUMBER OF REGISTERED	HIGHEST WAVES OBSERVATION MONTHS	WATER DEPTH (m)	-15.00 m BOTTOM CONTOUR DIRECTION	WAVES DIRECTION	
							MORE FREQUENT	HIGHEST WAVE
TRAMANDAI	10/72 to 9/73	8	335	April	-17.50	18^oN	100° to 120°	120°
S. FRANCISCO DO SUL	6/73 to 3/74	8	109	July		28^oN		
PARANAGUÁ	4/72 to 3/73	6	553	July/dec.	-16.00	38^oN	SE	SE
SANTOS	11/72 to 11/73	11	540	Jan/May July	-10.00	65^oN	SE	SE and S
	7/73 11/73	11	150	Jan/May July	-15.00	65^oN	SE	SE and S
MACAÉ	Jun/o 1977	10	120	August	-17.00	64^oN	SE	SE
TUBARÃO	6/71 to 4/72	10	387	Jul/Sept.	-9.00	40^oN	SE and S	S
PORTOCEL	9/73 to 11/76	7(P1) 8(P2)	3,800	July	-16.00	36^oN	SE and S	SE
RIO DOCE	10/72 to 5/73	7(P1) 8(P2)	187	Mar/May	-22.00(P1) -53.00(P2)	65oN	ENE and E	NE and S
ARACAJU	9/65 to 8/66	8	730	Jun/Aug	-20.00	35^oN	110° to 140°	130 to 140
MACEIO	8/72 to 6/73	7 to 10	2,000	Jun/July	-10.00	41^oN	140°	140° to 160°
RECIFE	7/76 to 2/77	10	400	Jul/Sept.	-10.00	4^oN	SE	102^oN

2 to 3 years. Up to the present moment around 9000 registers of 15 to 20 minutes are available, for 11 points from Tramandaí to Recife i.e. for a distance of approximately 3500 km. These registers represent periods of measurements around 10 to 15 months for each point during the years: 71/77. Besides these data, wave observation is being collected at two other points (Natal and Suape) for a 12 months period. This will provide 4000 new registers.

A wave observation station may be installed in Praia Mole - future Sea harbour near Vitória, ES. The more information is gathered, the better should be the understanding of the wave regimen. Regardless the present state of information, "Tendency of Agitation" in shallow water is predicted.in this paper.

BRAZILIAN COAST - SHALLOW WATER WAVES

Table 1 shows some wave characteristics in the Brazilian coast. All data with the only exception of Rio Doce data - measured in almost deep water condition (-53.00 m), may be considered as shallow water data. Among all field measurements, only Rio Doce's didn't include the Winter period. Generally, the highest waves come from SE direction, with exception of Rio Doce, where NE and S waves, of periods of 7 and 8 sec have not yet suffered refraction. All these wave height and direction data were statistically treated, and accumulated frequencies were obtained for each zone and for NE, E, SE and South wave directions, and plotted on log-paper (H_s and H_{max}). For 30 and 50 years of recurrence period H_s and H_{max} were estimated and the "Tendencies along the Brazilian Coast - Shallow Water" - figure 1 enclosed, were obtained for NE, E, SE and South directions. In case of NE direction, there is a tendency in the agitation to decrease from the area of Portocel towards Maceió. For E direction, the waves are higher in the South coast, decreasing and passing through a minimum, around Maceió and then rising a little more, in direction to Recife. The wav waves which arrive to the coast with SE direction, have high values in the region of Tramandaí, decreasing until passing by a minimum in Maceio and rising again a little towards Recife. The South waves have the tendency of decreasing from Santos to Maceió.

THE BRAZILIAN COAST-DEEP WATER WAVES AFTER SHALLOW WATER MEASUREMENTS

For each one of the 11 wave record stations, refraction studies have been carried out in order to extrapolate for deep water the shallow water parameters obtained. It was observed that refraction causes the waves crest coming from NE and East to spill clockwise (+), for all the 11 points. The spilling is counterclockwise (-) for SE direction with the exception of

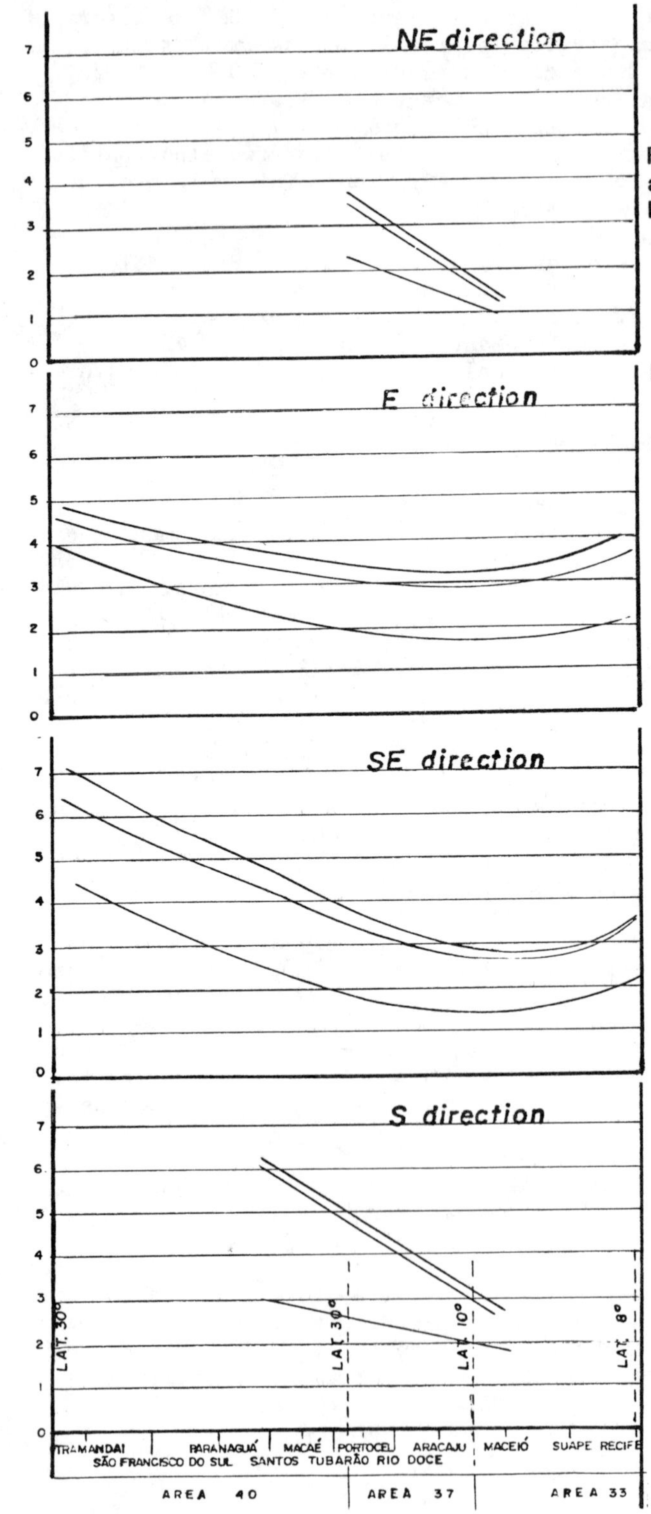

Figure 1a. Tendencies along shallow water-Brazilian Coast.

H_s(m) Observed
(30 and 50 years recurrence period)

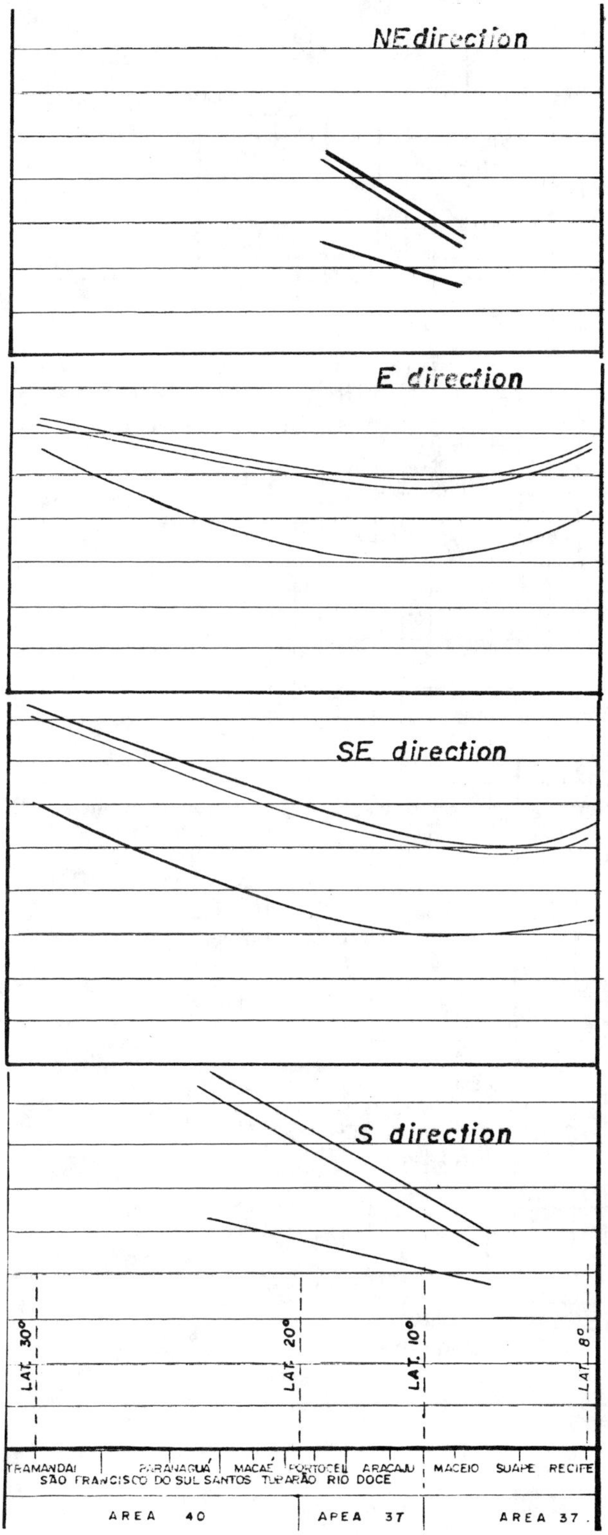

Figure 1b. Tendencies along shallow water-Brazilian Coast.

H_{max}(m) Observed (30 and 50 years recurrence period)

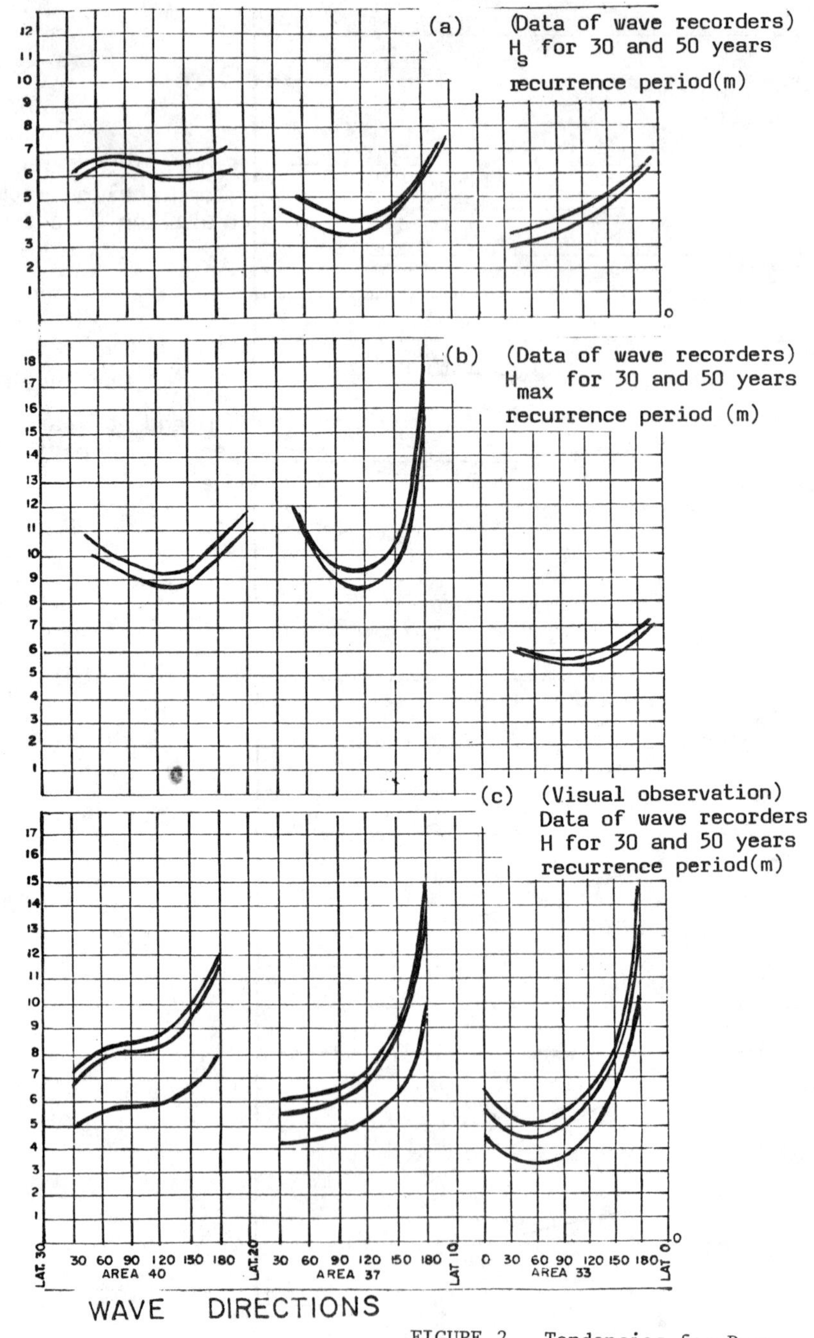

FIGURE 2 Tendencies for Deep Water Waves

Santos and Tubarão. For South direction, the spilling is counterclockwise for all the coast.

To compare these results with the visual observation data ("Ocean Wave Statistics"), they were grouped (see figure 2-a and 2-b) in function of the areas MS 40, 37 and 33, taking for each one of the NE, E, SE and South directions, the average of the heights obtained for the points in each area. The values obtained were plotted to give the "Deep Water Wave Height Tendencies of the Brazilian Coast". Generally, the highest waves for all directions are found on area MS 40, followed by areas 37 and 33.

THE BRAZILIAN COAST DEEP WATER WAVES - VISUAL OBSERVATIONS DATA (OWS)

The visual observation data from "Ocean Wave Statistics" have been treated similarly as were treated the shallow water data and "Tendencies for deep water waves" were obtained for period of recurrence of 1 - 30 and 50 years, as shown in figure 2-c.

CONCLUSIONS

Within certain limitations, the curves obtained for H_s and H_{max} in shallow waters can be used for structural designs in coastal regions. For the case of deep water, the obtained curves for H_s 30, H_s 50, H_{max} 30 and H_{max} 50 could be useful in projects of "off-shore" structures.

The study should be continued to include,

- the large data to be obtained from the Natal, Suape, Macaé and Praia Mole wave record stations
- the influence of wave periods.
- the assessment of homogeneous wave regimes in certain coastal regions.

WAVE OSCILLATION PROBLEMS IN DEEP AND SHALLOW WATERS

C.A. Brebbia, S. Walker and M. Kavanagh

Southampton University, England.

1. INTRODUCTION

In this paper we present a unified approach to the solution of two important engineering problems i) wave diffraction in deep water and ii) harbour oscillations. The deep water case is important in the design of offshore structures. The shallow water case occurs mainly in coastal waters and its more interesting application is for the analysis of semi-enclosed bodies of water such as harbours. The method may also be applied to lakes. In both cases the linear or Airy wave theory is used.

Offshore structures in deep water are subjected to forces due to inertia and drag effects. When the dimensions of the structure become large, drag - which is proportional to diameter - decreases in importance, and inertia - proportional to the cross sectional area - is the predominant effect. The large cross sectional area will disturb the incident waves causing wave diffraction. The diffraction analysis assumes irrotationality, as drag forces are neglected, this allows us to use the Helmholtz equation as the governing equation for the problem, in terms of the reduced velocity potential. Here we indicate how the theory can be used in conjunction with finite elements by using the radiation condition on the external sea boundaries.

Many harbours are limited in their usefulness by seiching associated with excitation at one or more of its natural frequencies. For harbours of regular shapes the natural frequencies can be found by analytical techniques, otherwise one needs to undertake expensive model experiments or turn to numerical techniques. The finite element method is well suited to the study of harbours of arbitrary dimensions as the size (and depth) of elements can be varied at will. This paper describes how finite element techniques can be used to take into account the radiation of wave energy from the harbour mouth by utilisation of the radiation condition. In addition wave reflection due to the surrounding coastline can be incorporated in the model. Most theoretical models used up to now have concerned themselves with predicting the natural periods of the water body, but have ignored the effect of friction between the bed and the moving fluid. In order to obtain realistic estimates of displacements the friction effects need to be taken into account.

The usual design procedure for both cases is to estimate the likely period of waves in the locality and then analyse the response of the system in one or the other case (shallow or deep waters). The designer tries to mismatch the frequencies of the incoming waves and the fundamental frequency of the system. This assumption of a deterministic design wave is in fact a drastic oversimplification of a real situation. Waves are essentially random in nature and this necessitates a probabilistic approach. The wave record at a point is, in fact, the superposition of an infinite number of harmonic components of different amplitudes and frequencies. This record is normally represented by a wave spectrum for the location which will be peaked at the dominant period of the waves. These recorded spectra which vary from region to region of the sea according to the local characteristics of the waves are then to be used as input for the statistical analysis.

The statistical analysis is based on a transfer function given by the system response to simple harmonic wave excitations. The transfer function converts the wave spectrum into a response spectrum of forces such as diffraction forces or wave heights (such as those inside a harbour). Integration of the response spectrum with respect to the frequencies gives the variance of the variable under consideration and this allows one to predict the probability of a maximum value being exceeded.

2. DIFFRACTION PROBLEMS

Consider first the case of a body in deep water. In this case we assume that the water is of constant depth (or that the undulations of the sea bed are much longer than the waves, and at a depth of more than half of a wave length). Assuming that the fluid is incompressible and in irrotational motion we can define a velocity potential by

$$\vec{v} = \left(\frac{\partial \Phi}{\partial x}, \frac{\partial \Phi}{\partial y}, \frac{\partial \Phi}{\partial z}\right) = (v_x, v_y, v_z) \tag{1}$$

The incompressibility condition then gives

$$\nabla^2\Phi = \frac{\partial^2\Phi}{\partial x^2} + \frac{\partial^2\Phi}{\partial y^2} + \frac{\partial^2\Phi}{\partial z^2} = 0 \qquad (2)$$

For a fluid of constant depth d we may write $\Phi = \Phi(x,y,z,t)$ as

$$\Phi(x,y,z,t) = \phi(x,y)\, f(z) e^{i\omega t} \qquad (3)$$

where ϕ is the reduced velocity potential and we are assuming harmonic motion.

The linearised boundary conditions on the free surface and the sea bed are:-

(a) Free surface $z = 0$

(i) Kinematic Condition

$$\frac{\partial\eta}{\partial t} = \frac{\partial\Phi}{\partial z} \qquad \text{on } z = 0 \qquad (4)$$

(ii) Pressure Condition

$$\frac{\partial\Phi}{\partial t} = -\, g\eta \qquad \text{on } z = 0 \qquad (5)$$

η is the surface elevation

hence

$$\frac{\partial\Phi}{\partial z} = -\frac{\omega^2}{g}\,\Phi \qquad (6)$$

(b) Sea bed $z = -d$

Zero normal flux condition

$$v_z = \frac{\partial\Phi}{\partial z} = 0 \qquad (7)$$

Taking these boundary conditions into account we obtain

$$f(z) = \frac{\cosh\left[\kappa(z+d)\right]}{\cosh \kappa d} \qquad (8)$$

where κ is the wave number ($\kappa = 2\pi/\lambda$, λ is wave length). It is easy to show that

$$\omega^2 = g\kappa \tanh \kappa d \qquad (9)$$

For deep water

$$\omega^2 = g\kappa \quad . \qquad (10)$$

The governing equation for the reduced potential now becomes,

$$\frac{\partial^2 \phi}{\partial x^2} + \frac{\partial^2 \phi}{\partial y^2} + \kappa^2 \phi = 0 \tag{11}$$

which is Helmholtz equation in the two horizontal variables.

Consider the case of an incident wave described by the ϕ_I potential and travelling in the positive x direction

$$\phi_I(x) = -\frac{ig}{\omega}\eta(x) \tag{12}$$

or $$\phi_I(x) = -\frac{ig}{\omega} a_o e^{-i\kappa x}$$

from equation (5),
where a_o is the wave amplified.

The total wave potential is equal to the incident plus the diffracted fields, i.e.

$$\phi = \phi_I + \phi_D \tag{13}$$

Hence we can substitute (13) into (11) which gives (taking into account that $\nabla^2\phi_I + \nabla^2\phi_I = 0$),

$$\nabla^2\phi_D + \nabla^2\phi_D = 0 \tag{14}$$

On a circle sufficiently far from the body (radius r) we can apply the following radiation condition

$$\frac{\partial \phi_D}{\partial r} + \frac{1}{c}\frac{\partial \phi_D}{\partial t} = 0 \tag{15}$$

where c is the wave celerity ($c = \omega/\kappa$).
For a harmonic wave this implies that

$$\frac{\partial \phi_D}{\partial r} + i\kappa\phi_D = 0 \tag{16}$$

Strictly speaking this boundary condition only applies if the circle is infinitely far from the body and sources producing the waves. In practice if the external sea boundary is three or four wavelengths from the obstacle reasonable results can be obtained with this boundary condition. An expression similar to (16) was deduced rigorously by Sommerfeld [1].

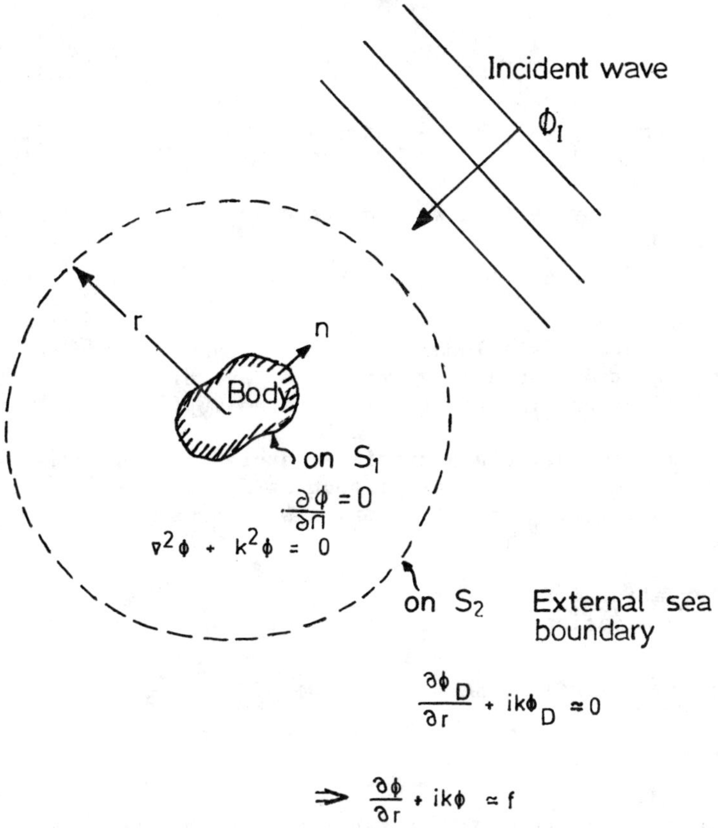

FIGURE 1 Definition Sketch for the Diffraction Problem

Using equation (13) we may re-write this condition in terms of the total potential to obtain

$$\frac{\partial \phi}{\partial r} + i\kappa\phi = \frac{\partial \phi_I}{\partial r} + i\kappa\phi_I \tag{17}$$

or
$$\frac{\partial \phi}{\partial r} + i\kappa\phi = f(r,\theta) \tag{18}$$

where f is a <u>known</u> function given in terms of the incident potential ϕ_I.

On the body surface we apply the usual zero normal flux condition

$$\frac{\partial \phi}{\partial n} = 0 \tag{19}$$

where n is a local coordinate measured into the fluid. For diffraction round a cylinder the governing equation and boundary conditions are illustrated in figure 1.

We can now write a variational expression in terms of ϕ and use a numerical technique such as finite elements to solve the problem. The above analysis is presented in more detail in reference |2|.

<u>Variational Statement</u>
The variational expression in terms of ϕ is

$$\iint \{\nabla^2\phi + \kappa^2\phi\}\delta\phi \, dA = \int_{S_1} \left(\frac{\partial \phi}{\partial n}\right)\delta\phi \, dS + \int_{S_2} \left(\frac{\partial \phi}{\partial n} + ik\phi - f\right)\delta\phi \, dS \tag{20}$$

where S_1 is the contour representing the boundary of the obstacle and S_2 is the external sea boundary. This expression can be integrated by parts to give

$$\iint \left\{ \frac{\partial \phi}{\partial x}\frac{\partial \delta\phi}{\partial x} + \frac{\partial \phi}{\partial y}\frac{\partial \delta\phi}{\partial y} - \kappa^2\phi \, \delta\phi \right\} dA + \int_{S_2} i\kappa\phi \, \delta\phi \, dS = \int_{S_2} f \, \delta\phi \, dS \tag{21}$$

Expression (21) will be used later in the finite element formulation.

If the force per unit depth on the obstacle is required, it may be computed by integrating the pressure on the obstacle, obtained from ϕ, by use of the linearised Bernoulli equation giving

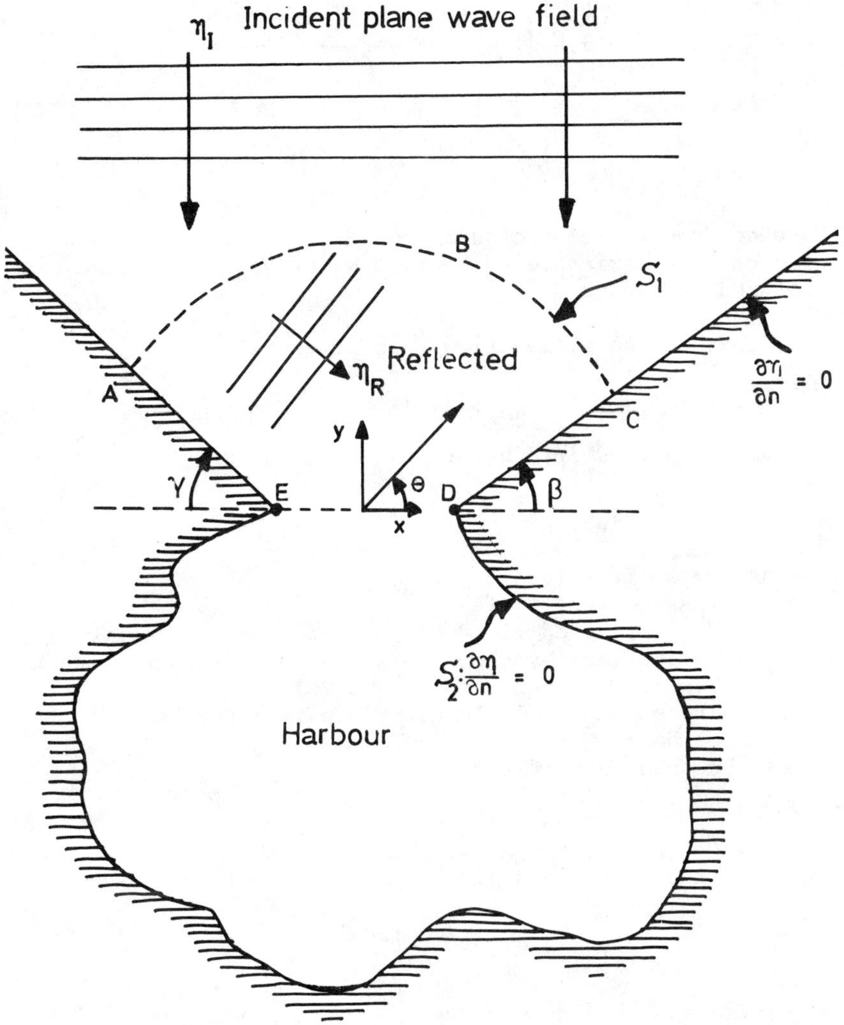

FIGURE 2 Definition Sketch for Harbour Oscillation Problem

$$F(z,\omega,t) = -i\omega\rho \int_{S_1} \phi \vec{n}\, dS \frac{\cosh|\kappa(z+d)|}{\cosh \kappa d} e^{i\omega t}$$

3. HARBOUR OSCILLATIONS

Consider now the case of shallow water, particularly the problem of a semienclosed body of water such as the harbour pictured in figure 2.

Note that the wave elevation η is given by

$$\eta = -\frac{i\omega}{g} \phi(x,y) \quad *$$

Substituting this into equation (11) we obtain

$$\nabla^2\eta + \kappa^2\eta = 0 \tag{22}$$

Let us now split η into 3 parts,

i) The incident field η_I.

ii) The reflected field calculated from plane wave reflections by the boundaries and called η_R (see figure 2, boundaries AG and CE).

iii) The scattered field η_S originating from inside the harbour.

Hence the total wave field is

$$\eta = \eta_I + \eta_R + \eta_S$$

with the incident field given by

$$\eta_I = a\, e^{-i\kappa r \cos(\theta-\alpha)}$$

The radiation condition we use on the sea boundary S_2 is now

$$\frac{\partial\eta_S}{\partial r} + i\kappa\eta_S = 0 \tag{23}$$

This gives

$$\frac{\partial\eta}{\partial r} + i\kappa\eta = \underbrace{\frac{\partial}{\partial r}(\eta_I + \eta_R) + i\kappa(\eta_I + \eta_R)}_{\text{known}} = f \tag{24}$$

The r.h.s. of this equation is a known function of the horizon-

*Note that now (shallow water case)

$$\omega^2 = g\kappa^2 d + \kappa^2 = \omega^2/gd$$

tal coordinates, the angles determining the local geometry and the angle of incidence of the wave field η_I. On the land boundaries S, we have $\partial\eta/\partial n = 0$. For the harbour oscillation problem the geometry and boundary conditions are shown in figure 2.

Variational Statement

We can now write (assuming an element of constant depth)

$$\iint (\nabla^2\eta + \kappa^2\eta)\,\delta\eta dA = \int_{S_1} \left(\frac{\partial\eta}{\partial n}\right)\delta\eta\, dS + \int_{S_2} \left(\frac{\partial\eta}{\partial n} + i\kappa\eta - f\right)\delta\eta\, dS \tag{25}$$

Integrating by parts

$$\iint \left(\frac{\partial\eta}{\partial x}\frac{\partial\delta\eta}{\partial x} + \frac{\partial\eta}{\partial y}\frac{\partial\delta\eta}{\partial y} - \kappa^2\eta\delta\eta\right) dA + \int_{S_2} i\kappa\eta\;\delta\eta\; dS = \int_{S_2} f\delta\eta\; dS \tag{26}$$

Note that this is an expression similar to (20).

4. FINITE ELEMENT MATRICES AND SYSTEM RESPONSE

Both expressions (20) and (26) can be written as

$$\iint \left\{ \frac{\partial u}{\partial x}\frac{\partial\delta u}{\partial x} + \frac{\partial u}{\partial y}\frac{\partial\delta u}{\partial y} - \kappa^2 u\delta u\right\} dA + \int_{S_2} i\kappa u\delta u\; dS = \int_{S_2} f\delta u\; dS \tag{27}$$

We now assume that the variable u can be approximated on each element by

$$u = \underset{\sim}{N}^T \underset{\sim}{U} \tag{28}$$

where N is an interpolation function vector and U are the nodal unknowns. Substituting (28) into (27) we obtain the following expression for an element

$$\delta\underset{\sim}{U}^T \left\{ \iint \left\{ \frac{\partial \underset{\sim}{N}}{\partial x}\frac{\partial \underset{\sim}{N}^T}{\partial x} + \frac{\partial \underset{\sim}{N}}{\partial y}\frac{\partial \underset{\sim}{N}^T}{\partial y} - \kappa^2\, \underset{\sim}{N}\,\underset{\sim}{N}^T\right\} dA \right.$$

$$\left. + \, i\int_{S_2} \kappa\, \underset{\sim}{N}\,\underset{\sim}{N}^T\, dS\right\} \underset{\sim}{U} = \delta\underset{\sim}{U}^T \left\{ \int_{S_2} \underset{\sim}{N}\, f\; dS\right\} \tag{29}$$

This expression can be written as,

$$\{ \underset{\sim}{K} - \kappa^2 \underset{\sim}{M} + i\kappa \underset{\sim}{M'} \}\underset{\sim}{U} = \underset{\sim}{F} \tag{30}$$

where

$$\underset{\sim}{K} = \iint \left(\frac{\partial \underset{\sim}{N}}{\partial x} \frac{\partial \underset{\sim}{N}^T}{\partial x} + \frac{\partial \underset{\sim}{N}}{\partial y} \frac{\partial \underset{\sim}{N}^T}{\partial y} \right) dA$$

$$\underset{\sim}{M} = \iint \underset{\sim}{N} \underset{\sim}{N}^T dA \qquad \underset{\sim}{M'} = \int_{S_2} \underset{\sim}{N} \underset{\sim}{N}^T dS \tag{31}$$

$$\underset{\sim}{F} = \int_{S_2} \underset{\sim}{N} f \, dS$$

We can now assemble the different matrices defined by (31) into the global matrices for the whole continuum. |See reference 3 |. This gives

$$\{ \underset{\sim}{\mathcal{K}} - \kappa^2 \underset{\sim}{\mathcal{M}} + i\kappa \underset{\sim}{\mathcal{M}'} \}\underset{\sim}{\mathcal{U}} = \underset{\sim}{\mathcal{F}} \tag{32}$$

It can now be seen that the radiation condition manifests itself as a damping term on the external boundary.

For problems such as harbour resonance it is possible to include an extra term representing frictional effects inside the finite element domain. The linearized version gives the matrix equation

$$\{ \underset{\sim}{\mathcal{K}} - (\kappa^2 - i\gamma\kappa) \underset{\sim}{\mathcal{M}} + i\kappa \underset{\sim}{\mathcal{M}'} \}\underset{\sim}{\mathcal{U}} = \underset{\sim}{\mathcal{F}} \tag{33}$$

where γ is an empirical coefficient.

γ can be related to the Chezy coefficient through the momentum equations. This gives,

$$\gamma = \frac{\lambda}{gd} = \frac{1}{dc^2} |v|$$

where $|v|$ is the magnitude of the velocity, c is the Chezy coefficient and d is the water depth. In order to estimate this we need to find the time and space average values of the ratio $|v|/d$.

We may write the inverse of the multiplier of the nodal

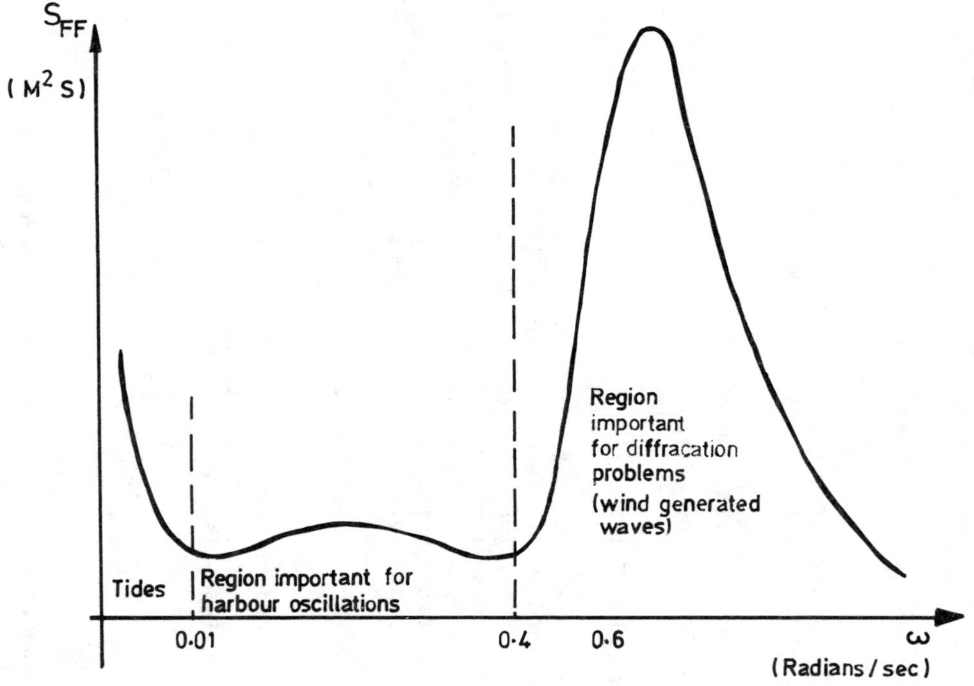

FIGURE 3 Typical Wave Spectrum (Logarithmic Scales)

unknowns vector $\underset{\sim}{U}$ as $\underset{\sim}{H}(\kappa)$ to obtain formally

$$\underset{\sim}{U} = \underset{\sim}{H}(\kappa)\underset{\sim}{F} \tag{34}$$

where $\underset{\sim}{H}$ may now be thought of as a complex transfer function and

$$\underset{\sim}{H}(\kappa) = \{\underset{\sim}{K} - (\kappa^2 - \gamma\kappa)\underset{\sim}{M} + i\kappa\underset{\sim}{M}'\}^{-1} \tag{35}$$

Equation (34) gives the response of the system to a harmonic excitation. At a point 'i' the displacement can be written as

$$u_i = \underset{\sim}{H}_i \underset{\sim}{F} \tag{36}$$

where $\underset{\sim}{H}_i$ is the ith row of the $\underset{\sim}{H}$ matrix.

5. SPECTRAL ANALYSIS

So far we have assumed that the incoming wave is harmonic with given period and amplitude. The problem is then one of a simple forced vibration and the solution is straightforward. Unfortunately this is not what happens in reality where the wave record is essentially the superposition of harmonic waves of differing amplitudes with random phase. The wave field is most commonly represented by the wave spectrum for the location. Figure 3 shows a typical plot of wave height spectral densities $S_{FF}(\omega)$ against angular frequency ω in radians per second.

If F(t) is a randomly varying quantity with zero mean then a basic property of the spectrum is that the mean squared value of F(t) or variance σ^2, is given by

$$\sigma^2 = \langle F^2(t) \rangle = \lim_{T \to \infty} \frac{1}{T} \int_{-T/2}^{T/2} F^2(t)\,dt = \int_o^{\infty} S_{FF}(\omega)\,d\omega \tag{37}$$

$S_{FF}(\omega)$ represents the way in which the harmonic components of F are distributed over the frequency range. The quantity of $\langle F^2(t) \rangle$ associated with a narrow frequency interval $|\omega,\omega+\Delta\omega|$ is simply $S_{FF}(\omega)\Delta\omega$.

It is clear that the response at a given point will be a random quantity and will in turn be described by a response spectrum $S_{UU}(\omega)$. To find the relationship between S_{FF} and S_{UU} let us first define a forcing vector F_o representing a unit actual value of the function which will be called F. Hence the height at a point 'i' can be written as

$$U_i = \alpha_i(\omega)F \tag{38}$$

(For instance F may be the incident wave height for a harbour).

Note that

$$\alpha_i(\omega) = \underset{\sim}{H}_i^T(\omega)\underset{\sim}{F}_o \tag{39}$$

To find the relationship between S_{FF} and S_{UU} we can compute the deviation of U and F by squaring equation (38). Taking into account the fact that U_i is a complex number.

Hence, we can write for a point 'i'

$$U_i^* U_i = \alpha_i^* F^* F \alpha_i \tag{40}$$

where the asterisk denotes complex conjugation.

Dividing equation (40) by T, the length of the record and letting $T \to \infty$ gives

$$S_{U_iU_i}(\omega) = |\alpha_i(\omega)|^2 S_{FF}(\omega) \tag{41}$$

as

$$S_{U_iU_i}(\omega) = \lim_{T \to \infty} \left\{\frac{1}{T} U_i^* U_i\right\}$$

and

$$S_{FF}(\omega) = \lim_{T \to \infty} \left\{\frac{1}{T} F^* F\right\}$$

See reference [2].

Equation (41) gives the relationship between the spectral density input function S_{FF} and the response spectrum $S_{U_iU_i}$ We can see that any particular ordinate of the $S_{U_iU_i}$ spectrum can be obtained by multiplying the corresponding ordinate by the modulus squared of the α_i function at that frequency. The ordinates of α_i are in fact the magnitude of the response at the point 'i' due to a unit amplitude forcing function for a given frequency ω. The values of α_i can be obtained using finite elements. Having obtained the spectrum of response we can integrate the area under it to obtain the mean squared value or variance σ^2. Note that

$$\sigma_u^2 = \langle U^2 \rangle = \int_0^\infty S_{UU}\, d\omega \tag{42}$$

for a process with zero mean. Furthermore, if the process is Gaussian or broad banded we can assign probabilities for the response exceeding various multiples -μ- of the standard deviation σ, i.e. the probability of the heights being within a certain $\pm\mu\sigma$ value is given in the following table.

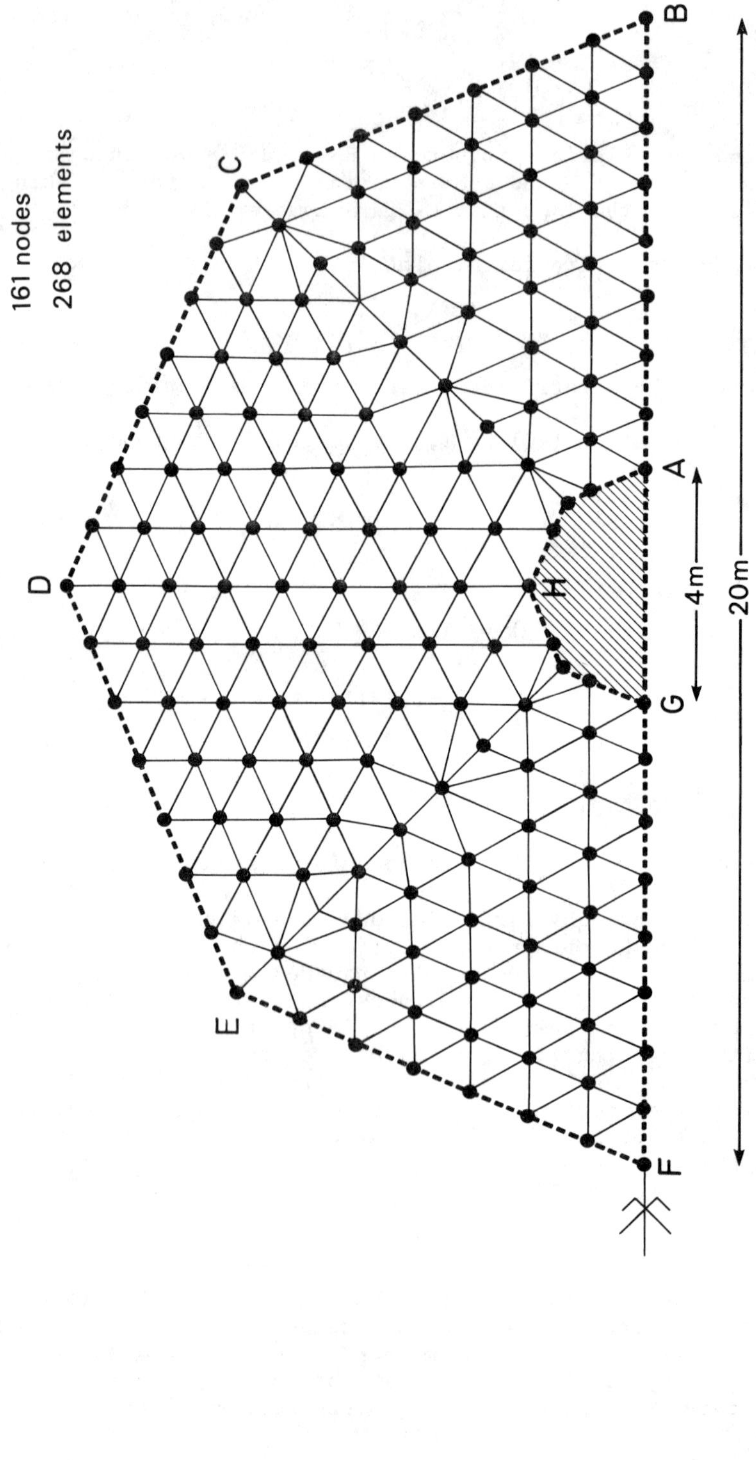

Figure 4a Finite Element Mesh

μ	Probability of $-\mu\sigma \le \eta \le \mu\sigma$	Probability of $\|\eta\| > \mu\sigma$
1	68.3%	31.7%
2	95.4%	4.6%
3	99.7%	0.3%

Alternatively if the input process is narrow banded (i.e. a storm or tsunami) we have a Rayleigh distribution, i.e.,

$$P(u) = \frac{u}{\sigma_u^2} \exp\{ - u^2/2\sigma_u^2 \} \tag{43}$$

for a given storm.

The expected maximum value of the response can be approximated by,

$$< |u|_{max} > = \sigma_u \{ \sqrt{2 \ln(T/T_m)} \} \tag{44}$$

where T_m is the mean period, given by

$$T_m = 2 \left\{ \frac{\int_0^\infty S_{uu} \, d\omega}{\int_0^\infty \omega^2 S_{uu} \, d\omega} \right\}^{\frac{1}{2}} \tag{45}$$

Example for Wave Diffraction

To illustrate the technique to be used, numerical results for wave diffraction round a fixed vertical circular cylinder were obtained. The mesh used for the solution of this problem is illustrated in figure 4a. As we are considering potential flow, we assume that the problem is symmetrical and that the contours FG and AB are streamlines and hence can be replaced by a solid boundary. In this way the cylinder is represented by the contour GHA. Linear elements were used in this preliminary example.

The wave to be diffracted by the cylinder was taken to be a harmonic plane wave travelling in the positive x direction. A subroutine was incorporated in the program to calculate the "radiation flux" f defined by equations (17) and (18). This function was then calculated at the mid-points of the element sides on the open sea boundary BCDEF. In this way wave energy flux away from the cylinder, as well as incident wave energy was represented in the matrix equations of the conventional finite element program.

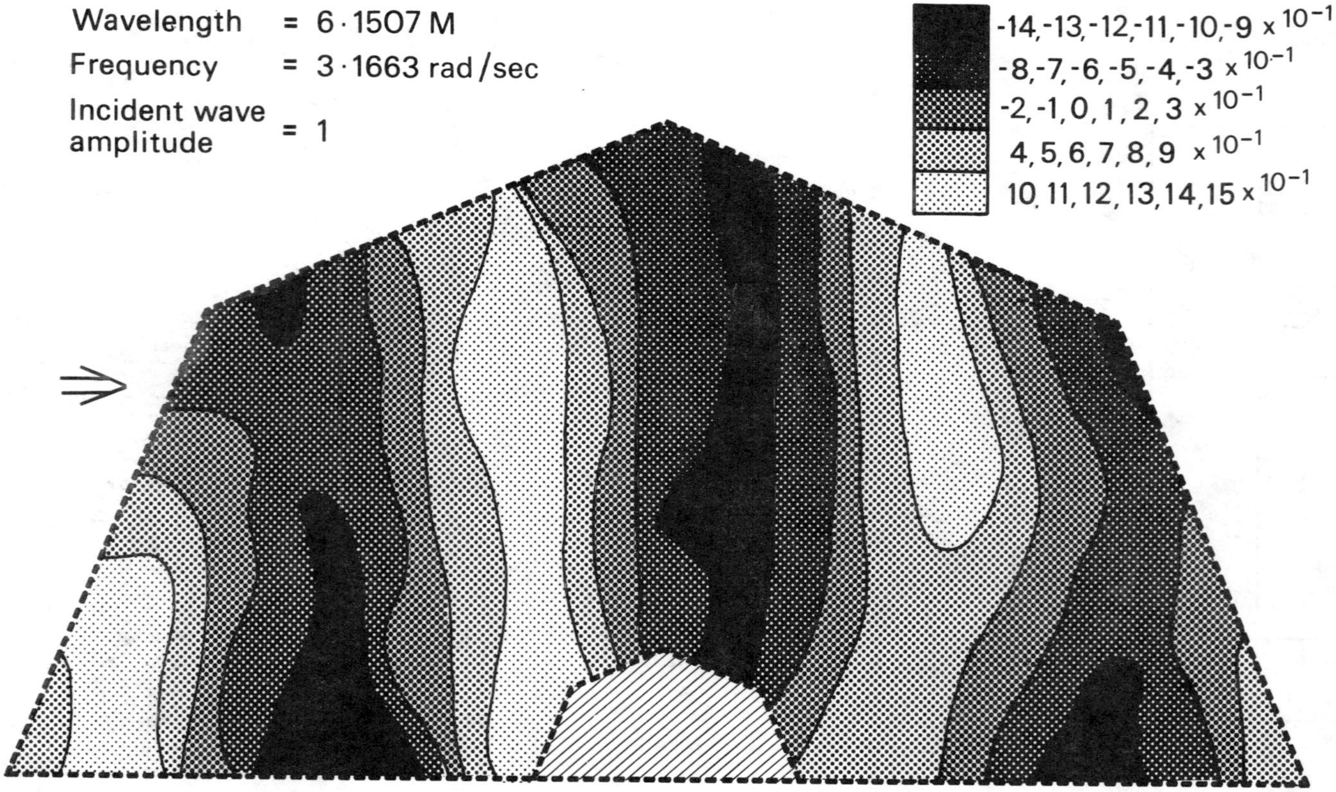

Figure 4b Surface elevations at time t=0

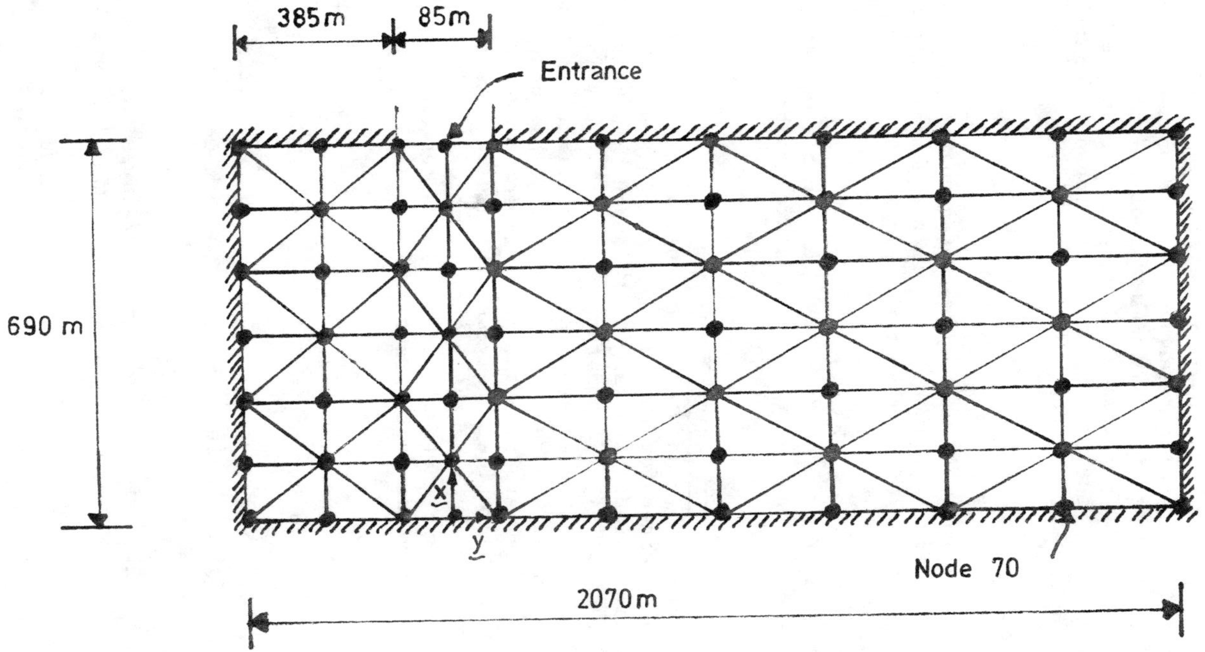

FIGURE 5   Mesh for Duncan Basin

The boundary condition applied on the contour FGHAG was the usual zero normal flux condition given by equation (7).

Figure 4b shows a contour plot of wave heights around the cylinder for an incident wave from the left of wavelength 6M. As a check on the accuracy of the results a very long wave of length 28M was assumed to be incident on the region without the cylinder. In these circumstances it was expected that the incident field would be virtually undisturbed by the presence of the boundary giving parallel vertical contours on the wave height diagram. The results were, in fact, found to confirm this assumption and were accurate to 3%.

Two important points were brought out by this example

(i) In order that the wave motion may be adequately represented by the linear finite element analysis the wave length must be at least six times longer than the maximum length of an element side.

(ii) For the radiation condition to be applicable on the points of the mesh boundary, these points need to be three or four wave lengths from the obstruction.

Example for Harbour Oscillations

Numerical results with, and without the use of the radiation condition were obtained for the case of the old Duncan Basin within Table Bay in South Africa. This harbour has been extensively studied [4], [5] and the shape of the Bay is such that some of the frequencies are greatly amplified, producing large oscillations within the Basin. This fact has been demonstrated by model experiments, harmonic analyses, seichograms and simple theoretical analyses, which give reasonable results here as the shape of the basin is approximately rectangular. The exact natural periods for a rectangular closed basin of constant depth are given by

$$T_{mm} = \frac{2\ell}{\sqrt{gd}\,(m^2 + \beta^2 n^2)^{\frac{1}{2}}} \quad , \quad \beta = \frac{\ell}{b} \tag{46}$$

where m and n describe the modal shapes in the two directions (m is associated with the length of the basin ℓ and n with the width b).

To simulate deterministically the effect of waves entering the harbour we discretised it into finite elements (figure 5). Along the contour BCDEFA we applied the usual zero normal flux condition $\partial\eta/\partial n = 0$. On the harbour entrance AB we applied the radiation condition for an incident plane wave of unit amplitude entering the harbour in the negative y direction, for the moment ignoring reflections from the neighbouring coastline.

Damping was not included in this part of the study. The matrix equation to be solved in this case is

$$\{ \underset{\sim}{K} - \kappa^2 \underset{\sim}{M} + i\kappa \underset{\sim}{M'} \} \underset{\sim}{U} = \underset{\sim}{F} \tag{47}$$

Solving equation (47) for $\underset{\sim}{U}$ gives a vector of the displacements in complex form at each node. Once the deterministic response is found it can be squared at the points under consideration to give the probabilistic transfer function for the resulting wave elevation at each node resulting from a wave of unit amplitude incident at the harbour mouth. One advantage of using the radiation condition is that the direction of this incident wave may be varied easily. The resulting elevation at node 70 for a unit incident wave is plotted in figure 6. The peaks correspond in position closely to the harmonic analysis carried out by Brebbia and Adey in reference [6], for a similar position in the harbour. The first peak represents the first significant period of about 11.45 minutes which is the mode corresponding to water flowing in and out of the basin and is called the "pumping mode". The second peak also corresponds to a simple motion the so-called "sloshing mode". Clearly the application of the radiation boundary condition on the harbour mouth is adequately representing the input of wave energy into the system as well as energy passing out of the harbour.

Another study was carried out on the same problem region without the use of the radiation condition but including the damping term representative of bottom friction. In this case the wave elevation at the entrance nodes was fixed equal to the amplitude of the incoming wave. The matrix equation to be solved is now

$$\{ \underset{\sim}{K} - (\kappa^2 - i\kappa\gamma) \underset{\sim}{M} \} \underset{\sim}{U} = \underset{\sim}{P} \tag{48}$$

The elements of the $\underset{\sim}{P}$ vector are formed by multiplying the known elevation values at the harbour mouth by elements of the original left hand side and transferring the result to the right hand side. The transfer function was obtained by squaring the elements of the response vector $\underset{\sim}{U}$ in the usual way.

The squares of the wave height response at node 70 is shown in figure 7 . Again the first significant modes appear at the expected frequencies.

The natural frequencies of harbours generally lie in the region between wind waves and tides, (i.e. approximately between frequencies ω of 0.01 to 0.4 radians/second). Usually the

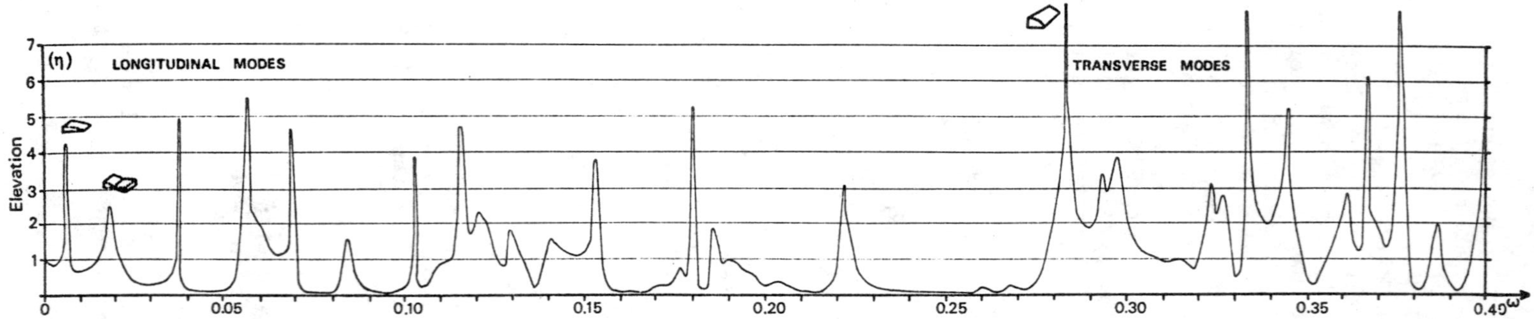

Fig.6 Maximum Surface Elevation at node 70 with Radiation but without Damping.

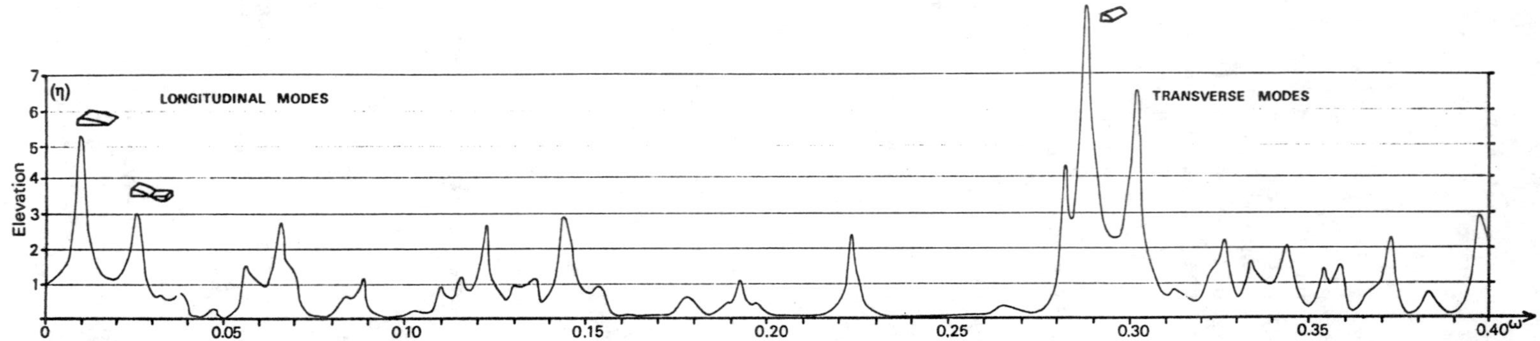

Fig. 7 Maximum Surface Elevation at node 70 without Radiation with Damping. $\gamma=0.0002$

energy of the oceans in this region is not considerable, but some phenomena can produce agitation which will damage the ships moored in harbours. Large displacements of the water occur for relatively small vertical motion.

Unfortunately, the spectrum corresponding to the Duncan Basin was not available, in consequence it was decided to use a white noise type spectrum with a cut-off at 0.4 radians/second and a magnitude 0.01 m^2s. Friction was considered in the form shown in equation (48). The value of the constant γ was taken to be 0.0002 (see references 7, 8 and 9).

Multiplying the ordinates of the diagram for the transfer function squared for point 70 (figure 7), by the ordinates of the (white noise) sea spectrum we obtained the response spectrum for the point under consideration. By integrating the area under the final figure we obtained the variance σ_u^2 .

$$\sigma_u^2 = 0.9039 \times 10^{-2} \text{ m}^2$$

For this case the maximum elevation to be expected within a 99.7% probability is

$$3.\sigma_u = 0.285 \text{ m}$$

The horizontal displacement spectrum can be obtained from the momentum equations. For harmonic motion they give

$$V_j = \frac{1}{\kappa^2 - i\kappa\gamma} \frac{\partial U}{\partial x_j} \qquad j = 1,2 \qquad (49)$$

where V_j is a maximum displacement such that

$$v_j = V_j \, e^{i\omega t}$$

Hence the displacement spectrum for a point i will be

$$S_{v_i v_i} = \frac{1}{\kappa^4 + \kappa^2\gamma^2} S_{u_{,j} u_{,j}} \qquad (50)$$

where $_{,j}$ denotes the derivative $\partial/\partial x_j$.

Once the spectrum of derivatives of wave height in the given direction is known, the variance of the displacements in that particular direction is given by

$$\sigma^2_{v_j} = \int_o^\infty S_{v_j v_j} \, d\omega \qquad (51)$$

Again we can define the maximum velocities which will be in the range $\pm\ 3\sigma_v$ within a certainty of 99.7%

This problem is discussed more fully in reference [23].

REFERENCES AND BIBLIOGRAPHY

1. Sommerfield, A. "Partial Differential Equations in physics" New York, Academic Press 1949 (pp. 256-344).

2. Walker, S. "Determination of the Wave Forces on Offshore Gravity Platforms, by Diffraction Theory and Spectral Analysis" University of Southampton, Department of Civil Engineering, Technical Report C.H. 1/77.

3. Brebbia, C.A. and J.J. Connor, "Introduction to Finite Element Techniques for Fluid Flow", Butterworths, London, 1976.

4. Wilson, B.W. "The Mechanisms of Seiches in Table Bay Harbor, Cape Town", Proc. 4th Conf. on Coastal Eng'g, 1953, Published by Council in Wave Research, 1954.

5. Wilson, B.W. "Research and Model Studies on Wave Action in Table Bay Harbor, Cape Town", Transactions of the South African Institution of Civil Engineers, Vol.1, No.6 and 7, Vol.2, No.5, June, July, 1959, May, 1960.

6. Brebbia, C.A. and R. Adey. "Circulation Problems", Proc. Finite Element Symposium, Atlas Computer Laboratory, Didcot, U.K. March 26-28, 1974.

7. Dorrenstein, R. "Amplification of Long Waves in Bays", Engineering Progress at the University of Florida Bulletin, Vol.15, No.12, 1961.

8. Harleman, D.R.F. and Ippen, A.T. "Friction and Energy dissipation in Long Wave Hydrodynamics", Proceedings of the Tsunami Meetings Associated with the 10th Pacific Science Congress. International Union of Ceodesy and Geophysics, Hawaii, August-September, 1961.

9. Shaw, R.P. and Lai Chi-Kwong, "Channel Friction and Shape Effects on Harbor Resonance", Proceedings A.S.C.E., Waterways and Harbors Div. Journal, August, 1974.

10. Raichlen, F. "Harbour Resonance", Estuary and Coastline Hydrodynamics", Edited by A.T. Ippen, Chapter 7, McGraw Hill, 1976.

11. Lee, J.J. and A.M. Raichlen, "Oscillations in Harbors with Connected Basins", Proc. A.S.C.E., Waterways, Harbors and Coastal Eng. Div., August, 1972.

12. Ippen, A.T. and Q.M. Fattah, "Wave Induced Oscillation in Harbors: The Effect of the Variation of Geometric Parameters on the Response of Rectangular Harbors", Ralph Parson Laboratory, Report No.70, M.I.T. June, 1964.

13. Ippen, A.T. and Y. Goda, "Wave Induced Oscillations in Harbors. The Solution for a Rectangular Harbor Connected to the Open Sea", M.I.T. Hydrodynamics Lab. Report No.59, July, 1963.

14. Eagleson, P.S. et al., "Hydraulic Model Study of Protective Works for Fleet Berthing Facilities in Coddington Cave, Narragansett Bay, R.I.", M.I.T. Hydrodynamics Lab., Report No.25, June, 1957.

15. Wilson, B.W., J.A. Hendrickson and J. Len. "Surge in the Southeast Basin, Long Beach Harbor, California", Proc. 13th Conf. on Coastal Eng., Vol.3, A.S.C.E., 1972.

16. Miles, J. and W. Munk, "Harbor Paradox", Journal of the Waterways and Harbors Division, A.S.C.E., Vol.87, No.WW3, Proc. Paper 2888, August, 1961, pp.111-130.

17. Lee, J.J. "Wave-Induced Oscillations in Harbors of Arbitrary Geometry", Journal of Fludi Mechanics, Vol.45, 1971, pp.375-394.

18. Lee, J.J. and F. Raichlen, "Resonance in Harbors of Arbitrary Shape" Proceedings of the 12th Conference on Coastal Engineering, Washington, D.C., 1970, Chapter 131, pp. 2163-2180.

19. Chen, H.S. and Mei, C.C. "Oscillation and Wave Forces in a Man-Made Harbor in the Open Sea", Tenth Sympsoium on Naval Hydrodynamics, Office of Naval Research, 1975.

20. Heaps, N.S. "Tidal Effects due to Water Power Generation in the Bristol Channel", Tidal Power (edited by T.J.Gray and O.K. Geshus), Plenum Press, N.Y. 1972.

21. Davenport, A.G. "Note on the Distribution of the Largest Value of a Random Function with Applications to Gust Loading", Proc. Inst. Civil Eng'g, Paper No.6739, 1964.

22. Kravtchenko, J. and McNown, J.S. "Seiche in Rectangular Ports", Quarterly of Applied Mathematics, Vol.13, 1955, pp.19-26.

23. Walker, S. and Brebbia, C.A. "Harbour Resonance Problems using Finite Elements", Advances in Water Resources, Vol.1, No.4, June 1978.

GENERALISED HYDRODYNAMIC FORCES ON VIBRATING OFFSHORE STRUCTURES BY WAVE DIFFRACTION TECHNIQUES

R. Eatock Taylor

University College London, England.

ABSTRACT

A unified method is presented for calculating the hydrodynamic loads associated with the wave excited dynamic response of an offshore platform. Wave forces and generalised added masss and damping terms are obtained, consistent with response of the structure in its modes of free vibration (including rigid body modes).

The method is based on the distributed singularity/boundary integral approach for ideal flow, which is first reviewed. The fluid/structure interaction problem is then solved, and results are given for a fixed and a floating structure to illustrate the generality of the approach when viscous effects may be neglected.

1. INTRODUCTION

Interest in the structural dynamics of offshore platforms has developed rapidly in recent years. This is partly because of the trend to place bottom supported platforms in ever deeper water (currently the tallest under construction is for 311 m water depth); partly because of the development of novel structures which may be relatively flexible (for example "compliant" towers, long wave energy devices etc.); and partly because of the appreciation that the in-service performance of more conventional offshore structures may be monitored by measuring their dynamic response to ambient environmental loading.

The analytical problem is one of fluid/structure interaction - hydroelasticity - (and in some cases foundation/structure interaction). A convenient approach to its solution is to separate so far as is possible the structural and the hydrodynamic analyses; and one way of achieving this is through a modal analysis of the structure. The principal modes of the structure may first be identified in the absence of the fluid. The hydrodynamic problem may then be solved, for boundary conditions on the submerged surface of the structure corresponding to motions in these modes. Finally the forced motion problem may be solved in terms of principal coordinates of the structure and associated generalised hydrodynamic forces.

The foregoing procedure is based on the assumption of linear superposition. It assumes that the motions are small (in structure and fluid), and that a linear solution can adequately describe the hydrodynamic problem. In what follows the fluid is taken to be incompressible and inviscid. The resulting procedures are therefore not directly relevant to offshore structures in which drag effects are significant, such as space frame platforms (for such cases an approximate linearisation would be needed). They are however immediately relevant to an extensive range of other types of offshore platform, including gravity platforms, large articulated columns, ships, barges and certain designs of tethered buoyant platform.

The emphasis of this paper is on solution of the hydrodynamic problem associated with the modal analysis of the structure. The first part is in the nature of a survey, describing how the generalised hydrodynamic boundary value problem, for an arbitrary body in the presence of a free surface, may be solved by potential theory using integral equations. This summarises the approach adopted by several previous investigators in wave diffraction studies. The paper then develops this approach in the context of wave excited structural dynamic response, leading to generalised wave forces and generalised added mass and damping terms. Finally some results for simple problems are presented, illustrating the versatility of the approach and satisfactory agreement with some results obtained by alternative methods.

2. THE BOUNDARY INTEGRAL APPROACH FOR FREE SURFACE FLOW PROBLEMS

It is assumed that this ideal flow hydrodynamic boundary value problem may be posed in terms of velocity potential functions. The essence of the approach is to obtain these by writing them in terms of a distribution of singularities on suitable boundaries of the fluid region. The strengths of the singularities are determined by the boundary conditions, and from these strengths the potentials themselves are evaluated (see for example Rouse [1]). In practice, solutions to the resulting integral equations are approximated by choosing a discrete set of singularities and by satisfying the boundary conditions at a discrete set of points. This leads to a linear set of matrix equations.

The method has received widespread usage, although not, it seems, in the calculation of hydrodynamic forces associated with the structural dynamics of offshore platforms. In the simplest application, outlined by Rouse, the singularity is taken as a simple source or doublet. The three dimensional source 1/R is used very widely for analysis of flow past bodies in an infinite fluid. It may also be used for free surface problems, provided that the singularities are distributed over all surfaces bounding the fluid domain of interest, including the free surface. This approach has been demonstrated by Bai and Yeung [2].

Alternatively the free surface water wave problem in an infinite half-space may be tackled by distributing over a closed surface, interior to the fluid domain, singularities which are wave potentials, satisfying all relevant boundary conditions except those on the body surface. This latter approach has been adopted by Lebreton and Margnac [3], Garrison and Chow [4], van Oortmerssen [5], Hogben and Standing [6], and Faltinsen and Michelsen [7]. The methods described by these authors are essentially similar, except that in the technique applied by van Oortmerssen the singularities are distributed on a surface inside the body. In the other cases, and in the approach used here, the surface defining the body is that on which the singularities are distributed.

2.1 Simple Singularities in Three Dimensions

The required singularity for the deep water case is a pulsating source in three dimensions, located at (ξ, η, ζ) and having velocity potential (which is the Green's function for the integral equations determining source strength)

$$\Phi(x, y, z, t) = \mathrm{Re}\left[\phi(x,y,z)\, e^{i\omega t}\right] \tag{1}$$

where

$$\phi(x, y, z) = \frac{1}{R} + \phi_o(x, y, z) \tag{2}$$

with

$$R^2 = (x - \xi)^2 + (y - \eta)^2 + (z - \zeta)^2 \tag{3}$$

It is to be understood that $\Phi(x, y, z, t)$ and $\phi(x, y, z)$ are also functions of ξ, η, ζ.

The Cartesian coordinate system is chosen with Oxy in the mean free surface, and z measured vertically upwards. The function $\phi_o(x, y, z)$ is harmonic in the whole region $z \leq 0$, including the point (ξ, η, ζ). It is chosen such that $\phi(x, y, z)$ satisfies boundary conditions on the free surface and the bottom, and the requirement that the waves at infinity progress outwards (the radiation condition).

These are

$$\frac{\partial\phi(x,y,o)}{\partial z} - \nu\ \phi(x,y,o) = 0 \tag{4}$$

$$\lim_{z\to-\infty} \frac{\partial\phi}{\partial z} = 0 \tag{5}$$

$$\lim_{r\to\infty} r^{\frac{1}{2}}\left(\frac{\partial\phi}{\partial r} + i\nu\phi\right) = 0 \tag{6}$$

where

$$r^2 = (x-\xi)^2 + (y-\eta)^2 \quad ; \ \nu = \frac{\omega^2}{g}$$

The solution, given by Wehausen and Laitone [8], is

$$\phi(x,y,z) = \frac{1}{R} + \frac{1}{R_1} + 2\nu\ \mathrm{PV}\int_o^\infty \frac{e^{\mu(z+\zeta)}}{\mu-\nu} J_o(\mu r)\,d\mu$$

$$+\ i2\pi\nu e^{\nu(z+\zeta)}\ J_o(\nu r) \tag{7}$$

with the interpretation that the principal value of the integral is taken, and

$$R_1^2 = (x-\xi)^2 + (y-\eta)^2 + (z+\zeta)^2\ . \tag{8}$$

For the purposes of comparison this potential may be expressed in alternative forms, by deforming the path of integration for example. This form however serves to indicate that at the low frequency asymptote ($\nu \to o$) the potential reduces to that of a simple source plus its image reflected in the free surface (c.f. Equation (4)). Using the alternative

$$\phi(x,y,z) = \frac{1}{R} - \frac{1}{R_1} + \mathrm{PV}\int_o^\infty \frac{2\mu e^{\mu(z+\zeta)}}{\mu-\nu} J_o(\mu r)\,d\mu$$

$$+\ i\ 2\pi\nu e^{\nu(z+\zeta)}\ J_o(\nu r) \tag{9}$$

it is seen that the high frequency asymptote corresponds to a simple source plus its negative image reflected in the free surface. Note that use has been made of the identity

$$\frac{1}{R} = \int_o^\infty \left[e^{\mu(z+\zeta)}\ J_o(\mu r)\,d\mu\right] \tag{10}$$

The corresponding pulsating source potential in water of depth d is given by Wehausen and Laitone [8] as

$$\phi(x,y,z) = \frac{1}{R} + \frac{1}{R_2} + 2PV\int_0^\infty \frac{(\mu+\nu)e^{-\mu d}}{\mu\sinh\mu d - \nu\cosh\mu d} \times \cosh\mu(z+d)\cosh\mu(\zeta+d)J_o(\mu r)\,d\mu + 2\pi i\,\frac{k^2-\nu^2}{(k^2-\nu^2)d+\nu}\cosh k(z+d)\cosh k(\zeta+d)J_o(kr) \tag{11}$$

where

$$R_2^2 = (x-\xi)^2 + (y-\eta)^2 + (z+\zeta+2d)^2$$

and

$$\nu = \frac{\omega^2}{g} = k\tanh kd \tag{12}$$

The low frequency asymptote, given by Garrison [9], is now

$$\phi(x,y,z) = \frac{1}{R} + \frac{1}{R_1} + \sum_{n=1}^{\infty}\left[\frac{1}{R_{2_n}} + \frac{1}{R_{3_n}} + \frac{1}{R_{4_n}} + \frac{1}{R_{5_n}}\right] \tag{13}$$

where

$$\begin{aligned} R_{2_n}^{\;2} &= r^2 + (z+\zeta+2nh)^2 \\ R_{3_n}^{\;2} &= r^2 + (z+\zeta-2nh)^2 \\ R_{4_n}^{\;2} &= r^2 + (z-\zeta+2nh)^2 \\ R_{5_n}^{\;2} &= r^2 + (z-\zeta-2nh)^2 \end{aligned} \tag{14}$$

This corresponds to the source and its image in the free surface, together with a series of reflections of this pair about the bottom and its reflections in the free surface.

The high frequency asymptote also given by Garrison [9], is

$$\phi(x,y,z) = \frac{1}{R} - \frac{1}{R_1} - \sum_{n=1}^{\infty}(-1)^n\left[\frac{1}{R_{2_n}} + \frac{1}{R_{3_n}} - \frac{1}{R_{4_n}} - \frac{1}{R_{5_n}}\right] \tag{15}$$

Again this could be interpreted as a series of reflections about the bottom, of the source and its negative image in the free surface.

An alternative to equation (11) is obtained by distorting the contour to integrate along the imaginary axis and around the poles given by $\mu = i\bar{\mu}_m$, where $\bar{\mu}_m$ are the positive real roots of

$$\bar{\mu}_m \tan \bar{\mu}_m d + \nu = 0 \tag{16}$$

The resulting expression is quoted by Wehausen and Laitone [8] in the form

$$\phi(x,y,z) = \frac{2\pi(\nu^2-k^2)}{(k^2-\nu^2)d+\nu} \cosh k(z+d) \cosh k(\zeta+d) \left[Y_o(kr) - iJ_o(kr)\right]$$
$$+ 4 \sum_{m=1}^{\infty} \frac{\bar{\mu}_m^2 + \nu^2}{(\bar{\mu}_m^2 + \nu^2)d - \nu} \cos \bar{\mu}_m(z+d) \cos \bar{\mu}_m(\zeta+d) K_o(\bar{\mu}_m r) \tag{17}$$

For the purposes of computation equation (17), truncated at a suitable number of terms, is generally preferable to equation (11). The exception is when r is very small, necessitating a very large number of terms in the series for satisfactory convergence and causing difficulties with the functions $Y_o(kr)$ and $K_o(\bar{\mu}_m r)$. In such cases the integral expression of equation (11) is used.

2.2 Approximate Solutions in Terms of Source Strengths

The above expressions give the velocity potential at the point (x,y,z) due to a simple singularity (a source) at the point (ξ, η, ζ), in the presence of a free surface. If such singularities are distributed over the surface S of a body, the resulting velocity potential may be obtained by integrating the contribution of each source over the total surface. The source potentials can be thought of as Green's functions which will henceforth be denoted $G(x,y,z;\ \xi,\eta,\zeta)$, where as in equation (2):

$$G(x,\ y,\ z;\ \xi,\ \eta,\ \zeta) = \frac{1}{R} + G_o(x,\ y,\ z;\ \xi,\ \eta,\ \zeta) \tag{18}$$

and G_o is everywhere regular in $z \leq 0$. If the strength of each source is $f(\xi,\eta,\zeta)$, then the resulting potential is

$$\phi(x,y,z) = \frac{1}{4\pi} \int_S f(\xi,\eta,\zeta)\ G(x,y,z;\ \xi,\eta,\zeta)\,dS \tag{19}$$

The source strengths are chosen to satisfy the boundary condition at the surface of the body. Thus if the normal component of fluid velocity out of the surface is

$$V_n = \mathrm{Re}\left[v_n \, e^{i\omega t} \right] \tag{20}$$

the boundary conditions on S may be shown to be (e.g. Rouse [1])

$$v_n(x,y,z) = \frac{1}{2} f(x,y,z) - \frac{1}{4\pi} \int_S f(\xi,\eta,\zeta) \frac{\partial G}{\partial n} (x,y,z;\xi,\eta,\zeta)dS \tag{21}$$

Strictly speaking, this condition holds for points just outside S. The first term on the right hand side arises from the $1/R$ contribution located at (x,y,z), one half of the flow going out from the surface. The second term gives the integrated effect at (x,y,z) of all the other sources distributed on the surface S.

In the numerical approximation to these integral equations, the surface S is divided into n_f small facets. Over the ith facet of area ΔS_i, the source strength is taken to be of the form

$$f(\xi,\eta,\zeta) = q_i \, \Delta S_i \, \delta(\xi - \xi_i, \eta - \eta_i, \zeta - \zeta_i) \tag{22}$$

where (ξ_i, η_i, ζ_i) are the coordinates of the centroid of the facet. Equation (21) may then be satisfied at the centroid of each facet, leading to n_f equations:

$$v_n(x_i,y_i,z_i) = q_i \left(\frac{1}{2} - \frac{\Delta S_i}{4\pi} \frac{\partial G_o}{\partial n} (x_i,y_i,z_i;\xi_i,\eta_i,\zeta_i) \right) - \frac{1}{4\pi} \sum_{\substack{j=1 \\ j \neq i}}^{n_f} q_j \, \Delta S_j \frac{\partial G}{\partial n} (x_i,y_i,z_i;\xi_j,\eta_j,\zeta_j) \tag{23}$$

In matrix form these may be written

$$\underset{\sim}{v}_n = \underset{\sim}{L} \, \underset{\sim}{q} \tag{24}$$

which may then be solved to obtain the source strengths $\underset{\sim}{q}$.

It should be noted that there is no guarantee that $\underset{\sim}{L}$ is a positive definite matrix. Thus, for example, in deep water at the high frequency aysmptote

$$L_{ii} = \frac{1}{2} + \frac{\Delta S_i}{4\pi} \frac{\partial}{\partial n} \left\{\frac{1}{R_1}\right\}$$

$$= \frac{1}{2} + \frac{\Delta S_i \, n_z}{16\pi\zeta^2}$$

For a horizontal facet pointing downwards at a distance b below the free surface the diagonal term L_{ii} will be zero if the facet area equals $8\pi b^2$.

The corresponding approximation to equation (19) is

$$\phi(x,y,z) = \frac{1}{4\pi} \sum_{j=1}^{n_f} q_j \, \Delta S_j \, G(x,y,z;\ \xi_j,\eta_j,\zeta_j) \tag{25}$$

When, however, the point (x,y,z) approaches the centroid of any one of the facets describing S, this expression may not be used directly because of the 1/R singularity. To evaluate the effect of the facet on itself, the 1/R source is spread uniformly over the facet instead of being concentrated at its centroid according to the distribution of equation (22). It may be shown that the potential at the centroid of a plane facet due to a simple 1/R source of unit strength spread uniformly over its area is given by

$$\hat{\phi}_i = \int_0^{2\pi} r_i \, d\Theta \tag{26}$$

where r_i is the distance from the centroid to the perimeter of the ith facet. The potential at the centroid due to all facets is therefore, from equation (18) and equation (25)

$$\phi(x_i,y_i,z_i) = q_i \left[\hat{\phi}_i + \frac{\Delta S_i}{4\pi} G_o(x_i,y_i,z_i;\ \xi_i,\eta_i,\zeta_i)\right]$$

$$+ \frac{1}{4\pi} \sum_{\substack{j=1 \\ j\neq i}}^{n_f} q_j \, \Delta S_j \, G(x_i,y_i,z_i;\ \xi_j,\eta_j,\zeta_j) \tag{27}$$

In matrix form these may finally be written

$$\underset{\sim}{\phi} = \frac{1}{4\pi} \underset{\sim}{G} \, \underset{\sim}{q} \tag{28}$$

This approximation is analogous to that developed by Hess and Smith [10] for flow past a body in an infinite fluid. However the formulation described by these authors is somewhat more complex. Rather than the two cases of a uniformly distributed source for the calculation of G_{ii}, and a point source for the calculation of G_{ij} $(i \neq j)$, a third type was introduced for neighbouring facets. This was a source plus a quadrupole at the centroid of the facet. This possibility of the use of multipole expansions of point singularities of various orders, to enhance accuracy, deserves further study.

Finally, to conclude this survey of an approximate technique for evaluation of generalised wave potentials, it should be noted that there is no certainty that the matrix $\underset{\sim}{L}$ always has an inverse. Difficulties would arise at the so-called "irregular" frequencies, as discussed by Hogben and Standing [11]. These are associated with the representation of surface piercing bodies by a distribution of sources. In practice the limitation does not appear to be severe, since the frequencies either lie outside the range of interest or may otherwise be circumvented, once recognised. It is however a problem which should be investigated further.

3. WAVE INDUCED DYNAMIC RESPONSE

The preceding development has described how velocity potentials may be approximated for the flow past bodies in the presence of a free surface. For any given boundary condition on the surface of the body the normal component of fluid velocity v_n is specified, and hence $\underset{\sim}{v}_n$ in equation (24) is prescribed. This in turn leads to a solution for the source strengths $\underset{\sim}{q}$ and ultimately to the potentials $\underset{\sim}{\phi}$.

We now consider particular cases of this boundary condition, corresponding to incoming waves interacting with the structure and to the resulting motions of the structure itself. Once these conditions have been specified and the corresponding velocity potentials have been calculated, the hydrodynamic loads on the structure may be formulated and the resulting equations of motion may be solved.

3.1 Wave Forces

The structure is assumed to be placed in a unidirectional sinusoidal incident wave having amplitude h, wave number k ($= 2\pi/\lambda$ where λ is the wavelength) and frequency ω. The velocity potential is a linear superposition of that due to the incident wave and that due to scattering by the body. Thus

$$\phi = \mathrm{Re}\left[(\phi_I + \phi_D)e^{i\omega t}\right] \qquad (29)$$

where ϕ_I refers to the incident wave, which in water of depth d is of the form

$$\phi_I = \frac{-gh}{\omega} \frac{\cosh k(d+z)}{\cosh kd} e^{ikx} \tag{30}$$

and g is the acceleration due to gravity. ϕ_D refers to the scattered wave, caused by the presence of the structure as a fixed body in the fluid.

Since there is no net flow into or out of the body

$$\frac{\partial \phi_D}{\partial n} + \frac{\partial \phi_I}{\partial n} = 0 \quad \text{on S.} \tag{31}$$

Hence with the interpretation

$$v_n = -\frac{\partial \phi_D}{\partial n} = \frac{\partial \phi_I}{\partial n} \tag{32}$$

equations (23) and (27) may be used to obtain ϕ_D in its discretised form.

The linearised pressure at a point on the submerged surface is

$$p(x,y,z,t) = \rho \frac{\partial \Phi}{\partial t} \tag{33}$$

where ρ is the density of the fluid.
Now the three components of force on the body, resulting from the incident wave, may be written

$$\underset{\sim}{F}(t) = -\int_S p \, \underset{\sim}{n} \, dS \tag{34}$$

where

$$\underset{\sim}{F} = \begin{bmatrix} F_x \\ F_y \\ F_z \end{bmatrix} ; \tag{35}$$

and

$$\underset{\sim}{n} = \begin{bmatrix} n_x(x,y,z) \\ n_y(x,y,z) \\ n_z(x,y,z) \end{bmatrix} \tag{36}$$

is the unit normal out of the surface. Similarly the moment about a reference point O (x_o, y_o, z_o) is

$$\underset{\sim}{M}(t) = \int_S p \, \underset{\sim}{\ell} \, dS \tag{37}$$

where

$$\underset{\sim}{M} = \begin{bmatrix} M_x \\ M_y \\ M_z \end{bmatrix} \tag{38}$$

and

$$\underset{\sim}{\ell} = \begin{bmatrix} (y-y_o)n_z - (z-z_o)n_y \\ (z-z_o)n_x - (x-x_o)n_z \\ (x-x_o)n_y - (y-y_o)n_x \end{bmatrix} \tag{39}$$

The three components of force and the three components of moment may be interpreted as the six generalised wave forces associated with the six rigid body motions of a body in three dimensions. These rigid body modes may be written in the form

$$\underset{\sim}{\psi}_r = \begin{bmatrix} \psi_{rx}(x,y,z) \\ \psi_{ry}(x,y,z) \\ \psi_{rz}(x,y,z) \end{bmatrix} \quad , \quad r = 1,2, \ldots 6. \tag{40}$$

where for example the mode of heave, translation in the z direction, is

$$\underset{\sim}{\psi}_3 = \begin{bmatrix} 0 \\ 0 \\ 1 \end{bmatrix} \quad ;$$

and the mode of pitch, rotation about an axis through the reference point O parallel to the y axis, is

$$\underset{\sim}{\psi}_5 = \begin{bmatrix} (z - z_o) \\ 0 \\ -(x - x_o) \end{bmatrix}$$

The six generalised wave forces associated with the rigid body modes, and expressed by equation (34) and equation (37), may therefore be concisely written as

$$F_r(t) = - \int_S p \, \underset{\sim}{\psi}'_r \, \underset{\sim}{n} \, dS \quad , \qquad r = 1, \ldots 6. \tag{41}$$

The form of equation (41) now suggests an alternative interpretation. If ψ_r represents any one of the n_o modes of vibration of an offshore structure discretised by n_o generalised coordinates (for example by a finite element idealisation), then

$$F_r(t) = - \rho \int_S p \, \underset{\sim}{\psi}'_r \, \underset{\sim}{n} \, dS \; , \qquad r = 1, \ldots n_o$$

gives the generalised force associated with that r^{th} mode. This may correspond to a rigid body mode or to a distortional mode. It is often convenient that these modes be defined as those of the structure oscillating freely in vacuo, as the fluid problem is then separated from the structural dynamic problem.

The final expression for the generalised wave excitation force associated with the rth mode is obtained by substituting equations (29) and (33) into equation (42). Thus

$$F_r(t) = \mathrm{Re}\left[-i\omega\rho \int_S (\phi_I + \phi_D) \underset{\sim}{\psi}'_r \, \underset{\sim}{n} \, dS \; e^{i\omega t} \right] \tag{43}$$

This may also be written in an alternative form following consideration of the motion problem.

3.2 Hydrodynamic Forces due to Motions of the Structure

In the discussion of generalised wave forces it was convenient to assume that the structure had free vibration modes $\underset{\sim}{\psi}_r(x,y,z)$ This column matrix contains the three components of displacement of the point (x,y,z) in the body, during oscillation in the rth mode (which may be a rigid body mode). Let us now assume that the structure is excited by sinusoidal waves of frequency ω so that it oscillates in each of its modes simultaneously. The response in the rth mode is defined by the rth principal coordinate P_r, such that the response of any point in the structure is

$$\underset{\sim}{\delta}(x,y,z,t) = \mathrm{Re}\left[\sum_{r=1}^{n_o} P_r \, \underset{\sim}{\psi}_r \, (x,y,z) e^{i\omega t} \right] \tag{44}$$

The resulting motions of the structure in the fluid induce hydrodynamic pressures on its surface S, and generalised hydrodynamic reaction forces may be defined in a manner similar to the generalised wave forces derived above.

The velocity potential corresponding to the resulting flow is defined by the boundary condition on S

$$- \frac{\partial \phi}{\partial n} = v_n = i\omega \sum_{r=1}^{n_o} P_r \underset{\sim}{\psi}_r \underset{\sim}{n} \quad \text{on} \quad S \ . \tag{45}$$

ϕ also of course satisfies the free surface and bottom conditions and the radiation conditions at infinity. It is convenient to define a spatial potential ϕ_s corresponding to deformation of the structure in its sth mode so that

$$\frac{\partial \phi_s}{\partial n} = - \underset{\sim}{\psi}'_s \underset{\sim}{n} \qquad \text{on } S. \tag{46}$$

The discretised approximation to ϕ_s may then be obtained from equation (27) using equation (23).

The pressure on the submerged surface due to its motions is

$$p(x,y,z,t) = \mathrm{Re}\left[- \omega^2 \rho \sum_{s=1}^{n_o} P_s \ \phi_s \ e^{i\omega t} \right] \tag{47}$$

In a manner analogous to the generalised wave force, a generalised hydrodynamic reaction force may be defined by

$$R_r(t) = - \int_S p \ \underset{\sim}{\psi}'_r \ \underset{\sim}{n} \ dS \ , \qquad r = 1, \ldots n_o \tag{48}$$

which by substitution of equation (47) leads to

$$R_r(t) = \mathrm{Re} \left[\omega^2 \rho \sum_{s=1}^{n_o} (P_s \int_S \phi_s \ \underset{\sim}{\psi}_r \ \underset{\sim}{n} \ dS) e^{i\omega t} \right] \tag{49}$$

With the use of equation (46), but with s replaced by r, this may also be written

$$R_r(t) = \mathrm{Re}\left[-\omega^2 \rho \sum_{s=1}^{n_o} (P_s \int_S \phi_s \frac{\partial \phi_r}{\partial n} \ dS) e^{i\omega t} \right] \tag{50}$$

The integral expression in equation (50) is generally complex, because ϕ_s is complex. It may be written in terms of its real and imaginary parts in the form

$$\rho \int_S \phi_s \frac{\partial \phi_r}{\partial n} \ dS = - \bar{M}_{rs} + i \frac{\bar{B}_{rs}}{\omega} \tag{51}$$

Then the expression for the hydrodynamic reaction forces becomes

$$R_r(t) = \text{Re}\left[- \sum_{s=1}^{n_o} (-\omega^2 \bar{M}_{rs} + i\omega \bar{B}_{rs}) P_s \, e^{i\omega t} \right] \tag{52}$$

The quantities $\bar{M}_{rs}$ and $\bar{B}_{rs}$ are recognised as generalised added mass and damping terms associated with response of the structure in its principal modes. It is clear that these hydrodynamic effects will generally introduce a degree of coupling between the principal coordinates associated with the modes of the structure vibrating in vacuo.

3.3 <u>Evaluation of Total Loads</u>

The total generalised force due to the incident waves and the resulting motions of the structure through the fluid is, by linear superposition

$$A_r^*(t) = \text{Re}\left\{ \left[\bar{F}_r - \sum_{s=1}^{n_o} (-\omega^2 \bar{M}_{rs} + i\omega \bar{B}_{rs}) P_s \right] e^{i\omega t} \right\} \tag{53}$$

(An added stiffness term may also be included if dynamic buoyancy effects are present, for example due to the motions of surface piercing members). The analytical expressions for M_{rs} and B_{rs} are given in equation (51). The expression for F_r given in equation (43) may be rewritten using equation (46): Thus

$$F_r(t) = \text{Re}\left[\bar{F}_r \, e^{i\omega t} \right] \tag{54}$$

where

$$\bar{F}_r = i\omega \rho \int_S (\phi_I + \phi_D) \frac{\partial \phi_r}{\partial n} \, dS \tag{55}$$

It may also be noted that this may be written in a form independent of ϕ_D by use of the Haskind relations (Newman [12]).

The final stage in the numerical evaluation of these loads is to compute the integrals of equation (51) and equation (55), in a manner consistent with the approximation employed to obtain ϕ_D and ϕ_r using sources at discrete points on S. It is assumed that the potential is uniform over each facet, and contributions from all facets are summed to obtain the integral over S. The general form is

$$I = \int_S \chi(x,y,z) \frac{\partial \phi_r(x,y,z)}{\partial n} \, dS \tag{56}$$

which is approximated (c.f. equation (25)) by

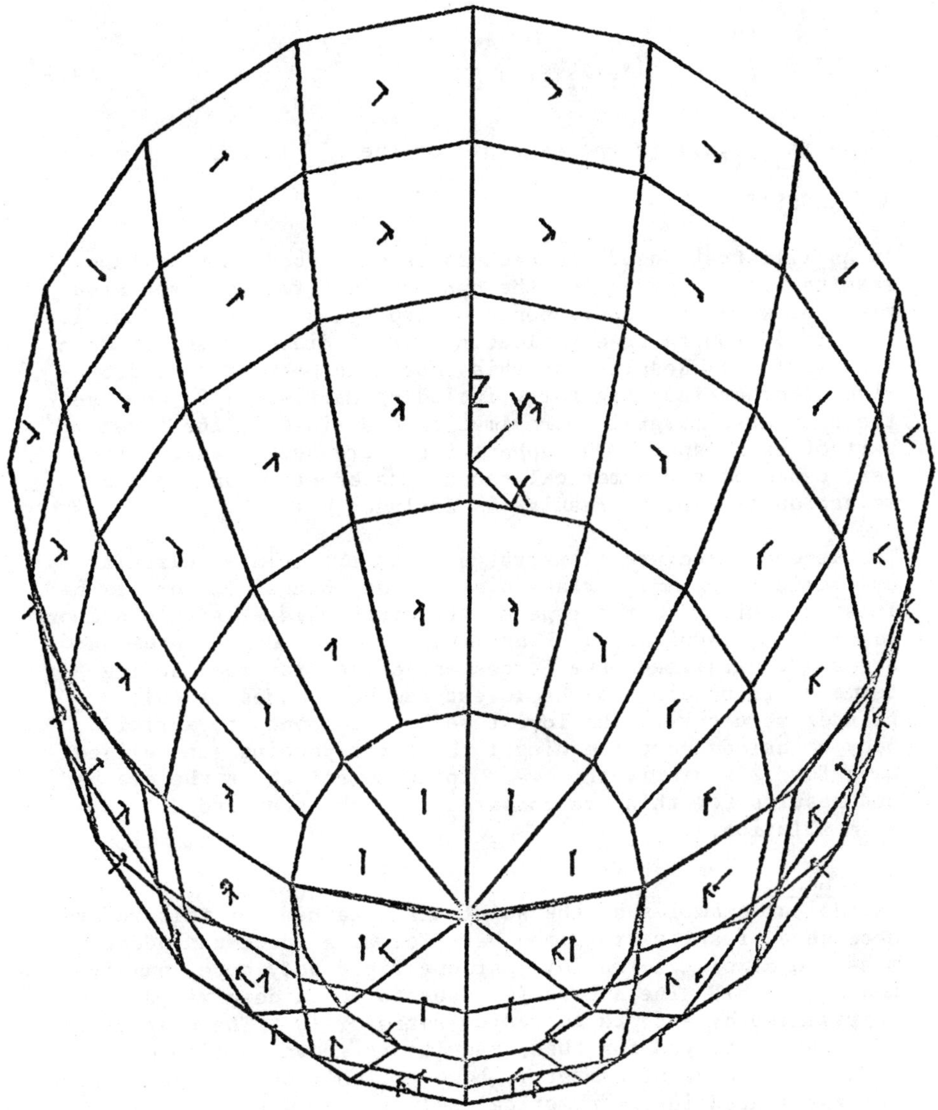

Fig.1 Disposition of facets idealizing submerged surface of sphere

$$I = \sum_{i=1}^{n_f} \Delta S_i \; \chi(x_i,y_i,z_i) \; \frac{\partial \phi_r(x_i,y_i,z_i)}{\partial n} \qquad (57)$$

where (x_i,y_i,z_i) is the centroid of the ith facet.

4. RESULTS

As an illustration of the techniques described in this paper, results will be given for the hydrodynamic forces associated with the wave excited response of two types of body. The first is a rigid sphere freely floating with a diameter in the mean free surface. Results for this case have been presented by several other investigators, including Garrison [13] who used the boundary integral approximation described in the first part of this paper. The sphere therefore serves as a useful test case for the numerical techniques adopted, and only a selection of typical results is included here.

The second structure illustrated is an articulated circular cylindrical column, surface piercing and hinged at the sea bed. This is similar to the type of structure used as single anchor leg mooring terminals or flare stacks. The results presented are the generalised wave forces and added mass and damping terms corresponding to the lowest few beam modes of this slender structure. The lowest mode corresponds to a rigid body rotation about the hinge: the corresponding generalised wave force is simply the overturning moment about the sea bed, and results for this are compared with the standard closed form solution.

4.1 Sphere

As typical examples of the solutions obtained for this reference case, results are given here for wave force and added mass and damping terms for just one rigid body mode, namely heave (r = 3). The sphere is taken to be in deep water, represented by a depth to radius ratio of 10. The disposition of facets employed for these results is shown in figure 1. As a consequence of symmetry, however, only one quadrant need be represented in the numerical solution.

The results are plotted against the non-dimensional frequency ka, where k is the wave number and a is the radius. Figure 2 shows the amplitude of the wave force $\bar{F}_3$, non-dimensionalised by the quantity $\rho g a^2 h$. The added mass $\bar{M}_{33}$ and damping $\bar{B}_{33}$ are shown in figure 3, non-dimensionalised by the terms ρa^3 and $\rho a^3 \omega$ respectively.

These figures show the results computed by the boundary integral technique described herein, and results given by Havelock [14] for infinitely deep water. Agreement is seen to

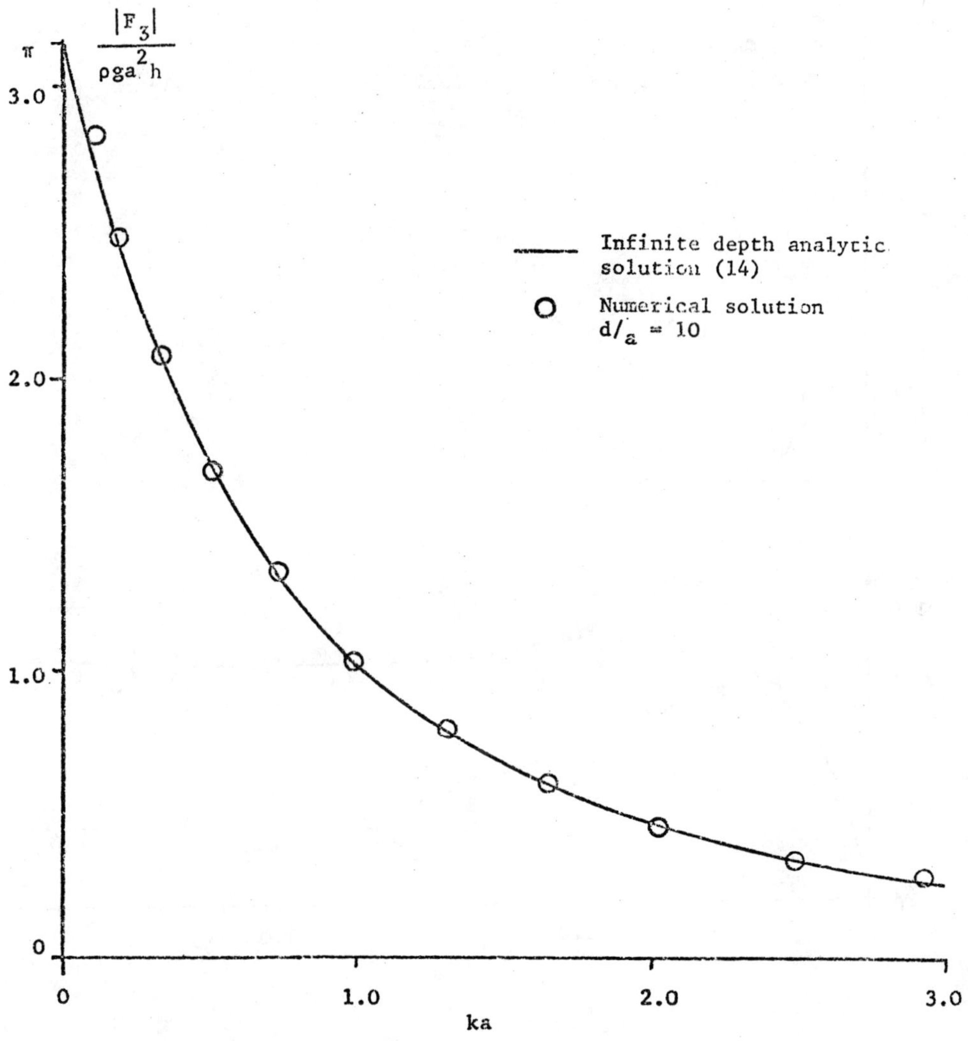

Fig. 2. Heave exciting force on sphere

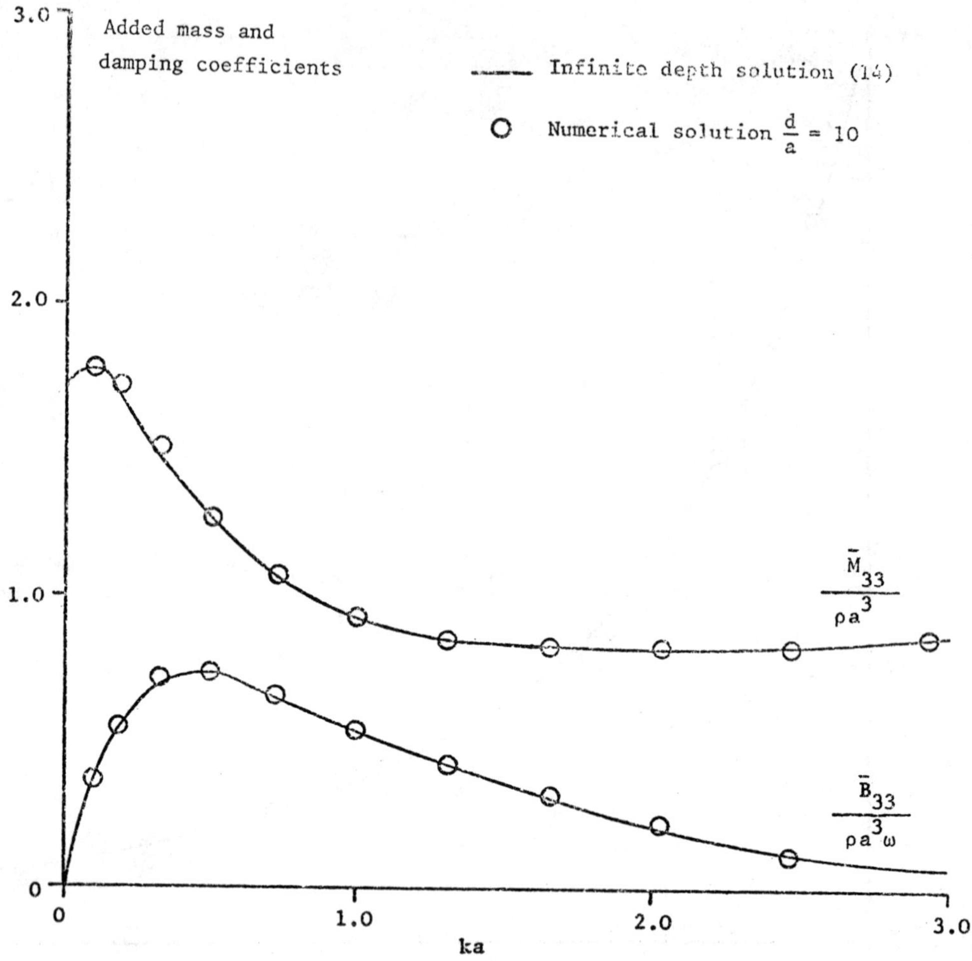

Fig. 3. Heave added mass and damping for hemisphere.

be satisfactory over the range examined, when a relatively coarse mesh of facets is employed. At low frequencies some of the discrepancies may be shown to be due to the finite depth representation in the numerical solution. For greater accuracy at high frequencies, corresponding to short wavelengths, a larger number of smaller facets would be used.

4.2 Hinged Cylinder

The ratio of cylinder radius a to water depth d is 1/10, and the arrangement of 84 equal facets is shown in figure 4. The first three modes of this structure illustrated in figure 5 are those of a uniform beam pinned at one end and free at the other, under the assumption that shear deformations and rotary inertia are neglected. The first is a rigid body mode whereas higher modes involve distortion and are therefore structural modes of vibration in the conventional sense. Each is normalised to give a unit displacement in the x-direction at the free surface.

The real and imaginary parts of the first two generalised wave forces are shown in figure 6. In addition to the numerical results computed herein, figure 6 also shows the overturning moment on a surface piercing circular cylinder computed from the diffraction solution of MacCamy and Fuchs [15]. Agreement between the approximate and analytical solutions is generally good, except at higher frequencies where accuracy of the 84 facet idealization decreases. The use of smaller facets at these short wavelengths has been found to yield significantly improved accuracy.

Generalised added mass and damping terms for the rigid body rotational mode (r = 1) and the first distortional mode (r=2) are shown in figures 7 and 8. The direct terms ($\bar{M}_{11}$ and $\bar{M}_{22}$ etc.) are in figure 7 while figure 8 contains the cross coupling terms ($\bar{M}_{12}$ and $\bar{B}_{12}$). Each of these quantities is non-dimensionalised by a reference generalised added mass in mode r defined as

$$\bar{M}^*_{rr} = \pi\rho a^2 \int_{-d}^{o} \psi^2_{rx}(o,o,z)dz \qquad (58)$$

This corresponds to the simple solution for motion of a circular cylinder transverse to its axis in an infinite fluid, for which the two dimensional added mass per unit length is $\pi\rho a^2$. Thus it is to be anticipated that $\bar{M}_{rr} = \bar{M}^*_{rr}$ when the flow is essentially two dimensional and free surface effects are negligible.

4.3 Conclusions

The results presented illustrate how the distributed singularity/boundary integral method may be employed to calculate hydrodynamic forces associated with the wave excited dynamics

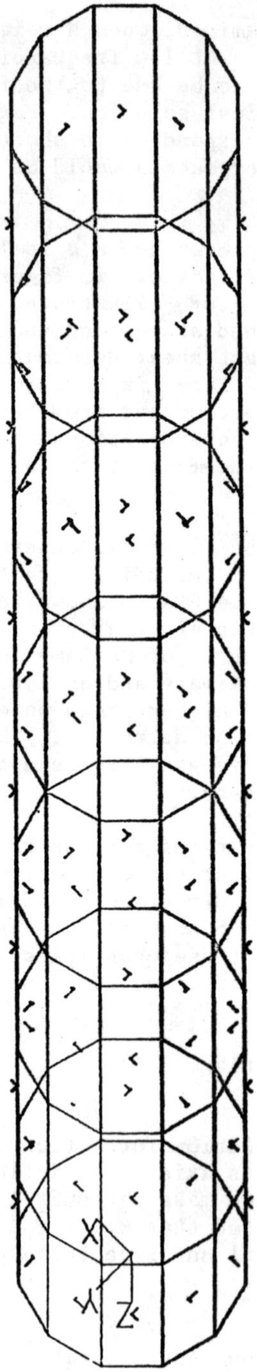

Fig.4 Disposition of facets idealizing submerged portion of cylinder

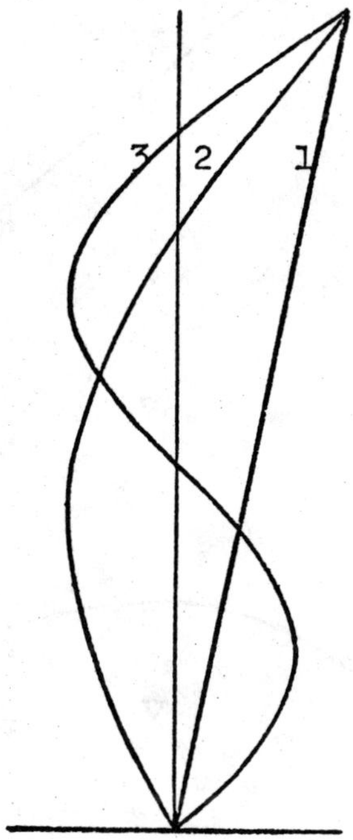

Fig.5 First three modes of pinned-free beam

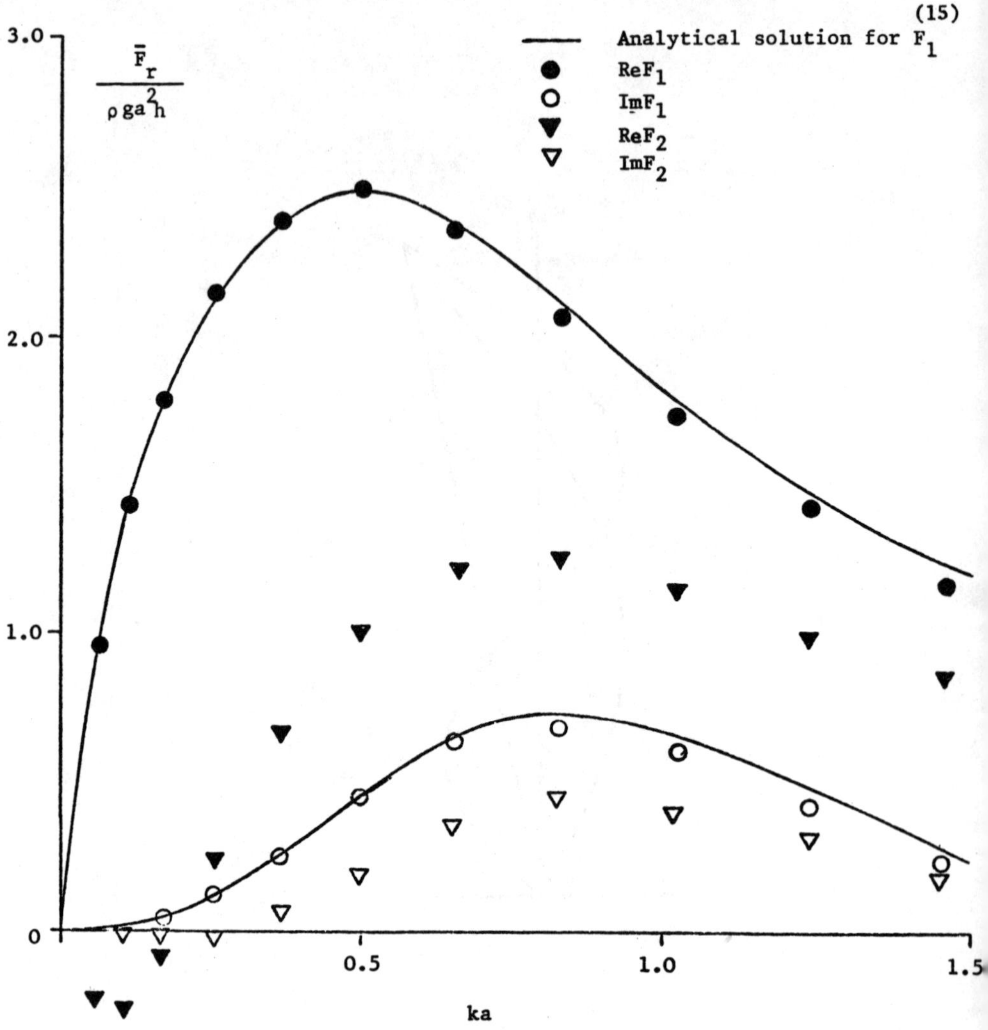

Fig. 6. Real and Imaginary parts of generalised wave forces on hinged cylinder. (a/d = 0.1)

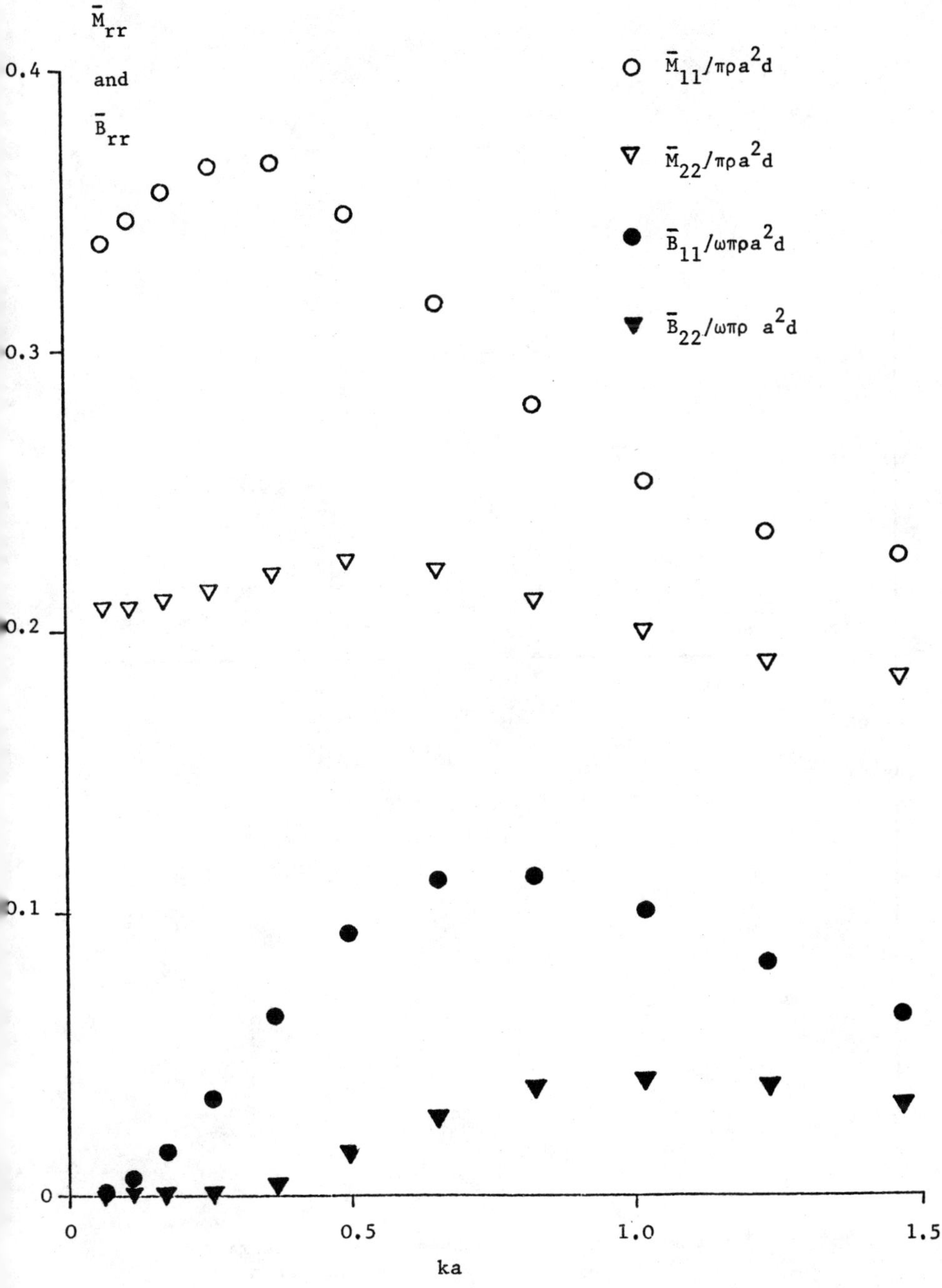

Fig. 7 Generalised added mass and damping terms for hinged cylinder: direct terms.

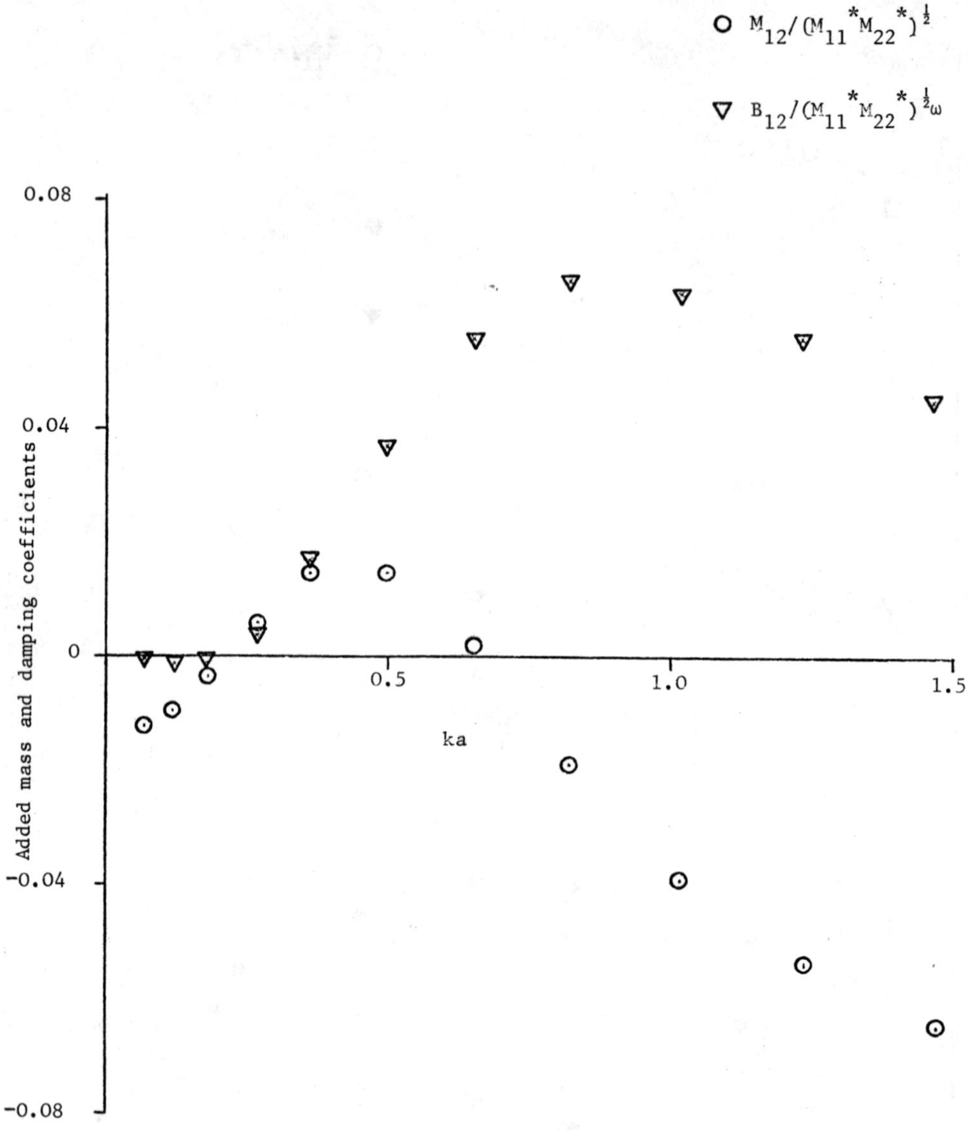

Fig. 8 Generalised added mass and damping terms for hinged cylinder: cross coupling terms.

of offshore structures. The solution is based on the assumptions of linearity and inviscid incompressible flow. But a wide range of practical problems may be analysed on this basis, including ship and barge dynamics, many types of tethered buoyant platforms, articulated columns and gravity platforms. The approach developed here provides a unified treatment for the assessment of wave forces and hydrodynamic added mass and damping terms, associated both with rigid body and vibrational modes of structures. The examples studied indicate that quite accurate results (as compared with alternative solutions) may be obtained with only a modest number of facets distributed on the submerged surface of these structures. Nevertheless further research is required to improve the efficiency of the method for complex structures, and to investigate such matters as error bounds and the problem of irregular frequencies.

ACKNOWLEDGEMENTS

This work was supported by the Science Research Council Grant GR/A 2095.8 to the London Marine Technology Centre. The work of Mr. J. B. Waite on the computer program is gratefully acknowledged.

REFERENCES

1. Rouse, H. Advanced Mechanics of Fluids, Wiley, New York, 1959.

2. Bai, K.J. and R.W. Yeung, "Numerical solutions to free surface flow problems", Tenth Naval Hydrodynamics Symposium, Office of Naval Research, 1974.

3. Lebreton, J.C. and A. Margnac, "Calcul des mouvements d'un navire ou d'une platforme amarrée dans la houle". La Houille Blanche, 5, 379-389, (1968).

4. Garrison, C.J. and P.Y. Chow, "Wave forces on submerged bodies", Proc. ASCE, J. Waterways and Harbors Div. 98, 375-392 (1972).

5. van Oortmerssen, G., "Some aspects of very large offshore structures", Ninth Naval Hydrodynamics Symposium, Office of Naval Research, 1972.

6. Hogben, N. and R.G. Standing, "Wave loads on large bodies", Proc. Int. Symposium Dynamics of Marine Vehicles and Structures in Waves, Univ. Coll. London. Inst. Mech. Eng., 258-277, 1975.

7. Faltinsen, O.M. and F.C. Michelsen, "Motions of large structures in waves at zero Froude number". Proc. Int. Symposium Dynamics of Marine Vehicles and Structures in Waves, Univ.Coll.London. Inst.Mech.Eng., 91-106, 1975.

8. Wehausen, J.V. and E.V. Laitone, "Surface Waves" in Encyclopaedia of Physics, IX, Springer, Berlin, 1960.

9. Garrison, C.J. and R.B. Berklite, "Hydrodynamic loads induced by earthquakes", Offshore Technology Conference, OTC 1554 (1972).

10. Hess, J.L. and A.M.O. Smith, "Calculation of potential flow about arbitrary bodies", Progress in Aeronautical Sciences, 8, 1-138 (1967).

11. Hogben, N. and R.G. Standing, "Wave loading on offshore structures - theory and experiment". Symposium on Ocean Engineering, Royal Institution of Naval Architects, London, 19-36 (1974).

12. Newman, J.N. "The exciting forces in fixed bodies in waves", J. Ship Research, 6, 10-17 (1962).

13. Garrison, C.J. "Hydrodynamics of large objects in the sea: Part I Hydrodynamic Analysis", AIAA Jnl. of Hydronautics, 8, 5-12 (1974).

14. Havelock, T. "Waves due to a floating sphere making periodic heaving oscillations". Proc. Roy. Soc. A, 231, 1-7 (1955).

15. MacCamy, R.C. and R.A. Fuchs, "Wave forces on piles - a diffraction theory", Beach Erosion Board Tech. Memo 69 (1954).

ASPECTS OF FLOW INDUCED OSCILLATIONS OF PILE STRUCTURES

F. Venancio Filho, Manuel A.G. Silva, R. Flores Coombs and A.C. Monteiro

PROMON Engenharia

1. INTRODUCTION

The dynamic behaviour of structural elements immersed in streams has been the subject of numerous studies and requires frequent analysis from designers. Wihtout aiming at an exhaustive search of illustrative situations, a reference can be made to trashrack gates in hydroelectric intakes, some types of girder-stiffened suspension bridges, cooling towers, poles, slender vertical vessels and ventilation stacks under the action of wind, flexible piles in flowing water and heat exchanger tube banks in nuclear reactor systems.

The present study is essentially addressed to pile structures built to support piers, bridge decks or similar superstructures and subjected to the flow of water induced e.g. by tidal motion.

A review of the importance that some structural, inertial and hydrodynamic parameters have on the response is made. An evaluation of the amplitude of the oscillations caused by in line harmonic excitation is performed. An assessment of the influence of the ratio of the mass M of the superstructure over the mass of the piles on the response is presented and interpreted. The effects that vortex shedding may have on the serviceability of the superstructure are analysed. The results of the analysis are summarized in tables and/or graphically displayed.

2. DYNAMIC RESPONSE OF IMMERSED PILES

A brief account of the main aspects that concern the dynamic response of piles subjected to water flow is given below, in order to facilitate the presentation of the results of the analysis.

2.1 Cross Flow Oscillations

The steady flow of water past a clamped cylinder may cause lateral oscillations designated, here and in the sequel, as cross flow oscillations. These vibrations are the result of vortex shedding at a frequency f(cps) that approaches a natural frequency of the cylinder.

The frequency f is linearly related to the speed V of the water, the proportionality being measured by the Strouhal number S. For Reynolds numbers corresponding to situations of practical importance, the frequency of the cross flow harmonic excitation is

$$F = S \frac{V}{D} \tag{1}$$

where D is the diameter of the cylindrical pile; the remaining symbols have been defined and the number of Strouhal has a magnitude nearing 0.2.

The value of the speed of the water for which a finite motion exists is called critical velocity V_{cr}. The critical velocity associated to the fundamental frequency N of the piles makes $f \simeq N$ i.e. $V_{cr}/ND \simeq 5$. The parameter V/(ND) is called reduced velocity and, therefore, assumes a value near 5 when maximum response takes place in the direction of cross flow.

The case of structures supported by groups of piles represents an edtension that requires proper attention to other effects, as explained in 2.3. At the present state of knowledge, however, the phenomenon of cross oscillations is undesirable and its occurrence is prevented by means of devices like fins or shrouds, suppressing the need for studies of response to cross flow harmonic forcing functions.

2.2 In Line Oscillations

From the mechanism of vortex shedding it is known that the frequency of the in line component of the fluctuating force is twice the frequency of the cross flow oscillations.

The magnitude of this excitation was traditionally ignored in the studies involving wind because it is proportional to the mass density of the air, but its importance cannot be neglected when the fluid is water and the length of the cylinder reaches fairly high values.

The critical velocity for in line motion brings the frequency of the excitation near a natural frequency of the cylinder and, therefore, is inversely proportional to the square of the length of the pile. For large values of water depth, as required by modern tankers, the critical velocity assumes comparatively low values, but still associated with a non-negligible amplitude of the excitation, and therefore makes the pile structures vulnerable to in line oscillations.

This excitation was responsible for the problems at Immingham, has been the subject of recent experimental and analytical studies, and the major part of the present paper deals with aspects of it.

2.3 Group Effects and Hydrodynamic Coupling

While the study of isolated cylinders has received many contributions, there is relatively scarce information for the more general case of a group of piles responding simultaneously to the action of water flowing.

It has, nevertheless, been shown by Laird, Johnson and Walker [1] that neighbouring cylinders cause a reduction of the drag force when the cylinder under study is in the neighbour's wake, whereas the eddies shed by the leading neighbour also interfere heavily with the response. For a spacing between cylinders larger than 4 diameters these effects decrease to become negligible at about 8 diameters [2].

It is noted that, in general, the computation of the natural modes of a group of piles in water is difficult due to the coupling effect with the water. Equations of motion that take into account this effect have been derived by Mazur [3] and extensively used by Chen, e.g. [4], based on the concept of added mass.

In the study that follows the coupling effect is not introduced but an effort is placed at extending the concept of reduced velocity to pile structures and the "added mass", as understood by King [5], is introduced into the model that is analysed.

2.4 Fluctuating Drag Force

The in-line critical oscillations are associated with an exciting frequency f', of the shedding of complementary pairs of vortices, that is twice the frequency of the cross flow forcing function and there is experimental evidence of peak responses for still higher frequencies [6].

The reduced velocity for $f' = N$ assumes, then, a value of about 2.5 and the in-line critical response is attained for a velocity of water that is half the critical velocity for cross-flow oscillations.

Expressing the fluctuating drag force for a flexible cylinder that has N as a natural frequency by

$$P = P_o \sin(2\pi Nt) \tag{2a}$$

as explained, for instance, in [7] with

$$P_o = \frac{1}{2} \rho \, V^2 \, D \, C_D' \tag{2b}$$

where ρ is the mass density of the water and C_D' a drag coefficient that scales ($\frac{1}{2}\rho V^2 D$) so as to recover the actual maximum amplitude of response as experimentally known in similar cases.

Usually the value of C_D' is expressed in terms of dimensionless parameters, namely the reduced velocity V_R, the amplitude and the damping K_s defined by

$$V_R = \frac{V}{N \, D} \tag{3a}$$

$$\eta = \frac{A}{D} \tag{3b}$$

$$K_s = \frac{2m\delta}{\rho D^2} \tag{3c}$$

where the symbols that appear for the first time have the following meaning: A = actual amplitude of oscillation; m = mass of the pile per unit of length (see 3.3); δ = logarithmic decrement due to structural damping.

Based in earlier findings [7] the value of 0.01 for structural damping was assumed throughout, though in some cases the true value is expected somewhat higher.

A first mode analysis of the clamped - clamped beam permits the determination of the corresponding C_D' in the manner that follows. The maximum displacement can be expressed as the product of the static displacement y_{st}' by the dynamic amplification factor which, in this case, is (π/δ), δ being the logarithmic decrement. Making use of the exact expression for the first mode of the beam it is found that

$$y_{st} = \frac{\int_0^L p_o \, \phi_1 \, dx}{\omega_1^2 \, m \int_0^L \phi_1^2 \, dx} = \frac{0.8308 \, p_o}{4\pi^2 m N^2} \tag{4}$$

Noting that $y_{max} = \eta D = (\pi/\delta) y_{st}$ and replacing p_o by its value as given in (2.2) leads to

$$C_D' = 9.52 \left(\frac{ND}{V}\right)^2 \left(\frac{2m\delta}{\rho D^2}\right) \eta \tag{5a}$$

or

$$C'_D = 9.52K_s \ \eta/V_R^2 \ . \qquad (5b)$$

The numerical coefficient for the clamped - pinned pile was found equal to 10.2.

Based on this result, on the adoption of 1% critical damping (therefore $\delta = 0.01 \times 2\pi$) and $V_R = 2.5$, together with figures 55 and table IX of [7], the value obtained for C'_D was $C'_D = 0.09$.

The configuration that contains the maximum nodal displacement, both space and timewise, in the direction of flow is used to evaluate the stress resultants that would occur if it were a statically imposed configuration. The results found in this way serve as an estimate of the maximum stress resultants.

3. FACTORS INFLUENCING THE RESPONSE

3.1 Effects of Lateral Rigidity of Pile Structures

The avoidance of the harmful aspects of resonant oscillation for a single degree of freedom system may be achieved by designing a structure that has the natural frequency either sufficiently larger a) or smaller b) than the exciting . The extension to multi-degree-of-freedom systems is fairly simple, bearing in mind the existence of the several natural frequencies.

Case a) corresponds to rigid structures with a fundamental frequency that exceeds for all velocities of flow. In terms of pile structures a) is illustrated by the utilization of tripods or, at least, inclined piles whose lateral motion determines axial straining of these piles.

Case b) corresponds to structures composed of slender vertical piles supporting rigid decks and, eventually, other structures.

In the first case an approximate model showing a fundamental frequency above the specified minimum assures that the structure is safe, from the standpoint under analysis.

In the second case a more accurate analysis is necessary, including the determination of sufficient number of natural frequencies to avoid unpredicted higher mode resonance. Also, the response for water speeds V, below V_{max}, that may cause resonant-like conditions for $V_R < 2.5$ must be examined to assure that the latter is admissible. For laterally flexible structures, the sway modes should be associated with low frequencies so that they are excited by forces with low amplitudes. The fundamental mode of the piles, in b), sometimes called the bow string mode, has to be kept as "far" from being excited critically by high speed of water as feasible and this is more easily accomplished if the piles support heavy masses (see 3.2).

The greater economy of solution b), notwithstanding the higher challenge that its design poses, recommends further studies of its applicability.

3.2 <u>Importance of the Mass of Superstructure</u>
The increase of M with regard to the mass of the piles has an effect on the natural modes and frequencies, as well as on the response, that is presently analysed.

In qualitative terms the conclusion reached for the simple model described below can be extended to more complex pile structures, provided that the piles are sufficiently spaced and that their flexibility be much smaller than that of the deck that introduces M.

The model is composed of an individual pile, as shown in fig. 1, topped with a concentrated mass M whose inertia is accounted only for motion normal to the axis of the beam. The geometric boundary conditions prevent displacement at the base and rotation at both ends. At the top the equilibrium of the shearing and the inertia forces provides the fourth boundary condition.

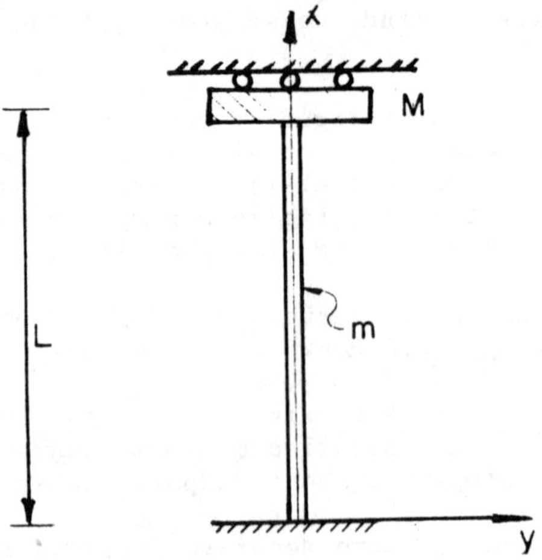

Fig. 1 - Model of Individual Pile With Attached Mass M

In terms of the dimensionless coordinate z = x/L the problem of the dynamic response is posed by

$$\frac{d^4y}{dz^4} + \frac{mL^4}{EI}\frac{d^2y}{dt^2} = \frac{pL^4}{EI} \qquad o < z < 1 \qquad (5)$$

$$y(o,t) = \frac{\partial y}{\partial z}(o,t) = \frac{\partial y}{\partial z}(1,t) = 0 \qquad (6 - a,b,c) \qquad (6a,b,c)$$

$$\frac{EI}{L^3}\frac{d^3y}{dz^3}(1,t) = M\frac{\partial^2 y}{\partial z^2}(1,t) \qquad (6d)$$

The characteristic equation, for free vibrations, is

$$\sinh\alpha\cos\alpha + \sin\alpha\cosh\alpha = \alpha R(1 - \cos\alpha\cosh\alpha) \qquad (7)$$

with

$$\alpha = m_t\omega^2 L^3/EI \qquad (8)$$

and

$$R = M/m_t \qquad (9)$$

where $m_t = mL$ designates the total mass of the pile and the remaining symbols are self explanatory.

The left hand side of equation (7) equalled to zero is recognised as the characteristic equation for the system with no attached mass, case on which the natural frequencies are well known. The right hand side is the product of R times the expression that, equalled to zero, leads to the natural frequencies of a clamped-clamped beam.

For large values of M the fundamental circular frequency may be approximated, if desired, by a series expansion of (7), neglecting higher order terms, that gives

$$\omega_1 \quad \sqrt{\frac{60}{2 + 5R}} \sqrt{\frac{EI}{m_t L^3}} \qquad (10)$$

whereas the second and higher natural frequencies are appropriately approximated by the first and higher natural frequencies of the clamped-clamped beam. The following table summarizes some results obtained by solving (7):

R	ω_1	ω_2	ω_3	Factor
0.1	4.995	28.030	69.767	
0.5	5.700	25.000	64.801	
1.0	2.955	23.939	63.432	$\sqrt{\frac{EI}{m_t L^3}}$
5.0	1.495	22.742	62.062	
10.0	1.075	22.562	61.870	
	22.373	61.670		

The results confirm that the second and higher modes correspond to the first and following modes of the clamped clamped beam and correct the statements in [5] and [8] that imply that all the modes of the system are approximated by the clamped clamped beam.

The reduction of the first fundamental frequency and the modification of the mode shapes due to the increase of R are also important in the interpretation of experimental results and the prediction of the response.

The velocity V for which the first mode is excited becomes lower and P_o, which is proportional to V^2, becomes so small for high values of R that the amplitude of motion is negligible. For higher values of V the second mode is excited and the magnitude of P_o is such that the motion assumes finite amplitude and the observed response corresponds, thus, to the second natural mode.

This interpretation is in agreement with and explains results found experimentally e.g. [9].

When the mass M increases, the question of eventual modification of the natural frequencies due to the axial load is also raised. For large values of the slenderness ratio, using safety factor of 1.92 (AISC) with respect to the buckling load, the decrease of the natural frequencies or, equivalently, of the critical velocity of flow would be close to 30%.

It is noted that, in practical design, the geometric inertia of the piles is determined by forces other than the axial force and the ratio P/P_{cr} is much smaller than 1.92 making this effect unimportant.

It is remarked that, when the translational mass effects are accounted in the vertical direction, the second mode, for R larger than a certain limit, corresponds to vertical motion and it is of no consequence for the study undertaken here.

3.3 Added Mass and Damping

The natural frequencies of piles immersed in quiescent water are lower than the values found when the vibrations take place in the air, due to the adherence of a mass of the water to the "solid mode" of motion of the pile. It has been shown [5] that the added mass to consider in order to account for this effect is independent of the flow and mode shape. In this investigation, the mass of a cylinder of water of same dimensions as the pile was added to the true mass of the latter, which includes the steel pipe plus the water inside, and distributed by the total length to define m.

As referred elsewhere, the structural damping is taken as 1% of critical and the need for an investigation of the effect of different coefficients remains. In particular the damping associated with the superstructure (for instance rubber supports, mechanical conveyors) may often reduce the response but its assessment could not be done with precision and was ignored.

3.4 Three Dimensional Effects

The study of a three dimensional structure, with its longitudinal axis normal to the direction of flow, was performed although the results are not included.

It is interesting to remark that the existence of very flexible longitudinal trusses determined the appearance of a low fundamental frequency for the global structure, within the range of cross flow oscillations.

The solution encountered consisted simply on changing the modulation of the structure by increasing the rigidity of the truss, see fig. 2, with the fundamental frequency of the revised structure assuming a value that was safe in terms of cross flow.

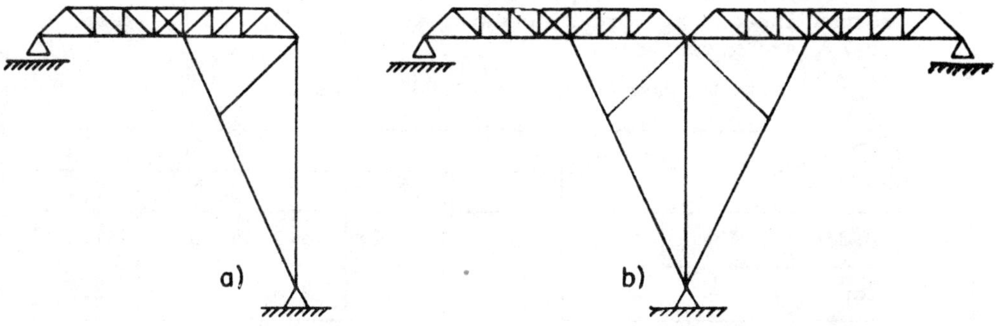

Fig. 2 - Simplified Sketch of Modulus of Metallic Truss, Before a) and After b) Revision of Structure

4. APPLICATIONS

4.1 Flexible Structure

The structure formed by two steel vertical piles, clamped at the base, supporting a horizontal, rigid platform that prevents flexural rotation of the top is geometrically defined by the model represented in figure 3 below.

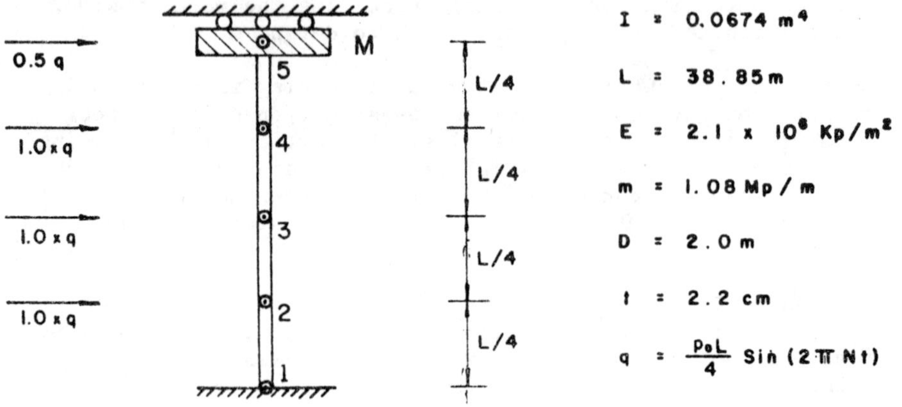

Fig. 3 - Model for Flexible Structures

The dimensions chosen for the model have the purpose of making the results directly comparable with those found for the rigid structure of 4.2 which is an actual model.

The mass is discretized at nodes 2 through 5 and the response is analysed for $M/m = 0.5$, 1.0 and 5.0.

The frequencies and mode shapes were obtained accurately and reported in 3.3.

The maximum lateral displacement, in cm, at 3 and 5, for forces induced by vortex shedding with $f' = N_1$ and N_2, respectively, are given below.

Maximum Displacements at 3 and 5

Point	M/m = 0.5		M/m = 1.0		M/m = 5.0	
	N_1	N_2	N_1	N_2	N_1	N_2
Mid-point (3)	12.8	39.9	8.2	45.1	2.1	51.2
Top (5)	24.7	14.2	16.0	9.6	4.2	2.6

The table shows, for instance,

a) Absolute maximum of displacements reached for second exciting frequency.
b) Steady decrease of importance of the response to first "resonant" condition with R increasing; at points 3 and 5, respectively, the sequence of displacements for successive R's is (12.8, 8.2, 2.1) and (24.7, 16., 4.2).
c) Steady increase of importance of response to second "resonant" condition at the point of maximum amplitude, (39.9, 45.1, 51.2), whereas at the top the clamped condition is being approached and, therefore, y_5 decreases.

Running a static analysis for the configurations that contain the maximum displacement, as explained before, and choosing the moments at base and top for the estimation of the maximum dynamic response it was found

Moment	M/m = 0.5		M/m = 1.0		M/m = 5.0	
	N_1	N_2	N_1	N_2	N_1	N_2
$10^3 \times M_1$	1.46	11.3	0.93	12.	0.23	12.8
$10^3 \times M_5$	1.21	11.4	0.52	12.2	0.23	12.8

and conclusions similar to the ones established above can be drawn.

The dynamic amplification factors for the displacement at 5 assume a monotonically decreasing value with increasing R: for $f = N_1 \rightarrow$ D.A.F. = 2.1, 1.4 and 0.4; for $f = N_2 \rightarrow$ D.A.F. = = 19.9, 12.9 and 3.4.

These results show that the response for the first mode excitation is considerably smaller than for the second mode. This in turn, indicates that excitations at low frequencies are admissible provided that the corresponding responses are within permissible range.

4.2 Rigid Structure

The model of the pile structures of the rigid type shown in figure 4 is one that was used in connection with part of a 2-dimensional study of the Ponta da Madeira
planned for construction in Northwestern Brazil. At Ponta da Madeira the water depth reaches 37 m and the velocity of tidal flow is expected to approximate 6 knots making the dynamic analysis necessary, more so after some preliminary work at the site confirmed the possibility of oscillations.

The analysis was conducted in the transversal and longitudinal

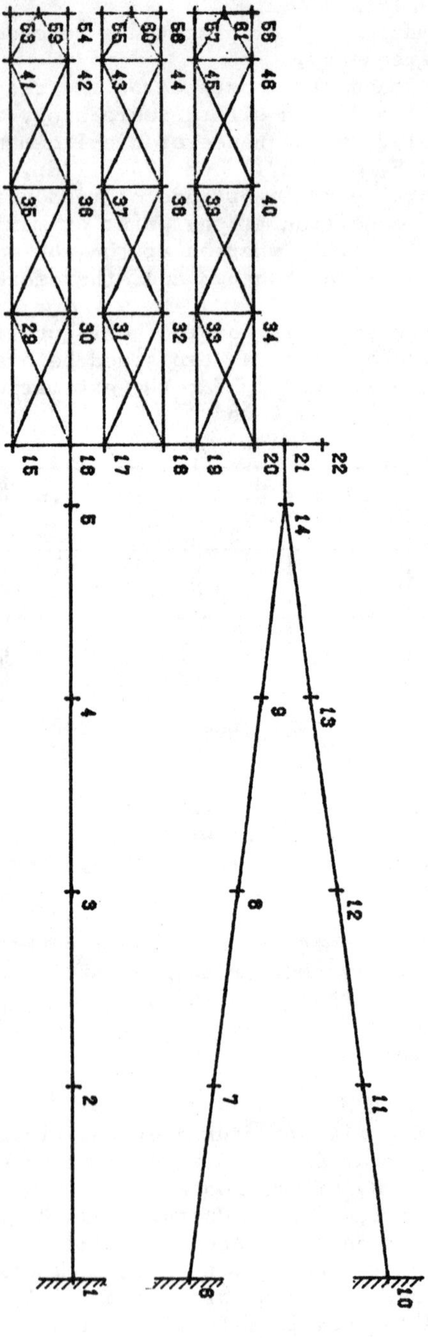

Fig. 4 - Geometric Model of Rigid Type Pile Structure

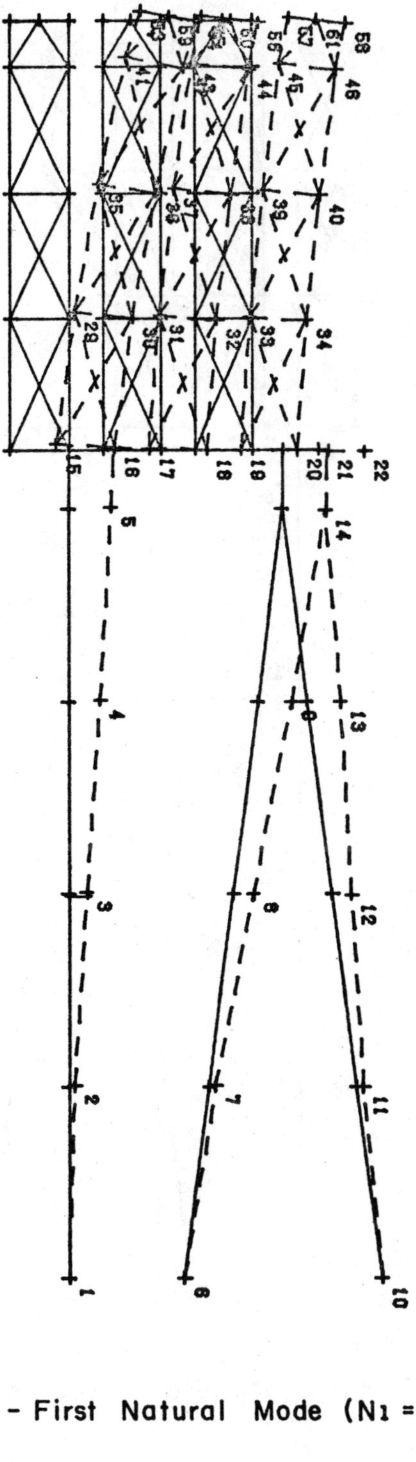

Fig. 5 - First Natural Mode (N1 = 0.694 cps)

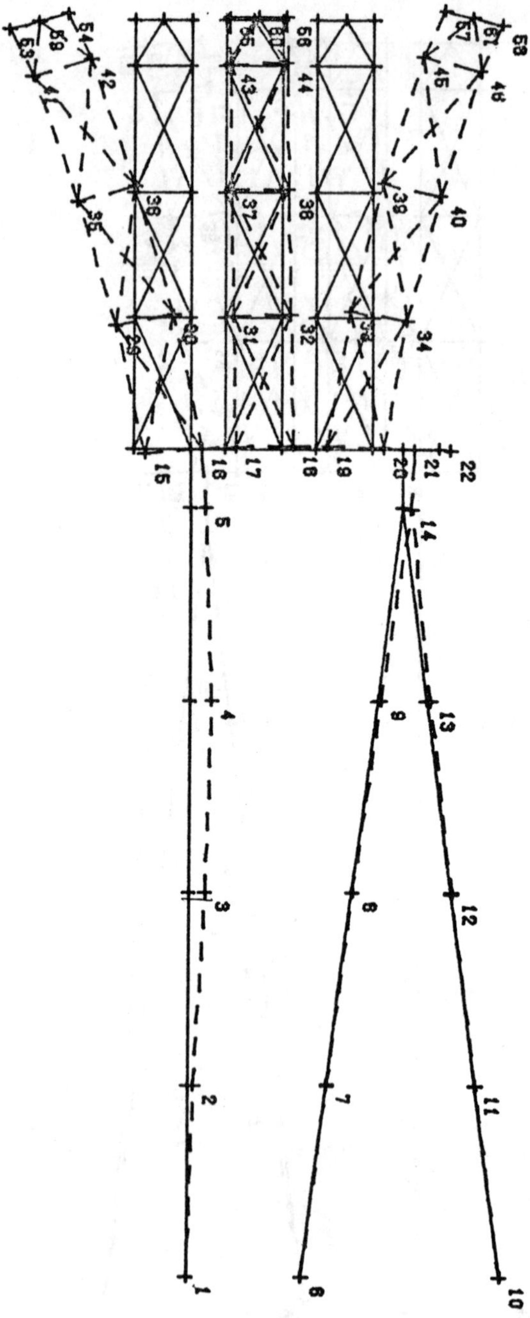

Fig. 6 - Second Natural Mode (N2 = 0.905 cps)

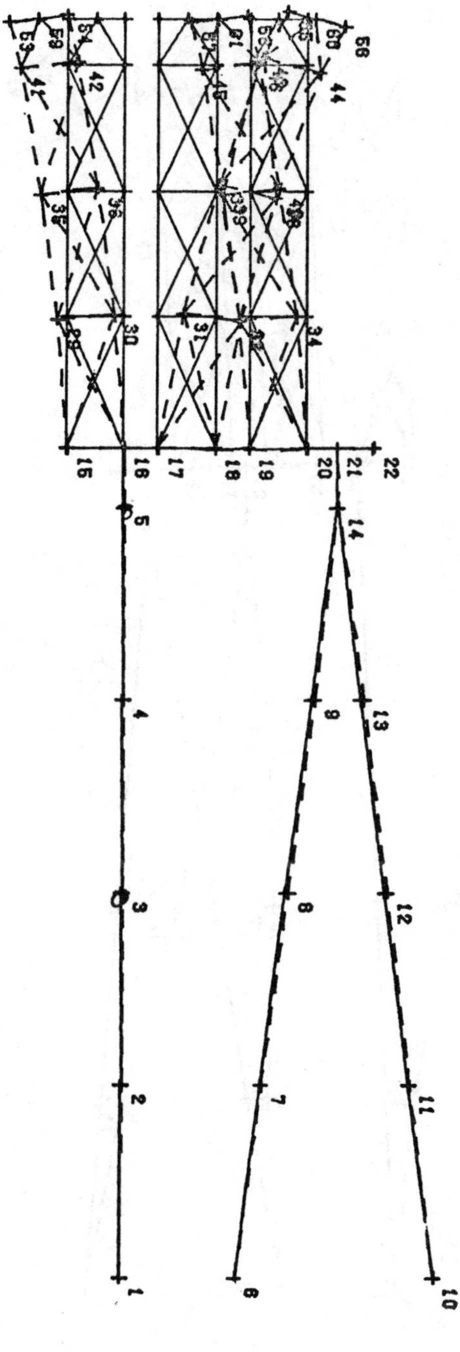

Fig. 7 – Third Natural Mode (N_3 = 0.985 cps)

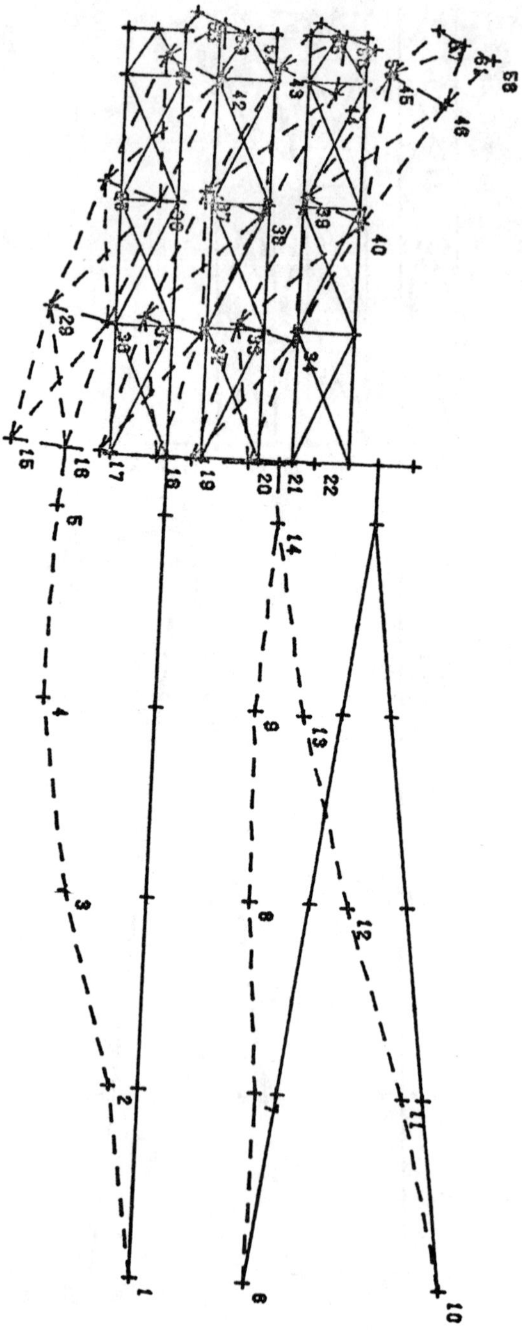

Fig. 8 - Fourth Natural Mode (N_4 = 1.355 cps)

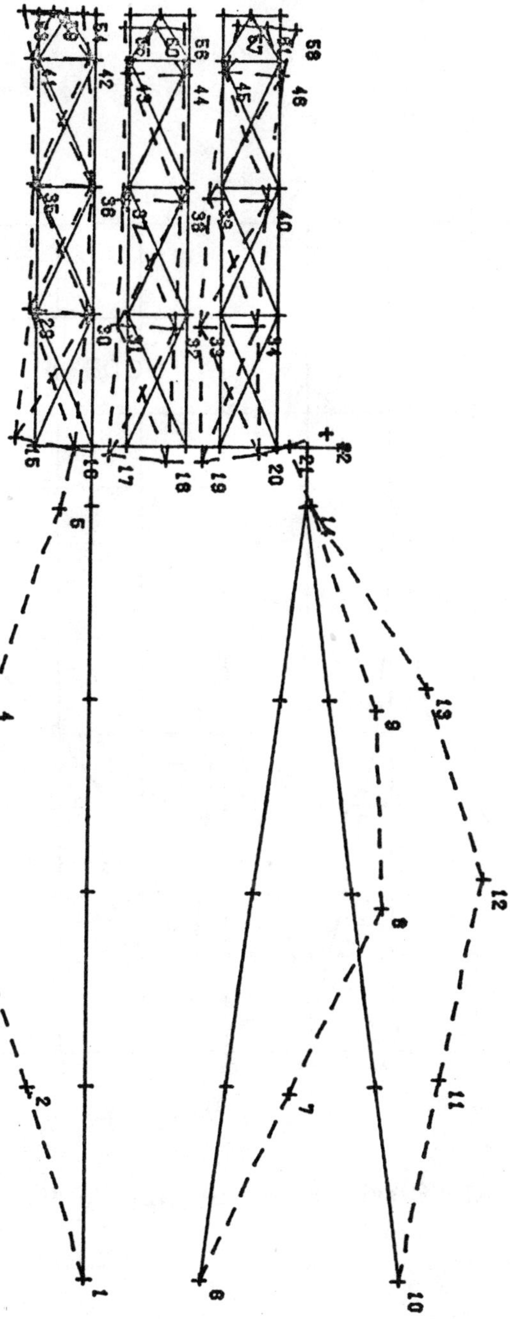

Fig. 9 - Fifth Natural Mode (Ns = 2.550 cps)

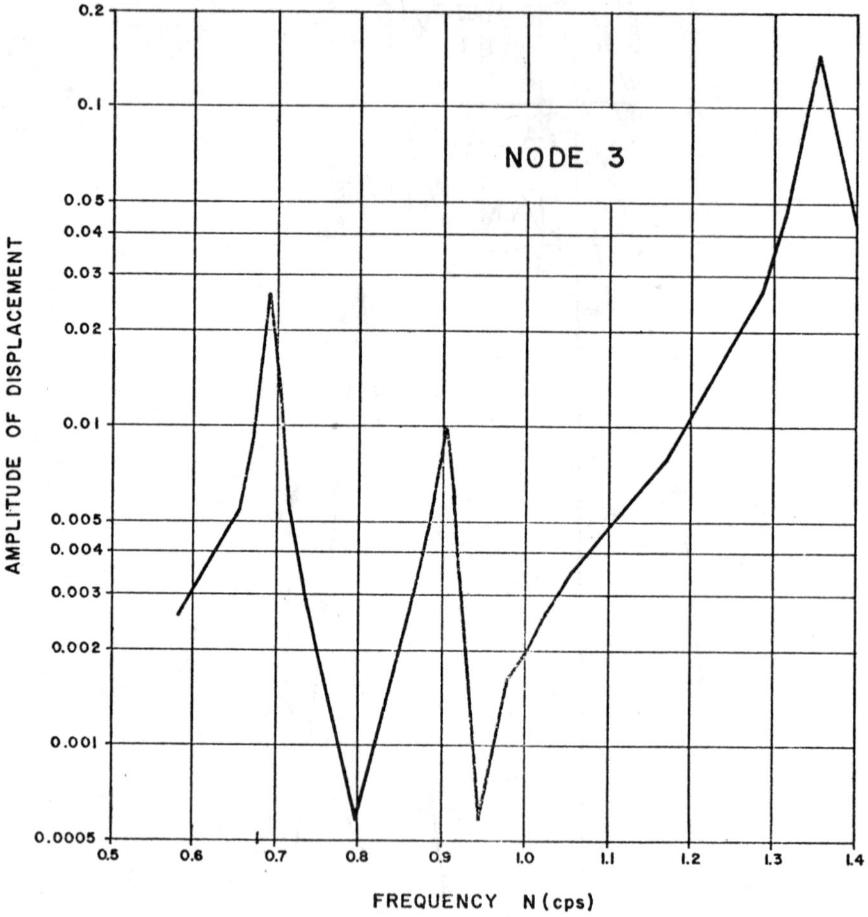

Fig. 10 - Displacement at Node 3 Versus Frequency

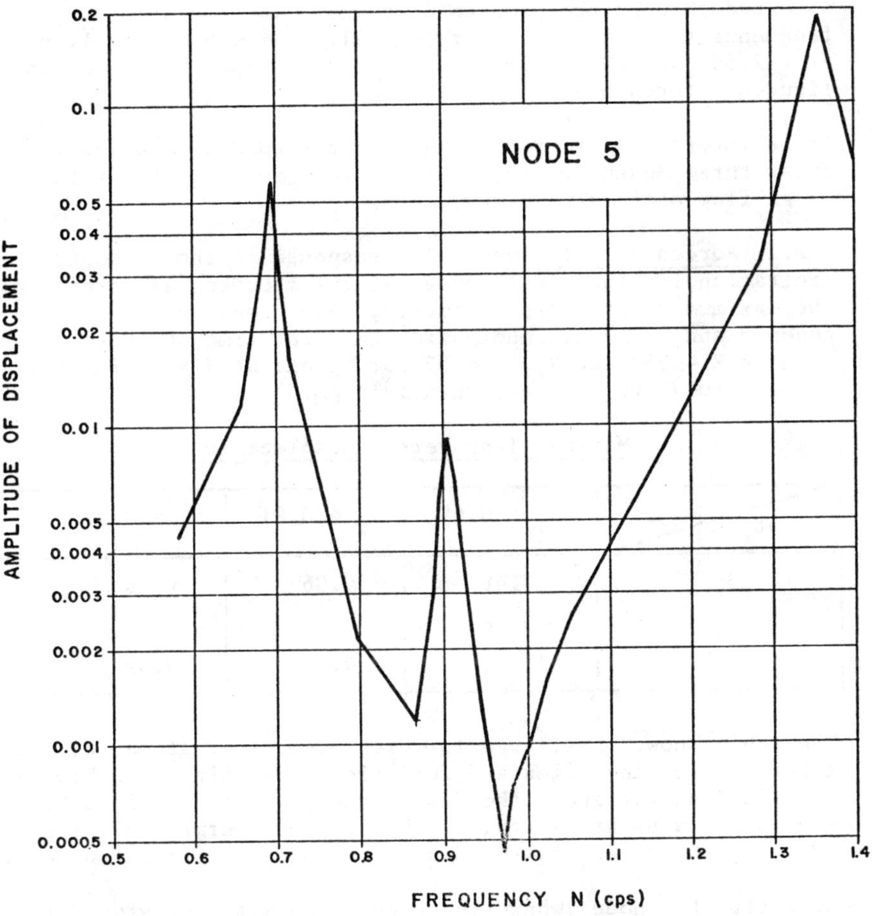

Fig. 11 - Displacement at Node 5 Versus Frequency

directions, both with plane and spacial models. The complicated structures necessary for the support ad functioning of the ore conveyors were modelled by equivalent beam and truss elements but the description of those steps exceeds and scope of this paper on which only the results obtained with the model of figure 4 are reported.

4.2.1 Free vibrations The natural frequencies and modes were obtained through the program Stardyne. The lowest five frequencies are $f_1 = 0.69$, $f_2 = 0.91$, $f_3 = 0.99$, $f_4 = 1.36$, $f_5 = 2.55$ cps. The corresponding modal shapes are shown in figures 5 through 9.

It is interesting to refer that, in the spacial analysis, the first three modes are longitudinal and are those for which cross-flow oscillations may occur.

4.2.2 Forced oscillations The response to the harmonic excitation for the first three natural frequencies determined the maximum lateral displacements, being reported those at nodes 3 and 5 and maximum maximorum (i.e. time and space wise) at node X = 53, for N_1, X = 57 for N_2 and at X = 55 for N_3. These values are tabulated next, in cm.

Maximum Transversal Displacement

Point	$N_1 = 0.69$	$N_2 = 0.91$	$N_3 = 0.99$
3	2.46	0.96	0.16
5	5.78	0.92	0.07
X	17.41	7.43	2.99

The table shows that, even for displacements that may be tolerable for the piles and for stresses within allowable range (considering fatigue), the displacements at some points of the superstructures assume values that are not permissible i.e. the concern of the analyst has to include the superstructure.

Selecting the nodes where the connection with the ground is made, plus node 5 for the evaluation of the stress resultants associated with the instantaneous configuration that includes $y_{53} = 17.41$ cm ($N_1 = 0.69$ cps) it is found:

"Maximum" Dynamic Stresses

Point	Moment Mpm	Shear Mp	Axial Force Mp
1	266.	13.1	262
5	97.	3.1	261
6	143	15.2	407
10	290	15.1	671

Taking the maximum values that would occur for a purely static response for forces p_o, the dynamic amplifications factors encountered are 10.6 for bending moment, 7.2 for shear force and 35.3 for axial force. These values are point-dependent and they serve only to appreciate the magnitude of the dynamic effects.

The curves that show the frequency dependence of the lateral displacements of nodes 3 and 5 are shown as Figs. 10 and 11.

5. CONCLUDING REMARKS

The study shows the importance that in line oscillations may assume in structures supported by very slender piles that are designed not to be subjected to cross flow oscillations.

The possibility of having to admit the presence of in line excitation near a fundamental frequency is shown to be tolerable, under certain limitations. The increase of the ratio R of masses has a beneficial effect in terms of the magnitude of the response induced by the flow.

The significant values that the response at points of the superstructure may assume, endangering its serviceability if not its safety, was also illustrated.

The need for further studies, both analytical and experimental, especially directed towards the importance and quantification of hydrodynamic damping, added mass, peak response at subcritical frequencies and mechanisms to prevent vortex shedding was also evidenced along the development of the present work.

ACKNOWLEDGEMENT

The authors acknowledge the permission from Amazonia Mineracao S.A. to release results connected with engineering studies, performed at Promon Engenharia, for the Terminal at Ponta da Madeira.

PART IV SOIL PROBLEMS

INFLUENCE OF THE SOIL CHARACTERISTICS IN THE BEHAVIOUR OF OF A FIXED STEEL PLATFORM

A.J. Ferrante, N.F. Ebecken, E.C. Prates de Lima and E.Chiang Valenzuela

COPPE-UFRJ and Petrobras-Dexpro/Denge

INTRODUCTION

Many different types of equipment are used to look for and to extract oil from marine oil fields. These include among others, drilling ships, jack-up rigs, semi-submersible platforms, subsea stations, tension leg platforms, articulated towers, buoy-type systems, and fixed platforms.

One of the most popular types, for production purposes, are the fixed platforms. Basically these can be concrete gravity structures and steel structures, although there is a wide variety of different types, and hybrid structures have also been used.

The fixed production platforms used in the Brazilian offshore oil developments are almost exclusively steel structures but three small concrete gravity platforms have also been built [1].

This paper is concerned with the structural analysis of the Petrobras PGA-06 platform, already installed in the Guaricema field. This is a piled steel structure, designed to operate in shallow waters of 32 metres depth.

The present study consists of two parts. In the first part the the following three different analyses were performed:
a) the structure was considered hinged at the sea bottom,
b) the piles were also included but the behaviour of the soil at the Guaricema field was represented by equivalent linear springs, and c) the soil behaviour was represented by non linear springs. The non linear soil characteristics were taken from P-Y curves, defined as indicated by the American Petroleum Institute Code [2].

Usually the task of getting boring samples and data about soil parameters is a very expensive and time consuming process. On the other hand the laboratory tests normally performed are rather simple and are applied to perturbed horing samples. This leads to inaccuracies in the geotechnical parameters required to estimate the pile behaviour, which is not very consistent with the sophistication level of the procedures used for structural analysis.

In some cases, due to difficulties encountered in the drilling operations, or because some wells have to be disregarded, a given platform needs to be relocated to a different area, having a soil with different characteristics. Additionally it has been found for the Brazilian coast that fine soils (usually mud) move with time, changing the topology and the batimetric characteristics of the sea bed, altering the conditions for pile design.

Due to the previous considerations, and in order to study the influence of different soils on the structural behaviour, in the second part of this work, the platform was also analysed for three additional types of soils, including that corresponding to the Robalo field, and two extreme cases indicated by the API code: a silty sand and a medium clay.

THE FIXED OFFSHORE STRUCTURES

It is generally considered that development of fixed offshore structures started in 1945, when Exxon installed a timber structure in 6 metres water depth, in the Gulf of Mexico [3]. After that, in 1947, a steel structure was installed in 10 metres water depth, and since then the number of fixed offshore structures built has grown very rapidly.

The fixed offshore structures can be classified basically into two kinds: the gravity platforms, and the piled steel structures. The gravity platforms are normally fabricated from prestressed reinforced concrete, and maintain their position due to their enormous weight. Lately some steel gravity structures have been also constructed.

By far the largest number of fixed offshore structures operating today are piled steel structures. For instance, by 1973 there were 1935 platforms of this type operating in the Gulf of Mexico [4]. By now that number has grown to more than 2600.

As the search for oil has progressed to deeper waters, it became necessary to build platforms of ever increasing dimensions. Thus the first platform operating in over 50 metres water depth was installed in 1960. By 1967 there were already platforms operating in more than 100 metres depths. In 1976 Exxon erected near Santa Barbara, in California, the Hondo platform, to operate in a 260 metres water depth. 1978 is the expected completion date for Shell's Prospect Cognac platform, which will be installed in 300 metres water depth, in the Gulf of Mexico [5].

These steel platforms are normally built as consisting of an upper deck of one or more levels, resting on top of a steel jacket, which is fixed to the sea bottom with the help of a set of driven piles.

From the point of view of the installations methods used for the jacket, these structures are classified into the following three types: a) launch type jackets, b) self floating jackets and c) multipart jackets.

The launch type jackets are transported on derrick barges. Once on location these are put in position by cranes, and finally piles are driven through their tubular legs, in order to fix the structure to the sea bed. The largest structure of this type is a steel jacket built for the Piper field, in the North Sea, weighing approximately 14000 tons, and operating in a 145 metres water depth [3].

Due to the size of the cranes which would be required in order to handle them, the heavier jackets have been built with self floating capabilities. In this case the jacket is towed to site and then is up-ended by means of a well controlled weight-buoyancy system, without having to use a derrick barge. Sometimes the buoyancy is obtained by floating units attached to the jacket, which are removed after installation and can be re-used. An example of a self floating jacket is the Ninian Southern jacket, to be installed in 141 metres water depth, in the North Sea. Its total height is 168 metres, and flotation is obtained through legs of 9.2 metres diameter [6].

The multi-part jacket concept was used in the case of the Hondo platform, the largest offshore structure built so far [5]. The jacket was built in two parts each lying on its side. Then they were transported and launched separately, and joined horizontally and welded together while afloat in deep sheltered waters. Finally, the full jacket was towed to site, put into the upright position, and installed. The full jacket weighing 12000 tons in place was fabricated in 17 months. More than 5000 tons of main and skirt piles were required. This 28 wells platform operates in a 260 metres water depth and is expected to reach a peak production of 28000 barrels per day.

Another structure of this type is the Prospect Cognac platform, currently being built. In this case the jacket is made of three separate parts. This platform will operate in over 300 metres of water, and will be able to accommodate 62 wells.

Its total weight will be 45000 tons.

The severe environmental conditions prevailing in the North Sea, including design wave heights of more than 30 metres, and more than 100 kilometres per hour winds, together with the hard soils encountered in the region, were the main reasons for the introduction of the concrete gravity structures. These structures remain in place due to their enormous weight, and provide very large deck area and large oil storage capacity. The first concrete gravity structure was installed in the Ekofisk field in 1973, in a 70 metres water depth. The largest structure of this type is probably the Ninian Central platform, currently being built by Howard-Doris. This structure consists of a 131 metres diameter base, supporting a central column of 55 metres diameter, atop which rests the deck area. The height of the structure is 170 metres and operates in 139 metres water depth. A total of approximately 600000 tons of concrete were required for its construction [7].

In addition to the oil platforms already operating, several others are being installed and constructed for the North Sea. By May of 1977 a total of 38 new offshore structures were in different stages of installation, and 6 concrete and 5 steel structures were under construction [8].

Lately some steel gravity structures were introduced. This is the case, for instance, of the four Loango platforms being built by Tecnomare to operate in the Congo Brazzaville, for water depths of the order of 90 metres [9]. The hard soils encountered in the area prevented the use of piled steel structures, thus making convenient the adoption of steel gravity structures. The Loango platforms consist of a hexagonal tower superimposed on a trinagular base, with foundation pads.

As the search for oil is taken to deeper waters, it can be foreseen that in the future platforms operating in water depths of more than 300 metres will have to be designed and constructed. Possibly, then, the concept of fixed structures will have to be abandoned, in favour of new solutions. The new generation of marine platforms may take the form of guyed or articulated structures, or tethered buoyant platforms.

OFFSHORE STRUCTURES IN BRAZIL

Brazil produces approximately 185000 barrels per day of oil, which represents around 20% of the total consumption. As the production of the land oil fields in Salvador, Bahia, is decreasing, the main effort is now directed to the development of marine oil fields. At present these already provide nearly

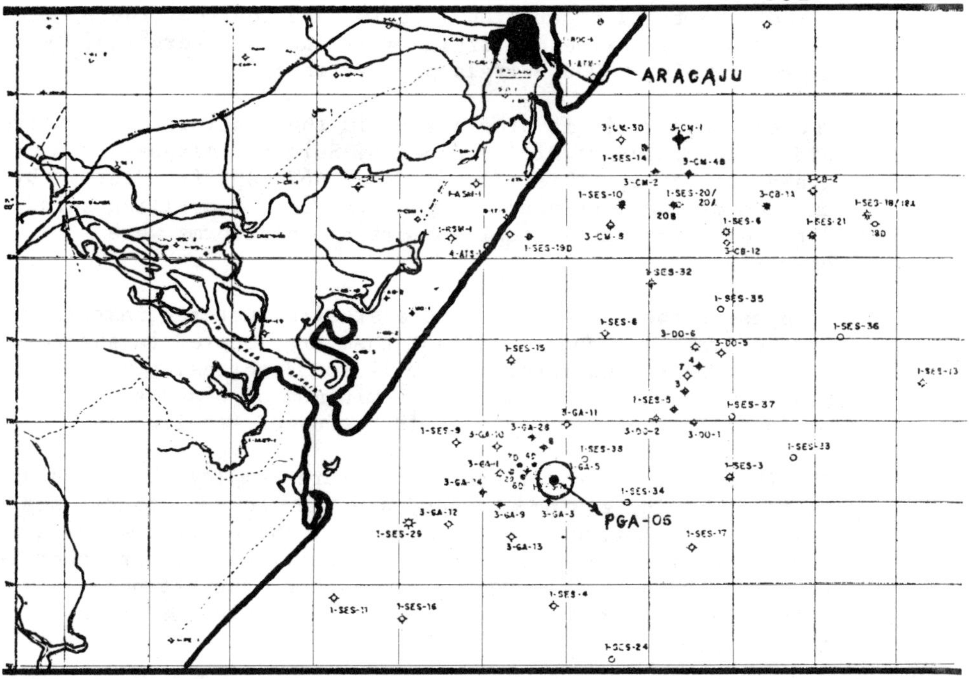
ARACAJU
PGA-06
3-CM-1
3-CM-30
1-SES-14
3-CM-48
3-CM-2
1-SES-10
3-CM-9
1-SES-19D
1-SES-32
1-SES-35
1-SES-8
1-SES-36
1-SES-15
1-SES-5
1-SES-37
3-GA-11
1-SES-9
3-GA-10
3-GA-28
1-SES-38
3-GA-5
3-GA-14
1-SES-34
3-GA-12
3-GA-9
3-GA-3
1-SES-29
3-GA-13
1-SES-17
1-SES-3
1-SES-4
1-SES-11
1-SES-16
1-SES-24
1-SES-21
3-CB-12
3-CB-2

FIG. 1

FIG. 2

30% of the total production. It is expected that the production from the marine oil fields will increase very rapidly to approximately 200000 barrels per day in 1978.

The search for oil is progressing in the continental platform along the states of Rio de Janeiro, Sergipe, Alagoas, Rio Grande do Norte, Bahia, Amapá, Ceará and Espírito Santo. By the end of the last year, 13 jack-up rigs, 2 drilling ships, 2 tender ships, and 5 semi-submersible platforms were operating.

At present there are 15 offshore steel structures installed in the Brazilian coast. All of them are located in shallow waters with depths ranging from 12 to 32 metres. These platforms, able to accommodate a minimum of 6 and a maximum of 15 wells, have from 700 to 1800 tons of steel weight.

The three main kinds of steel structures employed, depending on the drilling equipment used, are the jack-up, tender and self-contained structures, but most of the platforms built so far are of the first two types. In particular, this paper is concerned with the structural analysis of the Petrobrás PGA-06 platform for the Guaricema field, which is of the tender type.

Larger steel structures are currently being designed for the Campos basin. These include three piled steel structures for the Namorado, Garoupa and Enchova fields, which will operate in water depths of hte order of 160, 121 and 118 metres, respectively.

THE PETROBRAS PGA-06 PLATFORM

The Petrobras PGA-06 platform is the sixth to be installed in the Cuaricema field, with UTM coordinates: X = 8766150, Y = 714250 and a batimetric depth of 32 metres.

The Guaricema field is located south east from the coastal city of Aracajú, in the State of Sergipe, as shown in Fig. 1.

This platform was projected to operate with six wells. The drilling equipment to be utilized is the Delta Nine rig, with about 500 metric tons of weight, including the derrick, master skid, the main tanks, and the setback pipes. Due to the use of this equipment, the main deck of the platform is of 65 × 45 feet, rather than 45 × 45 as conventionally for small platforms. Still, the lower deck is of 45 × 45 feet dimensions, as normally for the production facilities.

The steel weight of the platforms is:

- Jacket	240 metric tons
- Decks	140 metric tons
- Piles	240 metric tons
Total	620 metric tons

The platform has a total height of 49 metres from the sea bed and a pile penetration of 55 metres. The design time was of two months, and the construction took four months. Fig. 2 shows this platform in the construction phase, in the Petrobras Salvador shop field.

This is a piled structure of the tender type for shallow water, where the rig is located on the main deck of the platform, but all other heavy equipment is on a support ship.

The structure was fabricated on its side, slid on to a barge and transported to site. Then it was put in the upright position, and lowered on to the sea bed. Then the piles were driven through the legs, and finally the deck frames and equipment were lifted on and welded.

These platforms are normally analysed under a number of different conditions, including loads acting when being transported, and when in service. In this case the interaction between the loaded structure and the reactions from the piles, may affect the stress levels in the components of the structure, particularly in its lower part. This type of effect has not been generally taken into account for smaller platforms, up to 20 metres of water depth, but for bigger platforms, and depending on the soil characteristics, this effect will have to be a determinant factor in the structural design.

THE ANALYSIS MODELS

Fig. 3 shows the space frame model adopted for the analysis of the platform structural system. This model includes 199 joints and 367 members. The overall dimensions are also indicated. Figs 3a and 3b show two plane views, parallel to the X and Y global axis respectively.

The structure was designed to withstand combination of the following loadings: structural equipment weight, buoyancy load, wave slamming, and wave action according to three different attack angles.

The wave data was taken from Glenn's one hundredth year design wave Report [10]. Its main parameters are the following:

- Chart depth	100.0 feet
- Highest astronomical tide	7.4 feet
- Storm tide	1.7 feet
- Total tide	9.1 feet

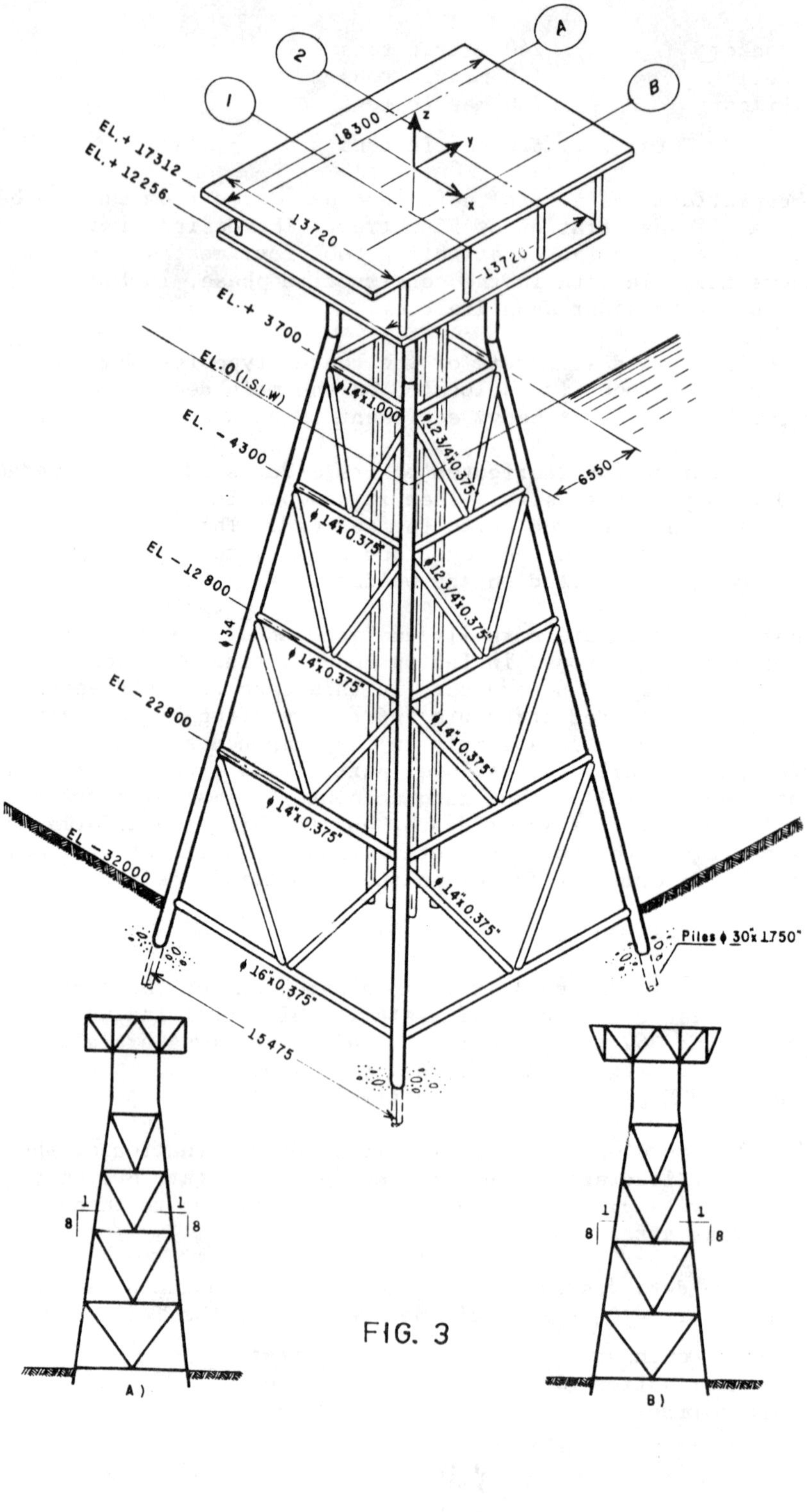

1
2
A
B
18300
13720
13720
EL.+ 17312
EL.+ 12256
EL.+ 3700
EL.0 (I.S.L.W)
EL. -4300
EL -12 800
EL -22800
EL -32000
6550
15475
ϕ 14"x1.000
ϕ 12 3/4"x0.375
ϕ 14"x0.375"
ϕ 34
ϕ 16"x0.375"
Piles ϕ 30"x 1750"
A)
B)

FIG. 3

- Still water depth	109.1 feet
- Height of maximum wave	41.4 feet
- Length of maximum wave	528.0 feet
- Period of maximum wave	10.5 seconds

The conductor pipes were also designed as load carrying structural components, in order to keep unchanged the preliminary design geometry of the jacket. However, these conductor pipes were rigidly jointed to the structure only at level (+) 3700, to avoid underwater welding at lower levels.

For this comparative study the wave loading in the X direction was assumed to be the only one acting on the structure. Due to the dimensions of the platform, the dynamic effects are not significant. Therefore the wave actions were transformed into a set of equivalent static loads. All the analyses to be presented in what follows were carried out for such loads.

The three different cases studied were:

1) Case A. Platform fixed at the sea bed. It was performed a linear structural analysis for static loads. This analysis will indicate the behaviour of the jacket independent of the soil characteristics.

2) Case B. In this case the piles and the conductor piles, penetrating 55 metres in the soil, were also included. The interaction between the piles, and the soil was presented by Winkler type springs, located every two metres, both in X and Y directions. The spring constants were obtained from the first part of P-Y curves. The analysis performed was again of linear static type. This case already includes the pile-soil interaction, and can be used for a preliminary design of the pile. Additionally, taken together with case C, can be used as a basis, in order to justify a more sophisticated non-linear analysis.

3) Case C. In this case the soil was also represented by springs, located every two metres, but non-linear load-displacement characteristics taken from the P-Y curves, were assigned to them. Notice, however, that in order to save computer time, the bars were defined as having a linear behaviour. The non linearity was only relative to the springs used to simulate the soil.

All this previous analyses were carried out using several languages of the LORANE system. This system developed jointly by the Posgraduate Program in Civil Engineering, of the Federal University of Rio Grande do Sul, and COPPE of the Federal University of Rio de Janeiro, includes a family of problem oriented languages for civil engineering applications.

The languages implemented so far are LORANE LINEAR [11] for linear static structural analysis; LORANE DINA [12] for linear dynamic structural analysis; LORANE NL [13] for static and dynamic geometrically non linear structural analysis; and LORANE HYDRO [14], for hydraulics applications.

The languages of the LORANE system are of general purpose kind, and can be applied to many different types of problems. In particular, several of them offer useful capabilities for off-shore structures analysis.

In the present case the LORANE LINEAR language was used for the analysis of cases A and B, both of linear static type.

The LORANE DINA language allows for free vibrations and deterministic and stochastic forced vibration analysis, capabilities which can be applied to study dynamic actions on larger off-shore structures, such as those due to waves. In addition it includes also the possibility of working with displacement dependent loads involving non-linear relationships. This was in order to perform the analysis of type C.

Several techniques for geometrically non-linear large displacements, and large deformations, analysis are part of the LORANE NL language. Even when its capabilities are considerably more powerful, they can also be used to perform the analysis of case C, using numerical schemes such as, for instance, Newton-Raphson.

SOIL CHARACTERISTICS FOR THE GUARICEMA FIELD

In the case of all previous platforms installed in this field drilling samples of the soil were not taken. For this platform such study was carried out using the diving bell system [15], to obtain boring samples up to a penetration of 55 metres.

The SPT showed that the first 11 metres corresponded to a very soft material, but after that a medium type sand was detected.

To evaluate the geotechnical soil parameters, the following laboratory tests were made:

1) Grain size analysis
2) Direct shear test
3) Vane test
4) Relative density
5) Humidity content

From these tests it was possible to prepare the overburden and shear stress resistance, versus depth, diagrams, shown in

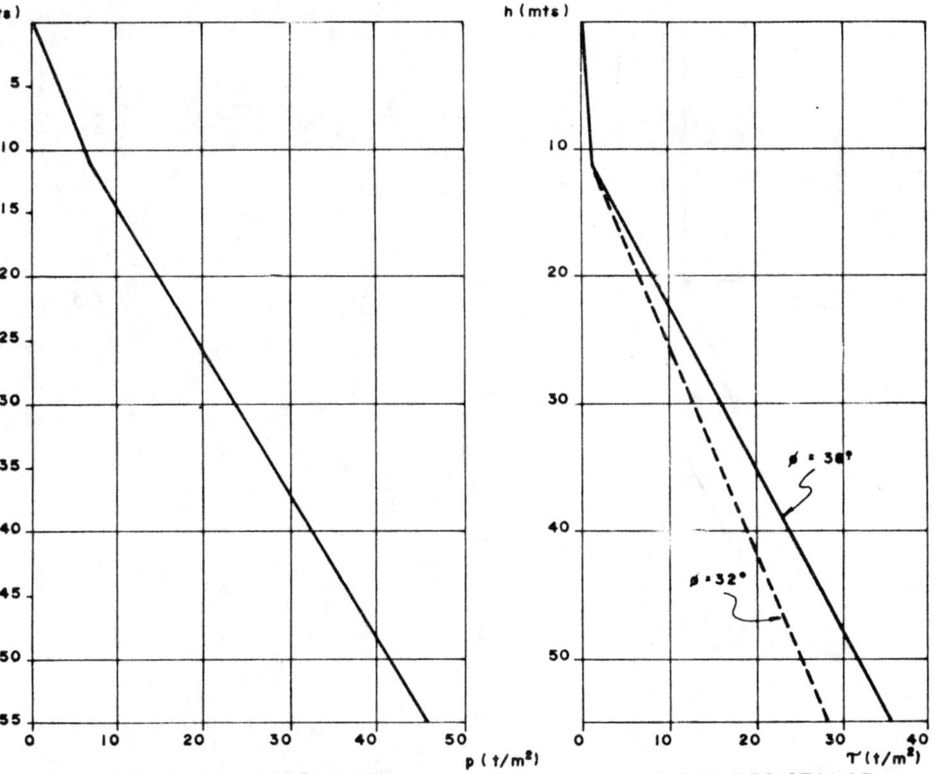

OVERBURDEN PRESSURE

SHEAR RESISTANCE

(a)

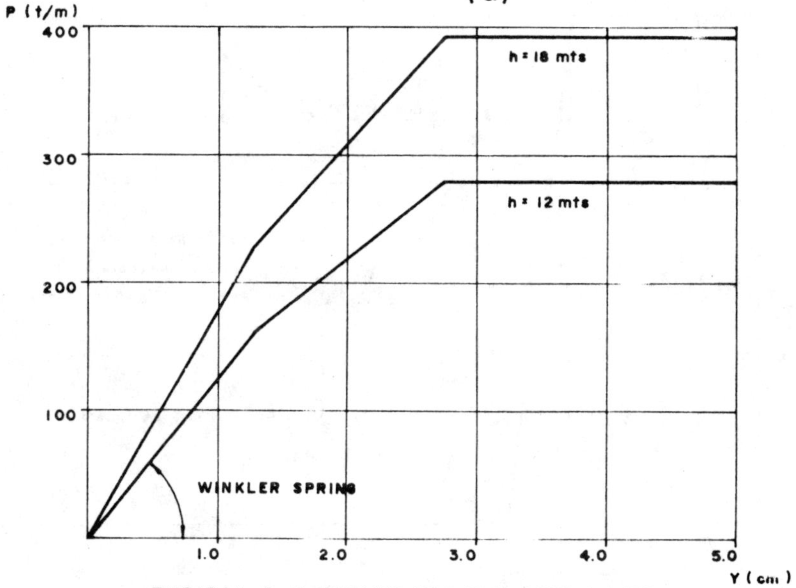

TYPICAL P-Y CURVES FOR THE SAND LAYER

FIG. 4

(b)

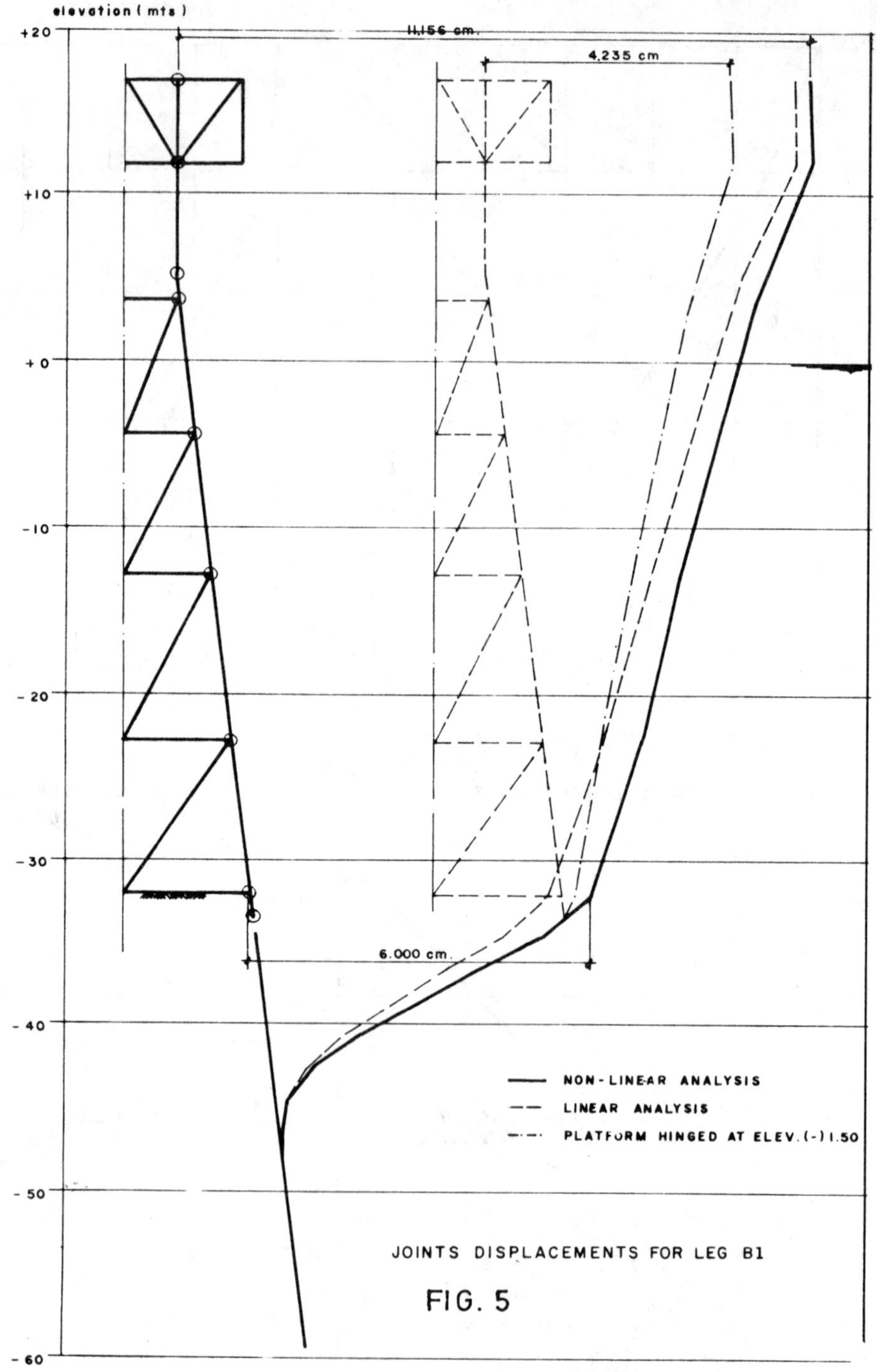

JOINTS DISPLACEMENTS FOR LEG B1

FIG. 5

Fig. 4a. The internal friction angle and the cohesion factor were also computed. All these parameters are enough in order to characterize the soil. The results were the following:

First layer. Material: Silty clay up to a depth of 11 metres
Submerged weight: 0.6 t/m^3
Angle of internal friction: 7°
Cohesion: 0.4 t/m^2

Second layer. Material: Medium sand from 11 metres up to a depth of 55 metres
Submerged weight: 0.9 t/m^3
Angle of internal friction: 30° (for maximum density)
32° (for minimum density)

To determine the lateral non-linear behaviour of the soil P-Y curves were computed, based on the previous data, as indicated by the American Petroleum Institute code [2]. Fig. 4b shows these curves for the sand layer.

OTHER SOILS CHARACTERISTICS CONSIDERED

In order to detect the influence of the soil parameters in the structural response, the following available data was used:

Robalo Field, Borings 1-2

First layer: Material: Silty sand up to a depth of 4.5 metres
Submerged weight: 0.9 t/m^3
Angle of internal friction: 30°

Second layer: Material: Clay up to a depth of 9. metres
Submerged weight: 0.8 t/m^3
Cohesion: 2 t/m^2
Angle of internal friction: 20°

Third layer: Material: Clean sand up to a depth of 30 metres
Submerged weight: 0.9 t/m^3
Angle of internal friction: 35°

Fourth layer: Material: Clay up to a depth of 44 metres
Submerged weight: 0.8 t/m^3
Cohesion: 2 t/m^2
Angle ot internal friction: 22°

Fifth layer: Material: Clean sand up to a depth of 55 metres,end of boring
Submerged weight: 0.9 t/m^3
Angle of internal friction: 35°

Theoretical Soils

Minimum soil parameters, based on disturbed boring samples and low values specified by the API Code, were used, with the following characteristics:

- Silty Sand. Submerged weight: 0.9 t/m^3
  Angle of internal friction: 20°
- Medium Clay. Submerged weight: 0.7 t/m^3
  Cohesion: 1.25 t/m^2
  Angle of internal friction: 15°

Fig. 4c shows the comparison of the soils profiles and the shear stress resistance versus depth diagram.

No rational theory for predetermining axial pile capacity for sand and clay in the elastic range is likely to be available for some time; observations of pile behaviour during driving could eventually be used.

Fig. 4d shows curves giving the ratio of load transfer to shear strength as a function of pile movement, adapted from Coyle Reese [16], and checked for highly plastic clays in the Pile-loading test, for the Morganza Floodway Control Structure, conducted by Mansur and Facht [17]. These curves can be used with caution, when no other data is available, to determine the elastic axial behaviour of the pile.

COMPARISON OF THE RESULTS

The results obtained are summarized in Figs 5,6 and 7 showing the lateral displacement in the direction of the wave, for the Guaricema field, the relevant bending moments and shear forces on the piles, also for the Guaricema field, and the lateral displacements computed for the four different soil conditions considered, respectively.

The displacements shown in Fig. 5 correspond to leg B1 (also see Fig. 3). The displacements for case A are shown relative to the displacement for case C at the sea bed. Comparison of cases B and C shows that consideration of non linear effects produces an increase of 15% of the displacement at the sea bed. Also the slope for case C is close to that for case A.

The bending moment and shear forces for the soil conditions corresponding to the Guaricema field, are shown in Fig. 6. For the bending moment the increase was 17% at the sea bed, and of 21% at 12.5 metres of pile penetration. In this last level the increase in shear force was 19%.

It was certified that the forces in the members of the jacket, were similar for cases A and C, even when the maximum lateral displacements at the top of the platform show an increase of

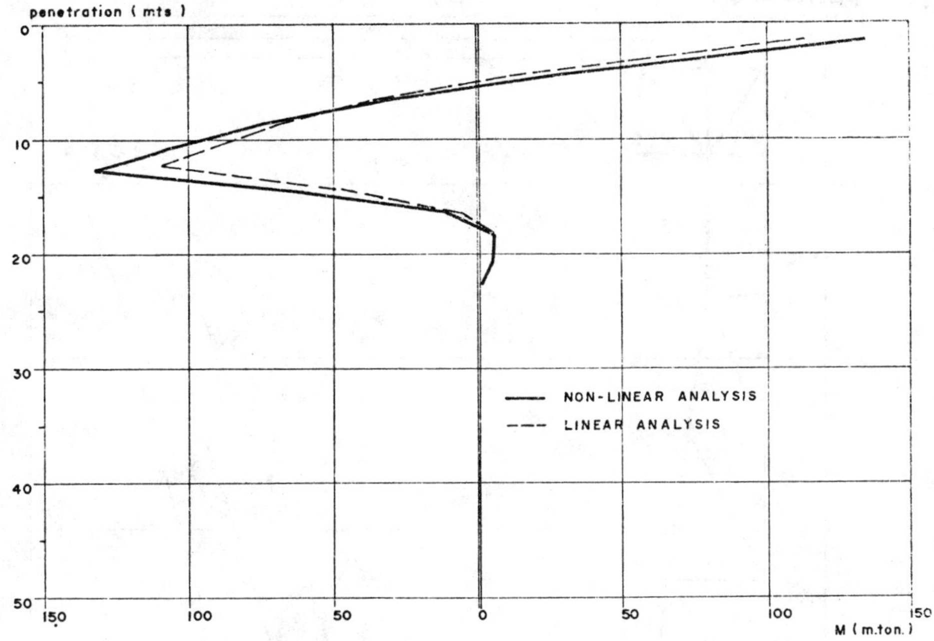

MOMENT DIAGRAM OF THE PILE FOR LEG B1

Figure 6A

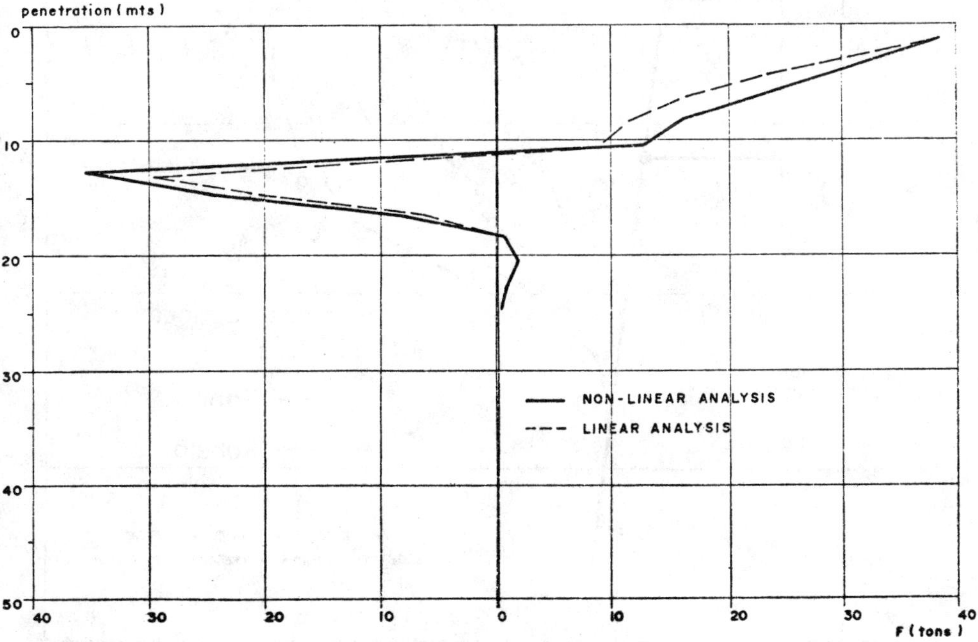

SHEAR DIAGRAM OF THE PILE FOR LEG B1

Figure 6B

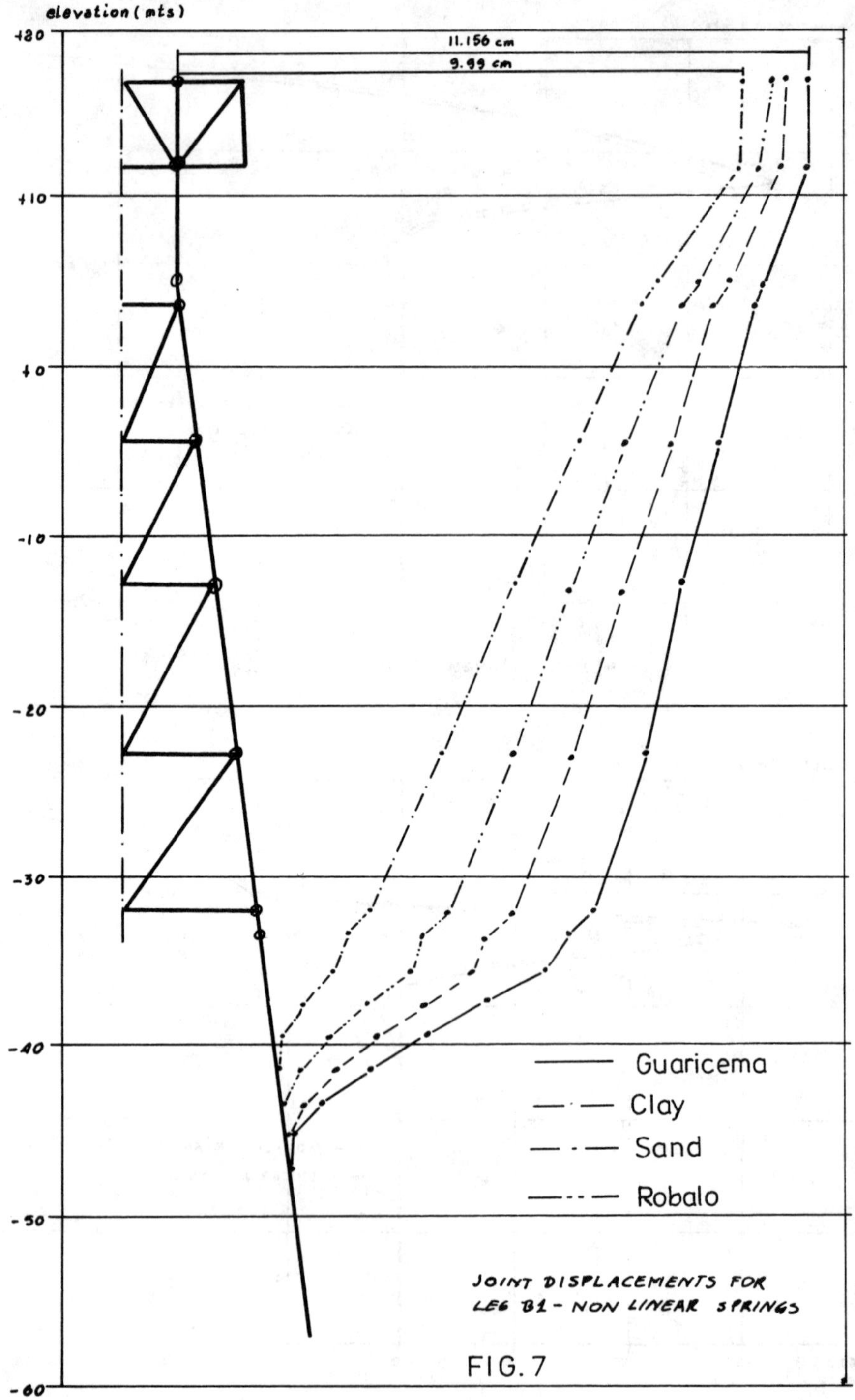
elevation (mts)
+20
+10
+0
-10
-20
-30
-40
-50
-60
11.156 cm
9.99 cm
Guaricema
Clay
Sand
Robalo
JOINT DISPLACEMENTS FOR
LEG B1 - NON LINEAR SPRINGS

FIG. 7

51%. This indicates that, for this type of platform, the analysis hinged at the sea bed will suffice for the design of the jacket.

The conductors, being carrying load members, take 51% of the total shear force, thus reducing the forces applied to the piles. Therefore, the non linear effects are smaller. A more rational design of the piles is thus achieved, making unnecessary the use of skirt piles or other similar schemes. This requires that the conductors be structurally designed, regarding both its dimensions and the quality of steel used. Without the conductors, the bending moments in the piles would have an increase of 90%.

Fig. 7 indicates that the soil conditions for the Guaricema field are even worse than the limiting cases for clay and sand given by the codes with extreme parameters. In particular, the soil at the Guaricema field is the worst to be found from Rio de Janeiro to Rio Grande do Norte, an area where most of the Brazilian oilfields are located.

CONCLUSIONS

In the first part of this paper a review is made of the fixed offshore developments, which have taken place so far. A reference is made to the Brazilian offshore oil field developments. In the second part the attention is focussed in the analysis of the Petrobras PGA-6 steel platform, considering different types of soil conditions.

The platform is initially analysed for the soil conditions prevailing at the Guaricema field where it was installed. Three different analyses are performed and their results are compared, including consideration of non linear soil behaviour Then the soil conditions are changed, and the cases of the Robalo field, a silty sand, and a medium clay, are taken into account.

The results obtained indicate that the deck and the jacket can be designed and analysed separately from the piles, performing a linear elastic analysis. Then they can be condensed, obtaining an equivalent stiffness matrix at the sea bottom. Finally the assembly of that condensed stiffness matrix and the piles is considered performing a non linear analysis leading to the design of the piles.

REFERENCES

1. Loze, B.J. and M.A. Bicalho, "Construction Developing of Concrete Off-shore Platforms for Oil Exploration in Brazil", International Conference on Off-shore Structures Engineering, COPPE-UFRJ, Rio de Janeiro, Brazil, September, 1971.

2. American Petroleum Institute, API-RP-2A Code, "Planning, Designing and Constructing Fixed Off-shore Platforms". Seventh Edition, January 1976.

3. Sjocrdsma, G.W., "Present and Future Development of Off-shore Structures", BOSS' 76, The Norwegian Institute of Technology, Trondheim, Norway, August 1976.

4. USDI, Bureau of Land Management. "Proposed 1973 Outer Continental Shelf East Texas General Oil and Gas Leox Sole", Draft Environmental Statement, Des 73-1, 1973, pp. 363-367.

5. Civil Engineering ASCE, Vol. 47, No.3, March 1977.

6. Off-shore Engineer, Number 6, June 1977.

7. Ocean Energy, IPC Industrial Press, Issue No.8, April 1977.

8. Off-shore Engineer, Number 5, May 1977.

9. Alloni, C., A. D'Agostino and R. Priarone, "Design, Analysis and Construction of the Loango Steel Gravity Platforms", Offshore Technology Conference, Dallas, Texas 1976.

10. Glenn, A.H. and Associates, "10 Year Storm Wind, Tide and Wave Characteristics: 100 Chart Depth , Offshore Aracajú, Brazil (11° South Latitude)."

12. Prates de Lima, L.C. "LORANE DINA: A Problem Oriented Language for Structural Dynamics", Ph.D. Thesis, COPPE-UFRJ, Rio de Janeiro, Brazil, 1977.

13. Ebecken, N.F,"Lorane NL: A Problem Oriented Language for Non-Linear Structural Analysis", Ph.D. Thesis, COPPE-UFRJ, Rio de Janeiro, Brazil, 1977.

14. Ferrante, A.J., "A General Purpose System for Computational Hydraulics", in Finite Elements in Water Resources, W.B. Gray, G.F. Pinder and C.A. Brebbia (Eds), Pentech Press, 1977.

15. Royal, S. and F. Spata, "Soil Research for the Installation of Fixed Platforms", Seminar on Engineering Technology on Sea Drilling and Production", Petrobras, August 1976.

16. Coyle and Reese: "Load transfer in deep foundations", Specialty Conference of the Geotechnical Engineering Division, American Society of Civil Engineering, Austin, Texas, June 11, 1977.

17. Mansur, C. and Fochet, J. Jr., "Pile Loading Tests, Morganza Floodway Control Structure", Transactions, ASCE, Vol. 121, 1956, pp.555-587.

BEHAVIOUR OF PILES FOR OFFSHORE STRUCTURES

J.L. Roehl and M. Gattass

Pontifícia Universidade Católica do Rio de Janeiro and
Escritório Técnico Guanabara Ltda.

1. INTRODUCTION

The behaviour of pile foundations in the analysis of offshore structures was considered, previously, in the same way as for general inshore structures. Later on, due to the high intensity of the lateral loads which are involved in such problems and to the special conditions of the sea floor soils, the behaviour of the soil-pile system had to be taken into account with more concern and other methods of analysis were introduced. This subject has received a large amount of investigation with the increasing demand of fixed offshore platforms for production fields.

In this problem there are two basic subjects which still remain in open discussion: the interaction in the soil-pile system and the convenient way to introduce this clearly non-linear effect in the analysis of the structure.

The establishment or choice of a relation to represent the interdependence of the soil pressure and the lateral displacement of the pile is, therefore, outside the objectives of this work. In a practical view of the question, one prefers to make use of the recommendations of the API RP 2A [1] which indicate Matlock's work [2] for clays and Reese study [3] for sands. Hence, the conclusions are restricted to the field of application of such recommendations; one has the merit, however, to follow experimental information very well divulged and routines already introduced in common design procedures.

Considering the second point, it is intended to facilitate its overpass by the consideration of the force-displacement relationships at the pile head; having them well understood and normalized, they can be used to match the response of the foundation and the structure, through the stiffness matrix of the pile-soil system referred to those displacements.

2. OBJECT AND SCOPE

This work intends to achieve the understanding of the behaviour of long piles, in soil of vari d properties, under lateral loading, through the static and cinematic quantities at the pile head. It is still intended to normalize conveniently those quantities according to the recommendation of the API for offshore platforms.

The present investigation was limited to isolated long piles, with linear behaviour under bending deformations; effects of inertia forces and axial loads are not taken into account; the soil reaction is defined by a nonlinear relationship according to API.

3. FORMULATION OF THE PROBLEM

The governing equation for a pile with bending stiffness, EI, driven in a soil giving a lateral reaction p(x,y) and subjected to a constant axial force P (Fig. 1) is as follows:

$$\frac{d^2}{dx^2}\left[EI\,\frac{d^2y}{dx^2}\right] + P\,\frac{d^2y}{dx^2} = p(x,y) \qquad (1)$$

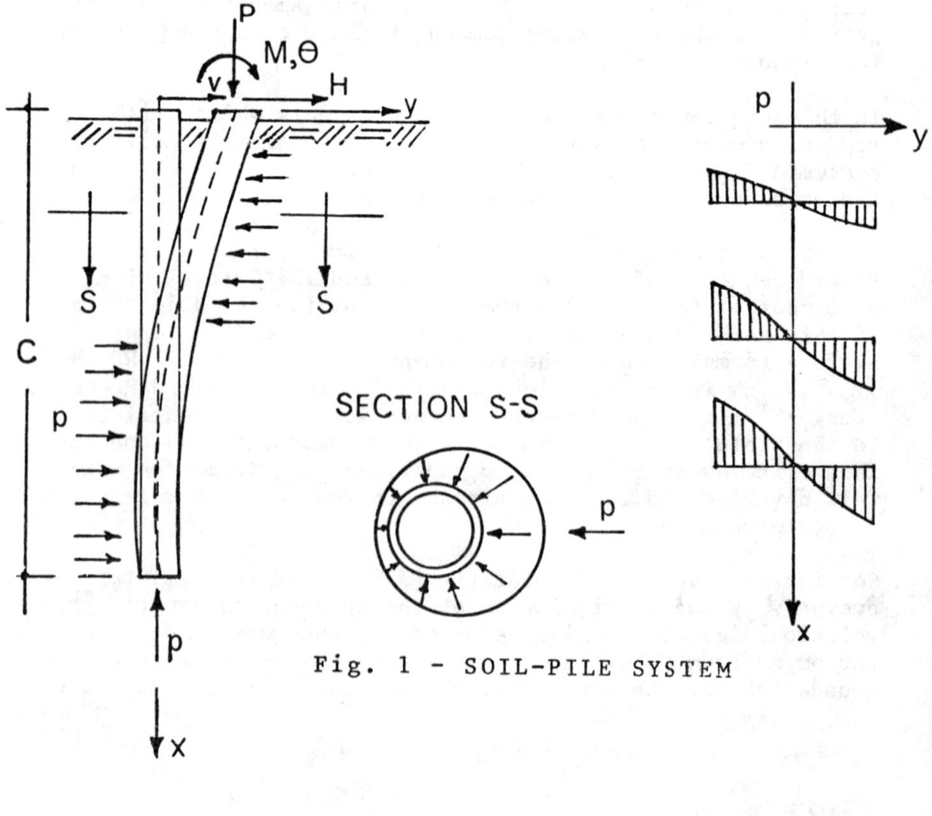

Fig. 1 - SOIL-PILE SYSTEM

It is largely accepted the expression $p = k.x^n.f(y)$ where k, n and f( ) depend on the characteristics of the soil and the pile. In general, f(y) is a nonlinear function and, moreover, difficult to make explicit what impedes one to find a general solution for equation 1.

As far as analysis is concerned it would be desirable to define the stiffness matrix at the pile head as

$$\begin{Bmatrix} H \\ M \end{Bmatrix} = \frac{EI}{T^3} \times \begin{bmatrix} k_{vv} & k_{v\theta}.T \\ k_{\theta v}.T & k_{\theta\theta}.T^2 \end{bmatrix} . \begin{Bmatrix} v \\ \theta \end{Bmatrix} \qquad (2)$$

in which, T is some characteristic length. However, this matrix changes with the displacement level.

To find a simple and workable way to define the elements of this matrix for any displacement level is the major object of this work. If one considers the variation of the soil reaction with the displacement y, as recommended by API, the nonlinearity of the problem allows one to state only that:

$$H = f_1(v,\theta)$$
$$M = f_2(v,\theta) \qquad (3)$$

being f_1, and f_2 nonlinear in v and θ.

In order to generalize a possible numerical solution of equation 1, it is considered for every soil-pile system a set of typical values v_t, θ_t, H_t, M_t and T with following the correlations:

$$\theta_t = \frac{EI}{T} \quad ; \quad H_t = \frac{EI}{T^3} v_t \quad \text{and} \quad M_t = \frac{EI}{T} \theta_t \qquad (4)$$

and one writes the forces and displacements at the pile head as:

$$v = v^*.v_t \ ; \ \theta = \theta^*.\theta_t \ ; \ H = H^*.H_t \ \text{and} \ M = M^*.M_t \qquad (5)$$

finally, equation 2 can be rewritten:

$$\begin{Bmatrix} H^* \\ M^* \end{Bmatrix} = \begin{bmatrix} k_{vv} & k_{v\theta} \\ k_{\theta v} & k_{\theta\theta} \end{bmatrix} . \begin{Bmatrix} v^* \\ \theta^* \end{Bmatrix} \qquad (6)$$

These typical values are characterized through a numerical approximation of the functions f_1 and f_2 obtained for a sufficiently large number of soil-pile systems.

Since API prescribes different procedures for granular and cohesive soils, it was necessary to consider separately each of these groups of soils.

The solution of equation 1 can be achieved by any of the available numerical techniques. In this work, a program is used in which an iterative procedure with secant stiffness on a finite differences formulation of the equation is employed. Such program was tested with problems of which solutions were available; the error becomes insignificant when the ratio of the interval size of discretization to the length of the pile is less than 0.1.

4. NORMALIZATION AND SPACE LIMITATION

a. Piles in granular soils - The API RP2A postulates for piles in sands the adoption of Reese's recommendations |3|. In that work the granular soils are characterized by three parameters: angle of soil friction on pile wall, ϕ', effective soil unit weight, γ' and initial soil modulus, K_1.

The set of soil characteristics chosen to represent the group of granular soils is shown in Table I.

Table I - Characteristic Granular Soils

Soil Type	ϕ'	γ' (t/m^3)	K_1 (t/m^3)
Very loose	25°	0.9	540
Loose	30°	1.0	1230
Medium	35°	1.0	2180
Dense	40°	1.1	3245

Initially, a 0.70 m diameter pile with $EI = 1.2 \times 10^5\ tm^2$ was used. Later on, two other piles were considered, $D = 0.50$ m with $EI = 2.4 \times 10^4\ tm^2$ and $D = 1.20$ m with $EI = 4.0 \times 10^5\ tm^2$. The length in every case was long enough to ensure no change in behaviour with an increasing in length.

Looking at equation 4 one verifies that if it is desired to normalize the different soil types it is required that the quantities v_t and T incorporate the major differential aspects of the soil-pile systems.

Cosnsidering the typical displacement, v_t, the pile diameter,

D, appeared to be the most indicated quantity, in the case of sands, because it is also used in the recommended p-y curves to normalize the horizontal displacements, y.

On the other hand, the length T has to rep·esent the reciprocal dependence of the soil and pile stiffnesses, e.g., it must characterize the relative stiffness of the soil-pile system. Hence, one makes it to be such to keep constant, in different systems, the ratio between the stiffnesses of the soil and of the pile, for a given level x* = x/T:

$$\frac{p_u/v_t}{EI/T^4} = c \qquad (7)$$

where p_u is the ultimate lateral bearing capacity of the sand. As p_u varies in different ways with the depth, it is necessary to find a simple and continuous function to substitute the recommended curve. A second degree parabola showed up to approximate very well the functions f_1, and f_2, equation 3, for the normalized representation of the granular soils. Then,

$$p_u = \frac{p_{u_t}}{x_t^2} \cdot x^2 \qquad \text{and } T \sim \sqrt[6]{\frac{EI}{\frac{p_{u_t}}{v_t \cdot x_t^2}}} \qquad (8)$$

where x_t is a typical depth for each sand. The other typical values in equation 4 are now defined starting from v_t and T.

b. Piles in cohesive soils - The behaviour of piles in clays is defined by Matlock's work [2] which has a similar orientation as Reese's in the case of granular soils. Three parameters are used to identify the cohesive soils: undrained shear strength, c_u, effective soil unit weight, γ' and strain which occurs at one-half the maximum stress on laboratory undrained compression tests of undisturbed soil samples, ε_c. The spectrum of clay soils used in this work is represented in Table II by the values of those quantities.

Table II - Characteristic Cohesive Soils

Soil Type	c_u (t/m^2)	γ' (t/m^3)	ε_c (%)
Medium Clay	2.5	0.5	2.0
Stiff Clay	5	0.7	1.0
Very Stiff Clay	10	0.9	0.5

In the above table, smaller ε_c values for increasing values of c_u, in order to introduce the more brittle character of stiff clays, are presented.

The typical values for this case are considered in a similar way as for sands:

for the typical displacements, v_t, is used a scale proportional to that of the p-y curves:

$$v_t = 100.\varepsilon_c.D \tag{9}$$

the relative stiffness is made as in equation 7, and the variation of p_u to approximate the functions f_1 and f_2 for the normalized soil-pile systems is taken as:

$$p_u = p_{u_t} \sqrt{\frac{x}{x_t}} \tag{10}$$

where $p_{u_t} = 9c_u$, for static loading, and $p_{u_t} = 0.72(9c_u)$, for cyclic loading. Then, the characteristic length, T, becomes:

$$T \sim 4.5 \sqrt{\frac{EI}{\dfrac{p_{u_t}}{v_t\sqrt{x_t}}}}$$

and the other typical values come out from equation 4.

c. Space Limitation - The space is limited by the values $|v^*| < 1$ e $|\theta^*| < 1$ and the set of points chosen for computation of the f_1 and f_2 values is given by $\{V\} \times \{\theta\}$ and

$\{V\} = \{\theta\} = \{$ ±1.0; ±0.9; ±0.8; ±0.7; ±0.7; ±0.5; ±0.4; ±0.3; ±0.2; ±0.1; ±0.08; ±0.06; ±0.04; ±0.02; ±0.01; 0$\}$

5. ANALYSIS OF BEHAVIOUR

The definition used for the typical values makes closely coincident the nondimensional representation of the force-displacement relationships for the various soil-pile systems; the largest discrepancies, in extreme cases, were less than 15% of the medium values. Fig. 2 represents those relationships for the sands in which there is no difference between the static and cyclic cases and it is also shown that the curves are very close for different soil types. In a similar

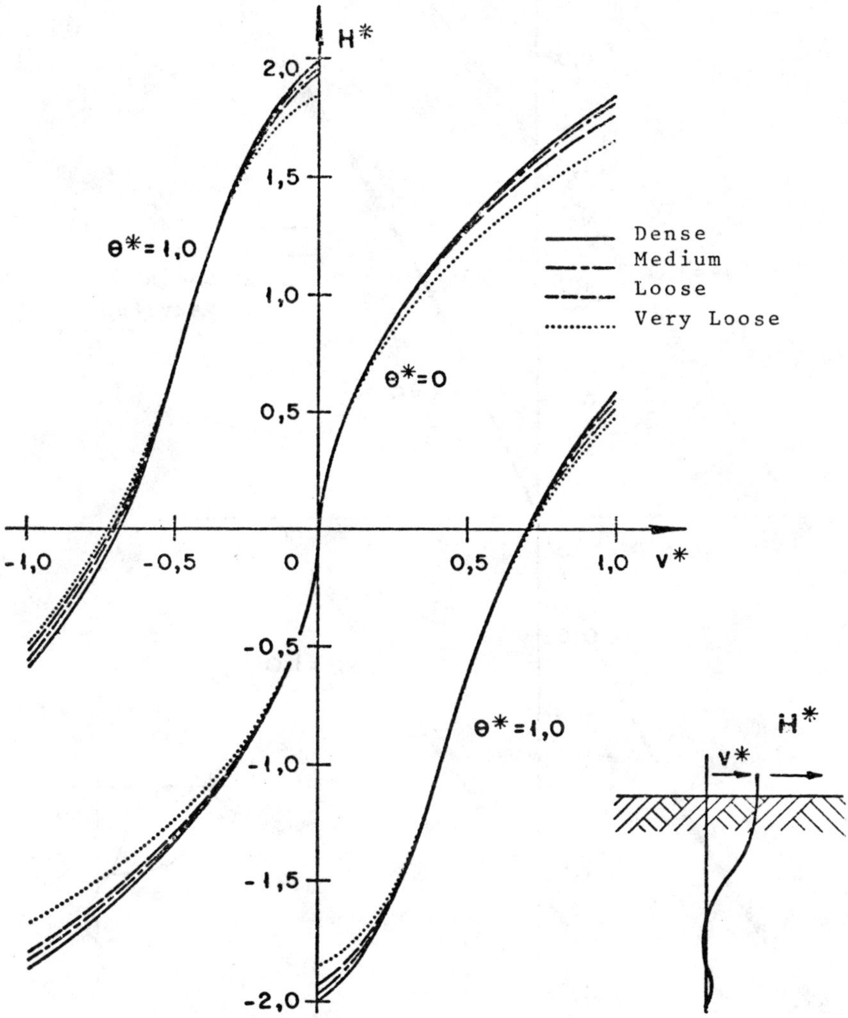

Fig.2a - FORCE-DISPLACEMENT FOR SANDS
STATIC AND CYCLIC LOADING

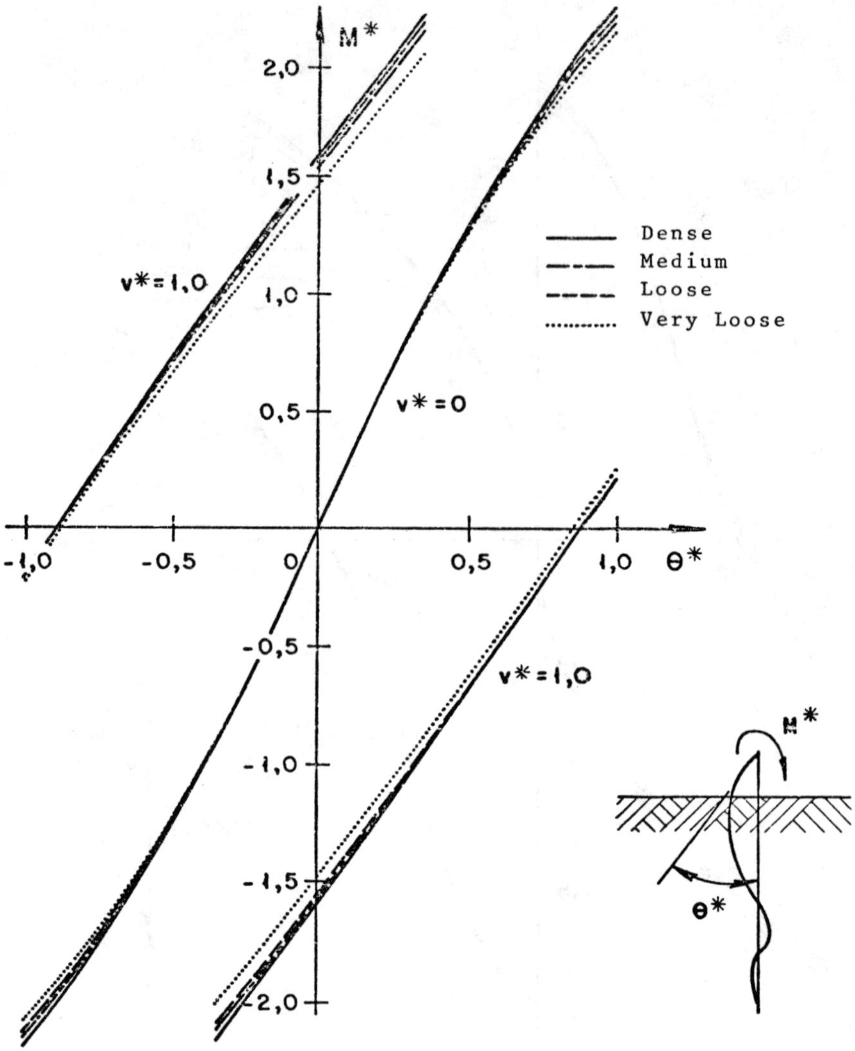

Fig.2b - MOMENT-ROTATION FOR SANDS
STATIC AND CYCLIC LOADING

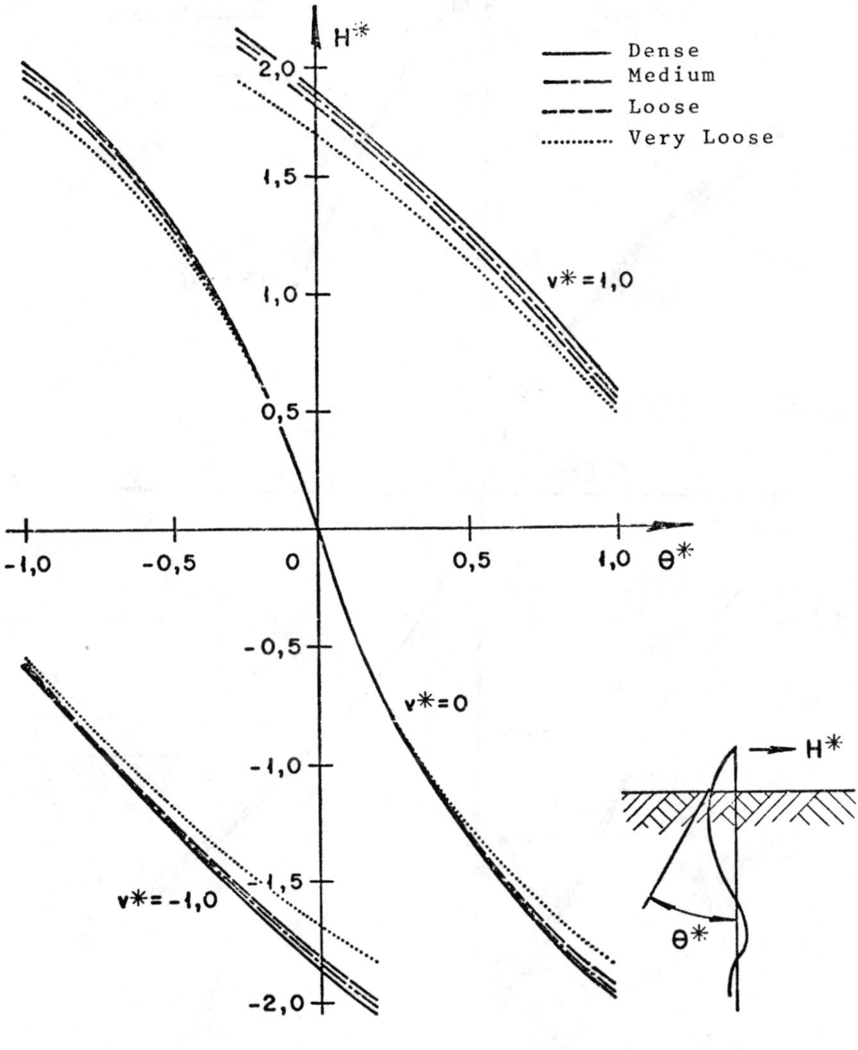

Fig.2c - FORCE-ROTATION FOR SANDS
STATIC AND CYCLIC LOADING

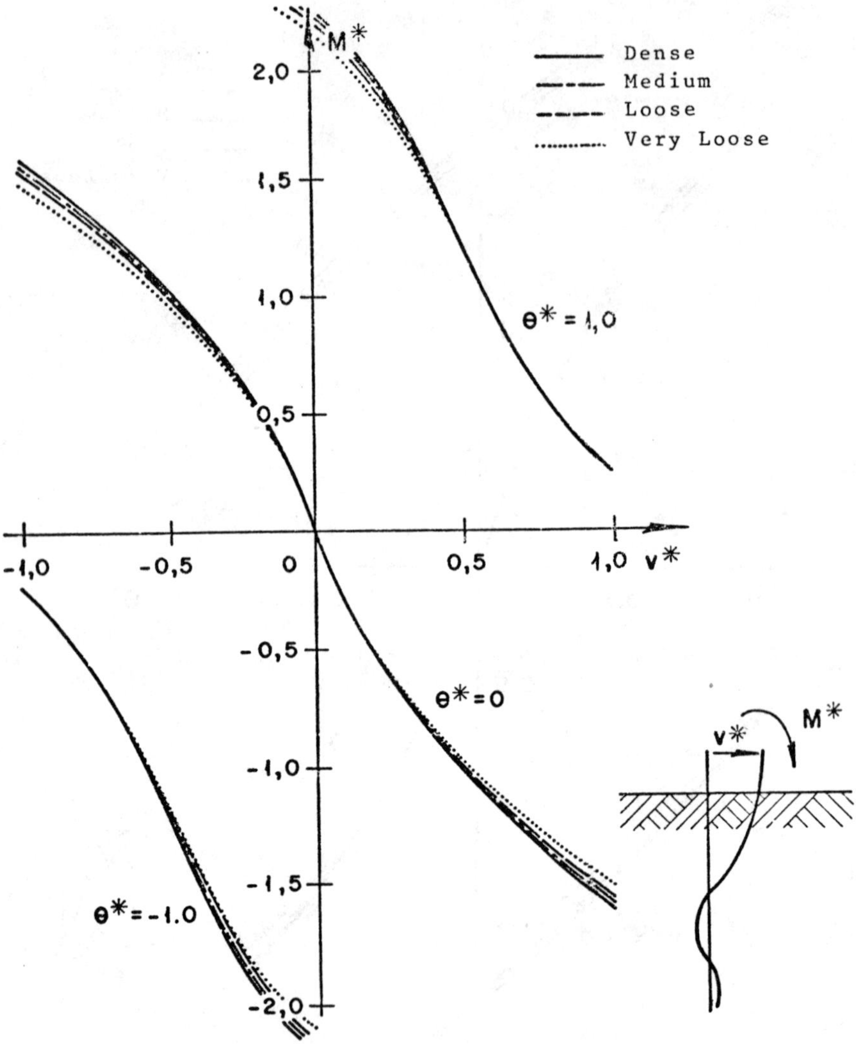

Fig.2d - MOMENT-DISPLACEMENT FOR SANDS
STATIC AND CYCLIC LOADING

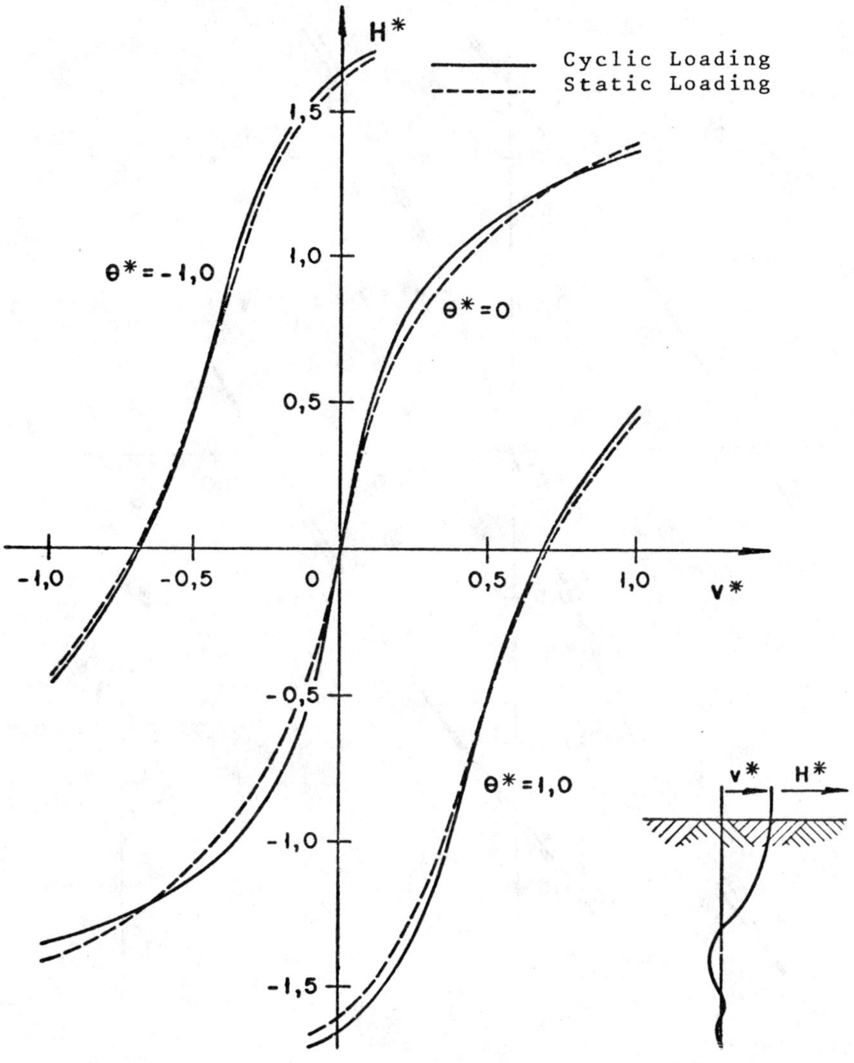

Fig.3a - FORCE-DISPLACEMENT FOR STIFF CLAYS

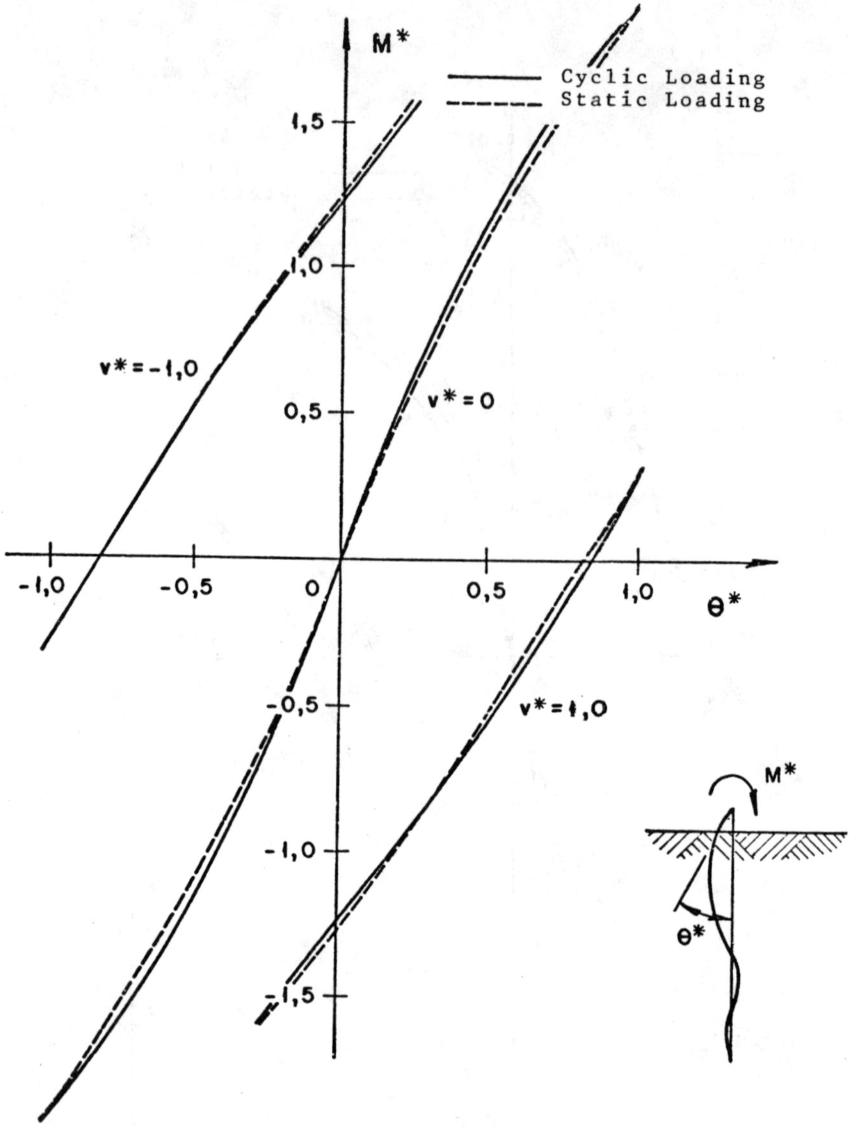

Fig.3b - MOMENT-ROTATION FOR STIFF CLAYS

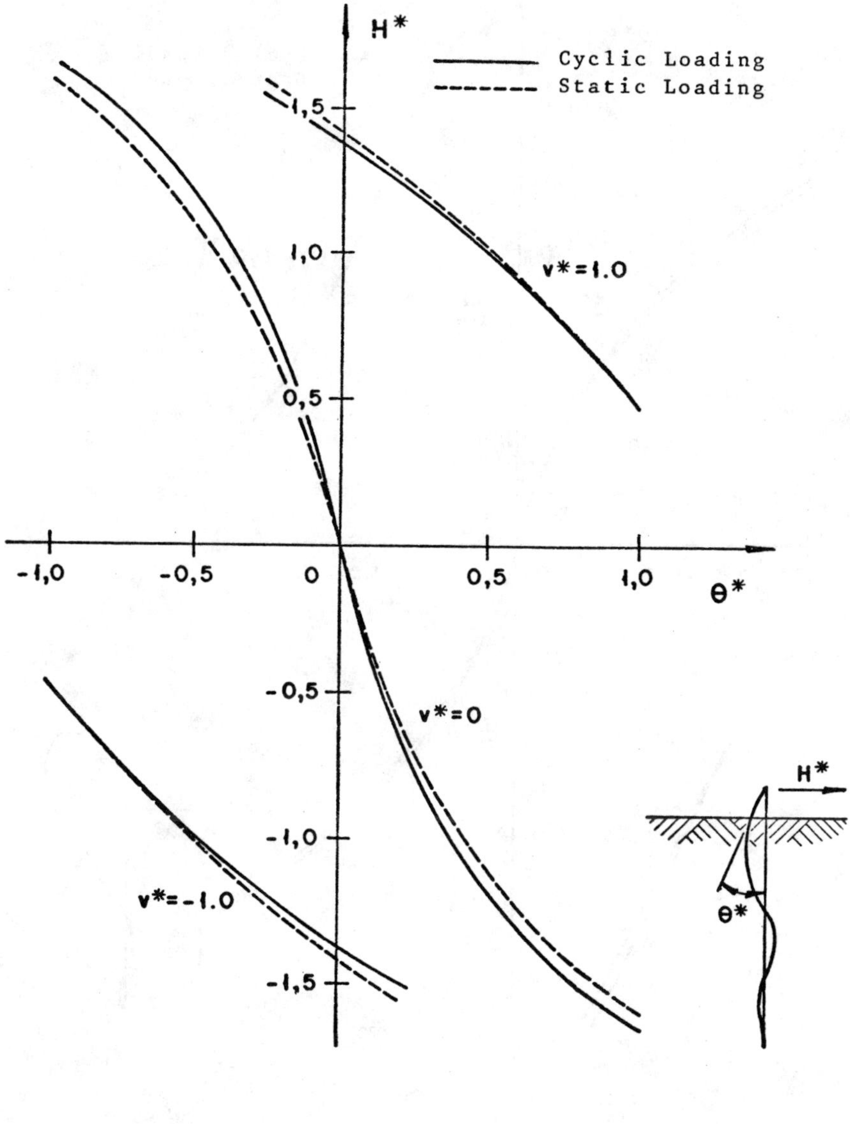

Fig.3c - FORCE-ROTATION FOR STIFF CLAYS

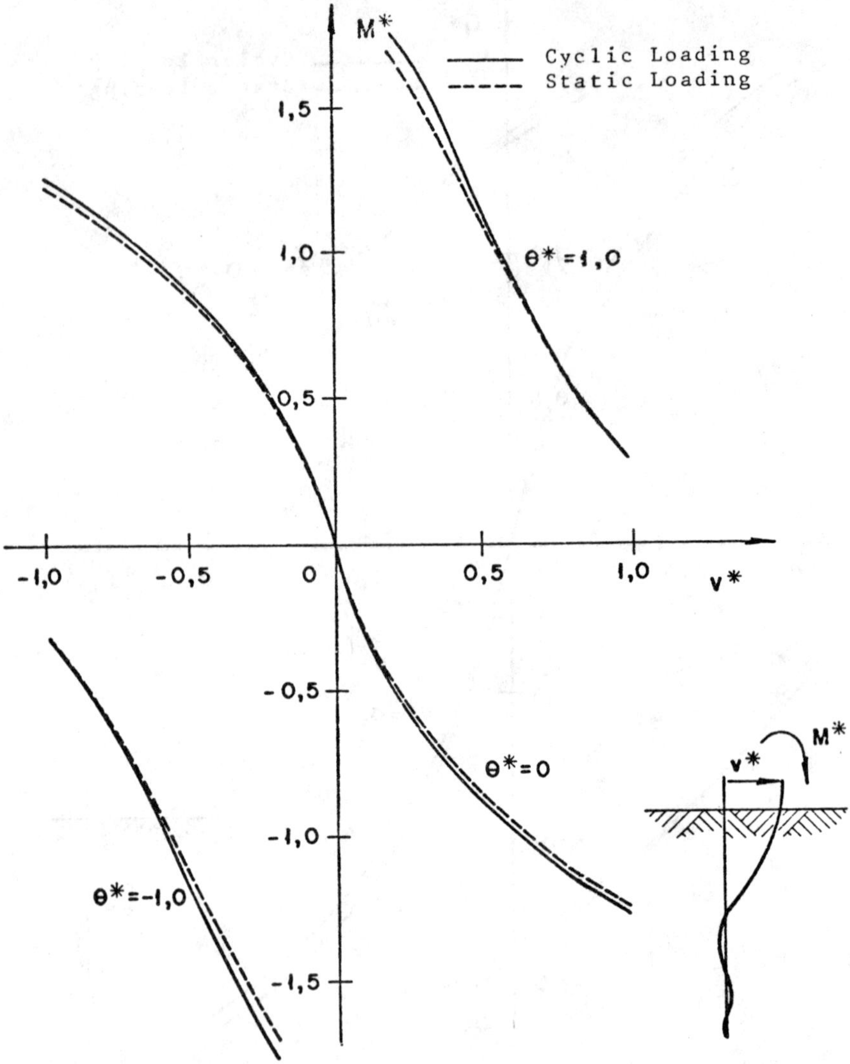

Fig.3d - MOMENT-DISPLACEMENT FOR STIFF CLAYS

manner, one finds in Fig. 3 that the results for the stiff clay soil-pile systems, under static and cyclic loading, show the same general behaviour identified for piles in sandy soils. The stiff clay curve is very good to represent any of the other cohesive soils considered; it is interesting to notice a small variation from the static to the cyclic loading case.

It is clear that the defferenciative aspects of the response, in the quantitative sense, are neutralized by the typical values and the remaining functions (H^*, M^*) make evident the qualitative trainds of the response.

From Figs 2 and 3 it is possible to make a few statements about the behaviour of the soil-pile systems.

- nonlinearity is more effective for H^*-v^* than for M^*-θ^* curves;
- there is no practical difference between the static and cyclic loading cases for the sandy soil-pile systems; for cohesive soils there is a small variation between the two cases;
- the rotations, θ^*, affect the H^*-v^* and M^*-v^* curves with a translation of the graph, more effective for the later than for the former relations;
- the influence of the displacements, v^*, over the curves H^*-θ^* and M^*-θ^* includes, apparently, a deeper modification which is more expressive over H^*-θ^*;
- the loss of stiffness in granular soils is almost the same as for the static loading case of the clay soils, for both H^*-v^* and M^*-θ^* curves;
- in cohesive soils, the H^*-v^* curve is lightly more linear for static than for cyclic loads.

6. CONCLUSIONS

The behaviour of vertical isolated long piles, under lateral loading, according to the practice recommended by the American Petroleum Institute can be synthetic and widely characterized if a convenient normalization is used.

The normalization recommended in this work approximates the response of the different soil-pile systems to one general behaviour identified by the force-displacement relationships at the pile head.

The soil stiffness can be represented by a polynomial of a degree between those of the functions that represent the ultimate lateral bearing capacity of the soil for small and large depths.

It is characterized a virtual length of the pile, proportional to a root of the ratio of the pile to the soil stiffness, which is very powerful to normalize the response quantities.

The proposed force-displacement formulation permits the analysis of an offshore structure with a reasonable approximation without the necessity to solve at every step the non-linear boundary value problem of the pile.

REFERENCES

1. "API Recommended Practice for Planning, Designing, and Constructing Fixed Offshore Platforms", Report PP-2A, American Petroleum Institute, Washington, D.C., Jan. 1976.

2. Matlock, H., "Correlations for Design of Laterally Loaded Piles in Soft Clay", Proceedings of the Second Annual Offshore Technology Conference, Houston, Texas, OTC 1204, Vol.I, pp.577-594, Apr. 1970.

3. Reese, L.C., W.R. Cox and F.D. Koop, "Analysis of Laterally Loaded Piles in Sand", Proceedings of the Sixth Annual Offshore Technology Conference, Houston, Texas, OTC 2080, pp.473-483, May 1974.

PART V DESIGN AND CONSTRUCTION

CODES FOR OFFSHORE STRUCTURES. DESIGN CRITERIA AND SAFETY REQUIREMENTS

Fernando Luiz Lobo B. Carneiro

COPPE, Universidade Federal do Rio de Janeiro, Brazil.

1. INTRODUCTION

In this paper a comparison is made of the criteria for the design and the analysis of the resistance for fixed offshore platforms, recommended by the following regulations or codes:

1.a) - DNV - Det Norske Veritas: Rules for the Design, Construction and Inspection of Fixed Offshore Platforms, 1974.

1.b) - BV - Bureau Veritas: Rules and Regulations for the Construction and the Classification of Offshore Platforms, 1975;

1.c) - API - American Petroleum Institute: Recommended Practice Code for Planning, Designing and Constructing Fixed Offshore Platforms, 1976;

1.d) - FIP - International Federation for Prestressing (Féderation Internationale de la Précontrainte): Recommendations for the Design of Concrete Sea Structures, 1974.

The first two regulations, DNV and BV apply for both steel and concrete structures (reinforced or prestressed); the third one, API, only for steel structures; and the FIP recommendations only for concrete structures. The design criteria are discussed based on the principles for treating the structural safety and serviceability adopted by the following draft of international code:

1.e) - CEB/SIRTUS - Comité Euro-International du Béton: International System of Unified Standard Codes of Practice for Structures (Système International de Règlementation Technique Unifiée des Structures), 1976.

The volume I of this draft, - "Common unified Rules for Different types of Construction and Material", whose provisions are to apply for both steel and concrete structures, prepared by six technical international organizations: CEB, (Comite Euro-International du Beton), CECM (European Council for Constructional Steelwork), CIB (European Council for Building Research), FIP (International Federation for Prestressing), IABSE (International Association for Bridge and Structural Engineering) and RILEM (The International Union of Testing and Research Laboratories for Materials and Structures).

This draft adopts the limit states design method, and identifies three levels at which the structural safety may be treated.

In this paper some suggestions are made related to the partial safety factors, adopted in the so called "Level 1" for the treatment of the structural safety (also known as the "semi-probabilistic process"), and to the requirements governing normal use and durability ("serviceability" requirements).

1.1 Semi-probabilistic process ("Level 1")
In the semi-probabilistic process the probabilistic aspects are treated specifically in defining the characteristic values F_k, of loads (or actions) and f_k, of strength of materials. For the verification of safety, in the so called ultimate limit state, the strengths f_k are divided by the partial safety factor γ_m ($\gamma_m = \gamma_s$ for steel and $= \gamma_c$ for concrete), and the loads (actions), or their effects, are multiplied by the partial security factors γ_f. These factors are derived whenever possible from a consideration of probabilistic aspects The values $f_d = f_k/\gamma_m$ ε $F_d = \gamma_f F_k$ are named design values (design strengths and design actions). The principles of the semi-probabilistic process are resumed in the following:

Partial safety factors:

For the strength of materials: $\gamma_m \begin{cases} = \gamma_c = 1{,}5 \text{ for concrete} \\ = \gamma_s = 1{,}15 \text{ for steel} \end{cases}$

Design strength: $f_d = \dfrac{f_k}{\gamma_m}$

For the actions: $\gamma_f = \gamma_{f1}\, \gamma_{f2}\, \gamma_{f3}$ (γ_{f1}, γ_{f2} and γ_{f3} to be defined later)

Design action: $F_d = \gamma_f\, F_k$

Stress (or stress resultants in bars): $S(F_d)$
(function of the actions)

Resistance of the structural elements: $R(f_d)$
(function of the strength of materials)

Safety condition (at the ultimate limit state)
$S(F_d) \leq R(f_d)$

or $S(\gamma_f F_k) \leq R\left(\frac{f_k}{\gamma_m}\right)$

(symbolically)

2. LIMIT STATES

The considered limit states can be placed in three categories:

- the ultimate limit states, which are those corresponding to the maximum load-carrying capacity of the structure;
- the serviceability limit states, which are related with its normal utilization and durability;
- the conditional limit states, which are ultimate limit states conditioned by accidental actions, whose probability of occurrence is very low and to which the structure will not be expected to resist without local damage, but there should be a reasonable probability that it will not collapse catastrophically.

2.1 Ultimate Limit States

2.1.1 Yielding (for steel structures only) The ultimate limit state is reached, in an elastic analysis, when in some point of the structure the yield stress of the material is attained, or, in the case of bars in flexure, when the moment of plastification is attained in some cross section (this does not imply that the structure is transformed into a mechanism). This limit state is generally considered in the analysis of steel offshore structures. When the partial safety factors γ_f are the same for all the loads, this limit state can be presented under the appearance of an allowable stress process, like in the DNV, BV and API codes; in this case the overall safety factor, equal to the ratio between the minimum specified yield stress (which is the characteristic strength f_{yk} of the steel) and the allowable stress σ_{all}, can be interpreted as the product of the partial safety factors $\gamma_s = 1.15$ and γ_f, defined in 1.1:

$$\gamma_s \gamma_f = \frac{f_{yk}}{\sigma_{all.}} \quad \text{then} \quad \gamma_f = \frac{f_{yk}/1.15}{\sigma_{all.}}$$

In the codes DNV and API the allowable stress in bending is increased by 10%; this is equivalent to admit a partial plastification of a cross section in the yielding limit state (for flexure with axial load it is given an interpolation formula). The items of the codes considered corresponding to this limit state are: DNV-4B200 and 4C200; BV-2.52, 21 and 22 and 2.53, 21; API-2. 18a, b and 19b.

If the partial safety factors γ_f are different for different types of loads, it will not be possible to define "allowable stress". In the DNV, BV and API codes the "allowable stress" for the case of "operating loads" is equal to $0.6f_{yk}$, and in the case of "operating loads plus extreme environmental loads" $0.8f_{yk}$. The corresponding γ_f factors are 1.45 and 1.09 (wit $\gamma_s = 1.15$).

2.1.2 Rupture (for concrete structures) This is the ultimate limit state corresponding to the rupture or excessive deformation of critical cross sections. This limit state is generally considered in concrete structures, on the base of the assumptions adopted by the International Recommendations CEB/FIP of 1970, maintained in the CEB/SIRTUS draft code: parabole-rectangle diagram for concrete, with a maximum compressive stress equal to $0.85f_{cd}$ (f_{cd} is the design strength of the concrete: $f_{cd} = f_{ck}/\gamma_c$); rupture shortening of the concrete variable between the limits 3.5×10^{-3}, in pure bending, and 2×10^{-3} in pure compression, with a linear interpolation; maximum elongation of the steel of the reinforcement 10×10^{-3}. The codes DNV, BV and FIP adopt these assumptions, with little modifications: DNV-5D207; BV-2.63, 41; FIP-R3.4.

2.1.3 Buckling (elastic or plastic instability) This ultimate limit state is defined in all codes considered, both for steel and concrete structures.

2.1.4 Transformation of the structure into a mechanism (ultimate strength plastic analysis) This ultimate limit state is mentioned specifically for steel structures, in the Norwegian code DNV-4B500 and 4C500 ("ultimate strength" plastic analysis). When the plastic hinges method is adopted, the structure is transformed into a mechanism after the plastification of a certain number of cross sections, and therefore, with a loading somewhat greater than the loading corresponding to the yielding ultimate limit state defined in 2.1.1. The International Recommendations CEB/FIP 1970 and the CEB/SIRTUS draft allow a limited application of this method for the concrete structures. A non-linear analysis (material non-linearity) is to be preferred, in the case of concrete structures.

2.1.5 Fatigue In opposition to the ultimate limit states previously mentioned, fatigue can lead to the rupture of one element of the structure in the service conditions, with the loads not multiplied by the partial security factor γ_f (in this case, $\gamma_f = 1$). The rupture derives from the cumulative damage due to repeated variations of stresses, during the period of planned life of the structure t_s (20 to 30 years in the case of offshore structures). The fatigue curves take into account the partial safety factor γ_m (γ_s for steel and γ_c for concrete).

2.2 Serviceability limit states

2.2.1 Local damage (cracking limit state, in the case of concrete structures) This limit is reached when some local damage occurs in the structure, opening a way to corrosion. In the case of concrete structures, this state is the cracking limit state. In the reinforced concrete structures, whose reinforcement bars present a low sensitivity to corrosion, the crack width is limited (in the CEB/SIRTUS draft, for the case of severe conditions of exposure, as in maritime environment, to 0.1 mm with a minimum concrete cover of 35 mm, and to 0.15 mm with a minimum concrete cover of 50 mm). In the prestressed concrete structures the prestressing reinforcement (heat treated or cold drawn wires, or cold worked bars, subjected to high permanent tension) is very sensitive to corrosion, specially to stress corrosion; in this case the uncracked limit state is required. Prestressed concrete structures must comply with the requirements of Class I (no tensile stresses) for frequent combination of loads, and of Class II (limited tensile stress, but no cracking) for infrequent (extreme) combinations of service loads. Class III (limited cracking as in reinforced concrete) should not be allowed, except for accidental actions. This point is very important, but the codes considered for offshore concrete structures present some discrepancies and this subject will be discussed later on.

2.2.2 Deformation Excessive deformation under service conconditions should not adversely affect the efficiency of the structure.

2.2.3 Vibration Excessive vibration should not cause discomfort or alarm, or adversely affect the efficiency of the equipment. The first natural period for the whole structure should be less than 4 seconds, to prevent resonance with wave forces, and measures should be taken to reduce vibration in the deck caused by machinery.

2.3 Conditional ultimate limit states

In certain accidental situations the structure could be locally damaged (local yielding or buckling in steel structures, localized ruptures or excessive cracking in reinforced concrete structures, limited cracking in prestressed concrete structures), but without a catastrophical collapse, resulting in loss of human lives or in strong pollution of the sea or coast. Examples of those situations are brutal collision of great tankers, earthquakes in non-seismic regions (or earthquakes of unexpected intensity, in seismic regions), hurricane explosions, and fire. Besides the factors normally taken into account for the analysis of the structural safety, we must consider, in the case of offshore structures, the facilities for sheltering the personnel in the platform itself or for a rapid evacuation in case of alarm, and the prevention of pollution.

3. ACTIONS AND SITUATIONS

3.1 Definition and classification of the actions

Actions are classified in:

- direct actions or loads (concentrated and distributed forces)
- indirect actions: imposed displacements (earthquakes, settlement of foundations). and restrained deformations (thermal deformations, shrinkage) or imposed deformations (prestress).

The actions can be natural (environmental), or resulting from the human activity (dependent from the existance of the structure, like its selfweight, or from its use, like operating loads).

According to their variation in time the actions are classified as permanent actions and variable actions, and according to their variation in space as fixed actions (when its distribution over the structure can be defined by one parameter, as is the case of stored fluids) and free actions (movable loads or loads which may have arbitrary distribution, as the distributed loads over the deck and the loads corresponding to stores and supplies). The actions which do not cause significant acceleration are classified as static actions, and those which cause significant acceleration, as dynamic actions. Many dynamic actions can be treated as static actions, when their dynamic effects are considered by the means of certain increases or impact coefficients. In the case of fixed offshore structures the environmental loads caused by wave actions are treated as dynamic actions, but the dynamic actions caused by the operation of the machinery are in general treated as static, with convenient increases. In the case of certain elements of the deck very much exposed to wind action, it should be necessary to analyse the dynamic effect of gusts, or von Karman vortices.

It will be later presented in this paper, a classification of the actions occurring in the offshore structures, according to these principles.

3.2 Situations

Besides the stages corresponding to construction, transportation and positioning of the structure, the following situations are considered during its planned life:

3.2.1 Permanent or frequent situations (normal conditions) In the case of offshore structures these situations are classified as normal or operating environmental conditions, because they are compatible with the normal activities of operating the structure.

The DNV and BV codes define the normal or operating environmental conditions as those for which the mean period of return of the environmental actions is one month (the probabilistic level, referred to waves, is $10^{-5\cdot 6}$ (DNV) or 10^{-5} (BV). This situation is considered in the design of concrete offshore structures, for the serviceability limit states.

3.2.2 Temporary or infrequent situations (extreme conditions) In the case of offshore structures these situations are classified as extreme environmental conditions. They should be considered for the serviceability limit states and, more specifically, for the ultimate limit states.

The DNV and BV codes define the extreme environmental conditions as those for which the mean period of return of the environmental actions is 100 or 50 years, respectively ("100 year" or "50 year" wave, in the case of the wave action on the structure). This corresponds to a probability level of $10^{-8\cdot 7}$ (DNV) or 10^{-8} (BV), referred to waves.

In the conditions of the North Sea, where the "100 year wave" is about 30 metres high, the extreme environmental conditions are not compatible with most of the operating activities in the structure. For the analysis of the structural safety, such activities are supposed partially discontinued, but the structure must resist without damage to the environmental actions and other loads. It is necessary to emphasize that the extreme environmental actions are not accidental actions. They are service loads, although infrequent, and they must be multiplied by the partial safety factor γ_f for analysis of the safety in the ultimate limit states.

In Brazil there are indications that the "100 years wave" is not more than 18 metres high. This corresponds in the North Sea to the "1 month wave" or to operating environmental conditions. It is possible then, that in Brazil all the operating activities should be considered as compatible with extreme environmental conditions.

3.3 Characteristic values of the actions The "characteristic values" of the actions are extreme values, and not the mean or frequent values. Although infrequent, these characteristic values can occur in service conditions, and the structure must be designed according with this possibility. As it was already observed, for the case of extreme environmental conditions, the characteristic values of the actions must not be confused with the accidental actions, which are very exceptional and unlike actions, causing partial damage to the structure, but not a catastrophic global collapse.

For verifications in the ultimate limit state the characteristic values must be multiplied by the partial safety factors γ_f.

3.3.1 Permanent loads In the case of permanent loads, whose statistical distribution is generally known, two characteristic values should be defined for the cases of a stabilising or non stabilising effect of dead-load. These values should correspond to the 5% or 95%, fractiles, in a normal distribution. In the case of the self-weight it is sufficient to take the mean value as characteristic value.

3.3.2 Variable loads

The characteristic values of the variable loads, which are extreme and infrequent values, are established, whenever possible, on the base of probabilistic aspects. The maximum and minimum values of the variable loads that can occur in service conditions are its characteristic values and zero, respectively.

A pragmatic method is recommended by the draft CEB/SIRTUS for assessment of the characteristic values of the actions. First the larger of the two following values is considered:

- nominal value, established by national codes based on wide experience, as the upper bound of an action, or as a value that should not be exceeded by users;
- value defined by a fractile between 22% and 33%, in an extreme statistical distribution, during the course of the life of the structure, corresponding to a mean period of return, from 2.4 to 4 times the planned life of the structure.

Further the adopted value should not be less than the result of dividing by the partial safety factor γ_{f1} the value corresponding to a probability of occurrence, in the course of the life of the structure, between 0.05 and 0.005. This value is calculated by an extrapolation of a theoretical extreme distribution law, and corresponds to a mean period of return of 20 to 200 times of the planned life of the structure.

The partial safety factor γ_{f1} takes into account the possibility of occurrence of actions greater than the characteristic value, and is taken equal to 1.4 for current structures, designed with the overall load factor $\gamma_f = 1.5$.

The statistics must include only the maximum value reached by each intermittent action during each application. In this case the mean period of return T is related to its probability of occurrence in a unit period by the formula

$$T = \frac{1}{\ell_n \left(\frac{1}{1-P}\right)}$$

For current structures, whose planned life is 50 years, the mean period of return adopted for the definition of the characteristic value F_k of an action is from 120 to 200 years. In the case of the offshore structures, whose planned life is only 20 to 30 years, it seems reasonable to adopt the mean period of return given by the DNV code (100 years) or by BV code (50 years).

The mean period of return of an action, equal to the characteristic action multiplied by γ_{f1}, which is 1000 to 10000 years for current structures, can be fixed between 500 and 5000 years for the offshore structures, according to the rule given above.

3.3.3 Waves   In the oceanographic statistics each "sea state", considered in general stationary during a period of time from half an hour to six hours, is represented by the "significant wave height" H_s (average of the highest one third of individual waves), and by the "average zero crossing period" $\bar{T}$.

In the long term statistics of sea states the distribution of the significant wave heights is generally described by a Weibull extreme distribution function:

$$Q(H_s) = 1 - P(H_s) = \exp\left[- \left(\frac{H_s}{H_o}\right)^{\gamma} \right]$$

from which

$$H_s = H_o \left[-\ell_n\ Q(H_s)\right]^{\frac{1}{\gamma}}$$

$Q(H_s)$ is the probability of exceeding the significant wave height H_s characterizing a "sea state"; H_o and γ are two parameters.

In each "sea state" the distribution of the individual wave heights can be described by a Rayleigh distribution:

$$Q(H) = 1 - P(H) = \exp\left(- \frac{H^2}{0.5H_s^2} \right)$$

Q(H) is the probability of exceeding the height H in a"sea state". The maximum expected height H_{max} in a "sea state" with N_c "zero crossing cycles" is given by the formula

$$H_{max} = H_s \sqrt{0.5 \ell_n N_c}$$

In a storm with 1000 cycles (during about three hours with an average period of 10 seconds), the maximum wave height will be

$$H_{max} = 1.86 H_s$$

It is then possible, on the basis of a long term statistics to define the wave height corresponding to a given mean period of return, like the "100 years wave".

Each "sea state" is represented by an "energy density" or "amplitude spectrum", taken as base for random dynamic analysis of the offshore structures. In general it is adopted the Pearson-Moskowitz spectrum. From this spectrum the maximum period can be evaluated at

$$T_{max} \cong 1.4\bar{T}$$

In the case of a "deterministic dynamic analysis", the scatter in the values of $\bar{T}$, and in consequence of T_{max}, makes necessary to investigate the dynamic response of the structure for many periods of the "design wave". The DNV code recommends to investigate three or more periods in the range from $\sqrt{6.5H}$ H in metres, to 20 seconds (from 14 to 20 seconds for a "design wave" of 30 metres, the "100 years wave" in the North Sea).

The "100 years wave", according to the DNV code, corresponds to a probability of exceeding $10^{-8.7}$ ($10^{8.7}$ waves in 100 years).

According to the BV code, the "50 years wave" corresponds to a probability of exceeding 10^{-8} ($10^{.8}$ waves in 50 years).

The "N years wave" corresponds to a probability of exceeding $Q = 10^{-n}$ given by:

$$-\log_{10} Q = n = 6.7 + \log_{10} N \qquad \text{(DNV)}$$

or

$$-\log_{10} Q = n = 6.3 + \log_{10} N \qquad \text{(BV)}$$

The "1 month wave", with a mean period of return of 1 month, corresponds then to n = 5.6 (DNV) or n = 5.3 (BV).

The relation between the stresses caused by waves and the wave heights can be represented by a flat quadratic polymonium which can be approached by a linear relation.

Without great error it can be assumed that the wave heights are proportional to the $\bar{n}$ values. Then the height of the wave with a mean period of return of 1 month will be:

$$H_{1month} = \frac{5.6}{8.7} H_{100years} = 0.64 H_{100years} \qquad \text{(DNV)}$$

$$H_{1month} = \frac{5.3}{8.0} H_{50years} = 0.66 H_{50years} \qquad \text{(BV)}$$

The "frequent value" of the wave heights (mean period of return of 1 month) is then about 65% of the "characteristic" or "extreme value" (mean period of return of 100 or 50 years), and the frequent and extreme values of the wave actions present the same ratio.

The height of "500 years wave" or "5000 years wave" will be only 1.1 to 1.25 times the "characteristic" wave. According to 3.3.2 the value of the factor γ_{f1}, which is 1.4 for current structures, can be reduced to 1.2, in the case of offshore structures. In consequence the partial safety factor γ_f can be reduced from 1.5 (current structures) to 1.3, in the case of offshore structures, as it is made in all the offshore codes considered. This is in good agreement with the CEB/SIRTUS draft.

3.3.4 Wind According to the BV code, in lack of observations, the following values will be adopted in the North Sea, for sustained wind speeds (maximum mean value over an hour), at 10 metres above the sea level:

extreme conditions: 36 m/s
normal operating conditions: 18 m/s

In this case the wind speed with a mean period of return 1 month is half the wind speed with a mean period of return 50 years; the frequent wind pressure is then 25% of the "characteristic" wind pressure.

The gust speeds (maximum mean speed value over 10 seconds) are 1.4 times greater.

The DNV code adopts the sample periods of 1 minute and 3 seconds, respectively, for the sustained wind speeds and for the gust speeds. The periods of return are the same adopted for waves, but numerical values for wind speeds are not given.

3.4 Characteristic values of the strength of materials

3.4.1 Steel The adopted characteristic value of the strength of steel is the minimum yield stress f_{yk} specified for the type of steel used, and warranted by the producer as the minimum 5% fractile in the statistical quality control.

3.4.2 Concrete The adopted characteristic value for the strength of concrete, is the 5% fractile from the statistical analysis of the results of compression tests in cylindrical test pieces:

$$f_{ck} = f_{cm} - 1.64s$$

where f_{cm} is the mean value and s the standard deviation.

The CEB/SIRTUS draft recommends to take 1.64s = 8 N/mm^2. (= 80kgf/cm^2).

4. CLASSIFICATION OF ACTIONS

On the base of the previous analysis, and as a result of comparing the codes considered, it is possible to establish the following classification of the actions that will be considered in the design of offshore fixed structures. For the environmental actions two types of values are defined: frequent values and characteristic values (corresponding to extreme conditions).

4.1 Direct actions

Symbols:

I PERMANENT ACTIONS AND HYDROSTATIC PRESSURE
- Characteristic values .... G_k
- self weight of structure
- weight of fixed equipment
- ballast
- external hydrostatic pressure
  (referred to mean sea level)

II ACTIONS CORRESPONDING TO THE USE OF STRUCTURE
- Characteristic values .... Q_k

IIa STATIC VARIABLE ACTIONS .... $Q_{k,est}$
- weight from drilling operations supplies
  (mud, cement, fuel, steel tubes)
- weight from stored products
- live load on the deck
- movable equipment

IIb ACTIONS DUE TO OPERATION OF EQUIPMENT
- dynamic actions treated as static means
  of convenient increases .... $Q_{k,op}$

- actions resulting from drilling operations
- actions resulting from lifting operations
- actions resulting from helicopter landing

III ENVIRONMENTAL ACTIONS
(frequent or normal values) .... $Q_{env.N}$

- Wave action:
maximum wave, mean period of return 1 month (probability level $10^{-5.6}$ according Det Norske Veritas, or $10^{-5.4}$ according Bureau Veritas) ($\cong 0.65\ Q_{env,E}$)
- Wind action:
wind pressure corresponding to maximum sustained wind speed, mean period of return 1 month ($\cong 0.25\ Q_{env.E}$)
- Action of currents (and ice)

IIIb EXTREME ENVIRONMENTAL ACTIONS
(infrequent or characteristic values; the operating activities are supposed to be reduced)
$Q_{env.E} = Q_{k,env.}$

- Wave action:
maximum wave, mean period of return 100 years according to Det Norske Veritas (probability level $10^{-8.7}$), or 50 years according Bureau veritas (probability level 10^{-8})
- Wind action:
wind pressure corresponding to maximum sustained wind speed, mean period of return 100 years, according Det Norske Veritas, or 50 years according to Bureau Veritas
- Action of currents (and ice)

IV INDIRECT ACTIONS
Temperature effects (restrained thermal deformations) .... $S_{k,temp}$

- temperature of the sea
- temperature of stored oil (thermal gradients in the reservoir walls)

V EFFECTS OF SEISM .... $S_{k,seism}$
- maximum level of seismic activity expected during the structure life, according to API

5. PARTIAL SAFETY FACTORS

5.1 Definitions

As it was already shown in 1.1 in the semi-probabilistic process the verification of the structural safety, in the ultimate limit state, can symbolically be expressed by the inequality

$$S(\gamma_f F_k) < R\left(\frac{f_k}{\gamma_m}\right)$$

where S means stress (or stress resultant in a cross section), and R means resistance.

The partial safety factor γ_m, takes into account imperfections of the structure, differences between the strength of the material in the structure and that obtained by testing standardized specimens, and inaccuracies in the assessment of the resistance of elements derived from the strength of the material, including inaccuracy in the dimensions. The following values are adopted:

for steel: $\gamma_m = \gamma_s = 1.15$

for concrete: $\gamma_m = \gamma_c = 1.5$

The partial safety factor γ_f is supposed to be the product of three factors: γ_{f1}, γ_{f2} and γ_{f3}.

The factor γ_{f1} takes into account the possibility of unfavourable derivations of the actions from the characteristics values and depends on the type of these actions.

The factor γ_{f2} is equal to 1 for the main action in a combination of actions of different types and is also equal to 1 for the permanent actions; for the accompanying variable actions in a combination of actions, it is less than 1 and takes into account the reduced probability of combination of actions, all at their characteristic values. The factor γ_{f3} takes into account the possibility inaccurate assessment of the action effects, including those derived from geometrical imperfections of the structure. The factor γ_{f3} does not depend on the type of the actions.

For the serviceability verifications the partial factors γ_{f1} and γ_{f3} are taken equal to 1 but a factor $\gamma_{f2} = \psi_1$, less than 1, is adopted to derive from the characteristic values of the actions their frequent values. In certain cases a factor ψ_2 is adopted to define a quasi-permanent value. In the case of environmental actions, considered in the design of offshore structures, the frequent values are directly defined by means of adoption of mean periods of return smaller than those adopted for the definition of the characteristic values. In this case the factor ψ_1 is not used, and $\psi_2 = 0$. This corresponds to adopt $\psi_1 \cong 0.25$ for wind action, and $\psi_1 \cong 0.65$ for wave action.

With the decomposition of the partial safety factor γ_f in three factors, the inequality above mentioned should be replaced by the following:

$$\gamma_{f3}\, S(\gamma_{f1}\, \gamma_{f2}\, F_k) < R\left(\frac{f_k}{\gamma_m}\right)$$

In the case of a linear static analysis (the action effects are proportional to the actions), the first inequality can be maintained:

$$S(\gamma_{f1}\, \gamma_{f2}\, \gamma_{f3}\, F_k) < R\left(\frac{f_k}{m}\right)$$

The factor γ_{f3} is generally taken as 1.08 to 1.12. The factor γ_{f1} is taken equal to 1.25 for permanent loads ($\gamma_f = 1.35$), and to 1.4 for the main variable load ($\gamma_f = 1.5$). As it was already commented, in the case of the wave actions the factor γ_{f1} can be reduced from 1.4 to 1.2, and γ_f from 1.5 to 1.3. It will be shown later than a further reduction for the γ_{f1} factor can be made, in the case of a rigorous random dynamic analysis.

5.2.1 Ultimate limit states The combination-of-actions effects for the ultimate limit states should be taken as:

$$S_d = \gamma_{f3}\, S\, |\gamma_{f1g}\, G_k + \gamma_{f1q}(Q_{1k} + \sum_{i=2}^{n} \gamma_{f2i}\, Q_{ik})$$

where Q_{1k} is the basic or main variable action. Each variable action is successively considered as a basic action, unless it is obvious that the combination resulting from it cannot be a critical one.

In the case of a linear analysis:

$$S_d = S|\, \gamma_g\, G_k + \gamma_q(Q_{1k} + \sum_{i=2}^{n} \gamma_{f2i}\, Q_{ik})$$

For current structures, according to CEB/SIRTUS draft, $\gamma_g = 1.35$ (G favourable) or 1.00 (G unfavourable), and $\gamma_q = 1.5$. The factor γ_{f2} is in general taken as equal to 0.6 or 0.7.

In the case of fixed offshore platforms the suggestion of this paper is to take for the extreme environmental loads $\gamma_q = 1.3$, and for the other variable loads $\gamma_q = 1.5$ and $\gamma_{f2} = 0.8$.

If the environmental loads are considered, in a combination of actions, as "accompanying loads", then the factor $\gamma_q = 1.3$ is maintained, but the infrequent or extreme value is replaced by the frequent or normal value, as it has been explained in 5.1.

5.2.2 Serviceability limit states

a) Infrequent combinations

$$S = S(G_k + Q_{1k} + \sum_{i=2}^{n} \psi_{1i} Q_{ik})$$

b) Frequent combinations

$$S = S(G_k + \psi_1 Q_{1k} + \sum_{i=1}^{n} \psi_{2i} Q_{ik})$$

As it was already commented, in the case of the environmental actions the products $\gamma_{f2} Q_k$ and $\psi_1 Q_k$ are replaced by the "frequent" values of these actions, directly defined by

$$\psi_1 Q_{env.k} = \psi_1 Q_{env.E} = Q_{env.N}$$

5.2.3 Conditional limit states (accidental combinations)

If Q_{acc} is the accidental action:

$$S_d = \gamma_{f3} S(G_k + Q_{acc} + \psi_1 Q_{1k} + \sum_{i=2}^{n} \psi_{2i} Q_{ik})$$

5.2.4 Random Dynamic analysis When a random dynamic analysis is made, using the spectral techniques and including the soil-structure and ocean-structure interactions, the factor γ_{f1} to be applied to the wave actions can be taken equal to unity as is the case in the conditional limit states. Appendix no. 1 of the CEB/SIRTUS draft, item 4.4, allows this assumption for the case of storms and earthquakes, when the event is "associated with further characteristics as spectral densities". The reduction of the factor γ_{f1} to unity should be conditioned to the replacement of the so-called "3σ criterium" by the assessment of the maximum value of a wave action effect on the basis of a Rayleigh distribution, and for a duration equivalent to at least 1000 cycles (about 3 hours). The so-called "3σ criterium" gives unsafe results, and in the case of its adoption the factor γ_{f1} should not be reduced to unity.

The following formula is recommended:

$$S_{max} = \sigma_S \sqrt{2\ell_n N_c}$$

where σ_S is the standard deviation of the action effect S.

For $N_c = 1000$ (3 hours), $S_{max} = 3.7\sigma_S$;

and for $N_c = 2000$ (6 hours), $S_{max} = 3.9\sigma_S$.

It must be emphasized that the factor γ_{f3} is maintained. Only the factor γ_{f1} is taken equal to unity. In this case, for the wave actions, $\gamma_f = \gamma_{f3} = 1.1$.

6. COMPARISON OF THE PARTIAL SAFETY FACTORS IN THE CODES CONSIDERED

This comparison is made in the tables I and II. As a suggestion, values of safety factors inspired in the CEB/SIRTUS draft are also given. For fatigue analysis, see 2.1.5.

7. REQUIREMENTS IMPOSED IN THE CASE OF CONCRETE STRUCTURES, FOR THE VERIFICATION OF SERVICEABILITY

These requirements are related with durability and their aim is to avoid or to make difficult the corrosion of the reinforcement. In table III three of the considered codes are compared with the CEB/SIRTUS draft. The main difference between these codes is related with prestressed concrete: the DNV and FIP codes accept, for the structural components not in contact with oil, the so-called "Class III" (limited cracking); the BV code recommends the "Class I" (uncracked concrete, no tensile stresses). In the suggestion presented in this paper, inspired in the CEB/SIRTUS draft, the "Class I" (no tensile stresses) is imposed for the frequent combinations (normal conditions) and "Class II" (limited tensile stresses, but no-cracking), for the infrequent combinations (extreme conditions).

All the codes considered impose severe conditions related to the thickness of concrete cover and concrete quality.

TABLE I Ultimate Limit State γ_f factors

(with $\gamma_s = 1.15$ and $\gamma_c = 1.5$)

	Situation	G_k	$Q_{k,est.}$	$Q_{k,op.}$	$Q_{env.N}$	$Q_{env.E}$	$Q_{Sism.}$
a) Concrete Structures							
Det Norske Veritas	a	1.4	1.6	1.6	–	–	–
	bE	1.1	1.3	1.3*	–	1.3	–
	bE	0.9	0.9	0.9*	–	1.3	–
Bureau Veritas	a	1.4	1.6	1.6	–	–	–
	a	0.9	1.6	1.6	–	–	–
	bN	1.2	1.6	1.6	1.4	–	–
	bE	1.2	1.2	1.2*	–	1.2	–
	bE	0.9	0.9	0.9*	–	1.3	–
FIP	bN	1.2	1.6	1.6	1.4	–	–
	bE	1.2	1.2	1.2*	–	1.2	–
	bE	0.9	0.9	0.9*	–	1.4	–
b) Steel Structures							
Det Norske Veritas							
Yielding	a	1.45	1.45	1.45	–	–	–
	bE	1.09	1.09	1.09*	–	1.09	–
Buckling	a	1.80	1.80	1.80	–	–	–
	bE	1.36	1.36	1.36*	–	1.36	–
Plastic analysis	a	1.50	1.50	1.40	–	–	–
	bE	1.30	1.30	1.30*	–	1.30	–
Bureau Veritas							
Yielding	a	1.45	1.45	1.45	–	–	–
	bE	1.09	1.09	1.09*	–	1.09	–
Buckling	a	1.96	1.96	1.96	–	–	–
	bE	1.46	1.46	1.46*	–	1.46	–

* operations compatible with extreme environmental conditions.

TABLE I (continuation) Ultimate Limit State γ_f factors (with $\gamma_s = 1.15$ and $\gamma_c = 1.5$)

	Situation	G_k	$Q_{k,est}$	$Q_{k,op}$	$Q_{env.N}$	$Q_{env.E}$	$Q_{Sism.}$
API							
Yielding	a	1.45	1.45	1.45	-	-	-
	bE	1.09	1.09	1.09*	-	1.09	-
Buckling	a	1.80	1.80	1.80	-	-	-
	bE	1.36	1.36	1.36*	-	1.36	-
Earthquakes (random dynamic analysis)	S	1.00	0.75	-	-	-	1.00
c) Steel and Concrete Structures							
Suggestion based in the CEB/SIRTUS draft							
Frequent combinations (normal environmental conditions)	bN	1.35	1.5	1.2	1.3	-	-
	bN	1.0	1.5	1.2	1.3	-	-
	bN	1.35	1.2	1.5	1.3	-	-
	bN	1.0	1.2	1.5	1.3	-	-
Infrequent combinations (extreme environmental conditions)	bE	1.35	1.2	1.2*	-	1.3	-
"Design wave method" (deterministic dynamic analysis)	bE	1.0	1.2	1.2*	-	1.3	-
"Spectral method"	bE	1.35	1.2	1.2*	-	1.1	-
(random dynamic analysis)	bE	1.0	1.2	1.2*	-	1.1	-
Earthquakes (random dynamic analysis)	S	1.1	0.8	-	-	-	1.1

*operating actions compatible with extreme environmental conditions.

Obs: The API code prescribes also an "accidental combination" corresponding to an "unexpected earthquake" with twice the "expected intensity".

TABLE II Serviceability Limit State γ_f factors (with $\gamma_s = \gamma_c = 1$)

	Situation	G_k	$Q_{k,est.}$	$Q_{k.op}$	$Q_{env.N}$	$Q_{env.E}$
Concrete Structures						
Det Norske Veritas						
	a	1	1	1	-	-
	bN	1	1	1	1	-
	bE	1	1	1*	-	1
Bureau Veritas						
	bN	1	1	1	1	-
FIP	bN	1	1	1	1	-
Suggestion based in the CEB/SIRTUS draft						
Frequent combinations	bN	1	0.8	0.8	1	-
Infrequent combinations	bE	1	0.8	0.8*	-	1

* operating actions compatible with extreme environmental conditions.

TABLE III Requirements Related to Durability (Concrete Structures) (Cracking Limit State)

Stresses in $N/mm^2 \cong 10\ kgf/cm^2$

		Structural components in contact with oil	Structural components in contact with oil	General	
	Situation	$P_{int.} \leq P_{ext.}$	$P_{int.} > P_{ext.}$	Reinforced Concrete	Prestressed Concrete
Concrete Structures					
Norske Veritas	a	$\sigma_{co} < 0$	$\sigma_{co} \leq -1.5$		
	bN	$\sigma_{co} < 0$	$\sigma_{co} \leq -1.0$	$\sigma_s \leq 160$	$\Delta\sigma_{ps} \leq 80$
	bE	$\sigma_{co} \leq 2.0$	$\sigma_{co} \leq -0.5$		
Bureau Veritas	bN	$\sigma_c < 0$	$\sigma_c < 0$	$w_{fiss.} \leq 0.2mm$	$\sigma_c < 0$
FIP	bN	$\sigma_c < 0$	$\sigma_c < 0$	$w_{fiss.} \leq 0.3mm$	$w_{fiss.} \leq 0.1mm$
Suggestion based in the CEB/SIRTUS draft					
Frequent combinations	bN	$\sigma_c < 0$	$\sigma_{co} \leq -1.0$	$w_{fiss.} \leq 0.1mm$	$\sigma_c < 0$
Infrequent combinations	bE	$\sigma_c \leq f_{ctd}$	$\sigma_{co} \leq -0.5$	$w_{fiss.} \leq 0.1mm$	$\sigma_c \leq f_{ctd}$

Obs: 1) σ_c = stress in concrete, under service conditions
σ_s = stress in steel, under service conditions
$\Delta\sigma_{ps}$ = stress variation in the prestressing steel
w_{fiss} = maximum width of cracks
f_{ctd} = design tensile strength of concrete ($\geq 2.0\ N/mm^2$)

DESIGN, CONSTRUCTION PRINCIPLES AND SETTING OF ONE TYPE OF CONCRETE GRAVITY PLATFORM INSTALLED ON OIL FIELDS IN THE NORTH SEA

C. G. Doris do Brasil

The concrete gravity platforms described in the following pages are partly constructed onshore in a dry dock then completed afloat in a protected area called a wetdock. They may be equipped, prior to being towed to the final site, in such a way that they are ready to operate a short time after they arrive on the final site. Towing will last one week approximately and the on-site installation may be completed in a few hours, which is remarkable considering the large dimensions of the platforms and the high sea conditions.

DESCRIPTION OF PLATFORMS OF C.G. DORIS DESIGN

The platforms which have been constructed or are presently under construction in the North Sea, have been designed as follows:

A circular raft foundation, designed as a rigid unit, including a series of radial and circumferencial skirts.

A floating vessel. This enclosure comprises a solid vertical wall, lobate or circular in shape, which is stiffened by a certain number of radial diaphragms and which rests on the raft foundation. Its diameter is calculated so that the structure will float under the useful loads exerted during the various construction, towing and on-site installation phases. The so achieved volume may be used for ballast and crude oil storage in cases in such a function is specifically required.

A second enclosure designed to dampen the effects of the waves. It is either placed above or around the first enclosure. A perforated wall envelopes the upper part of the structure and reduces the horizontal wave induced forces and corresponding

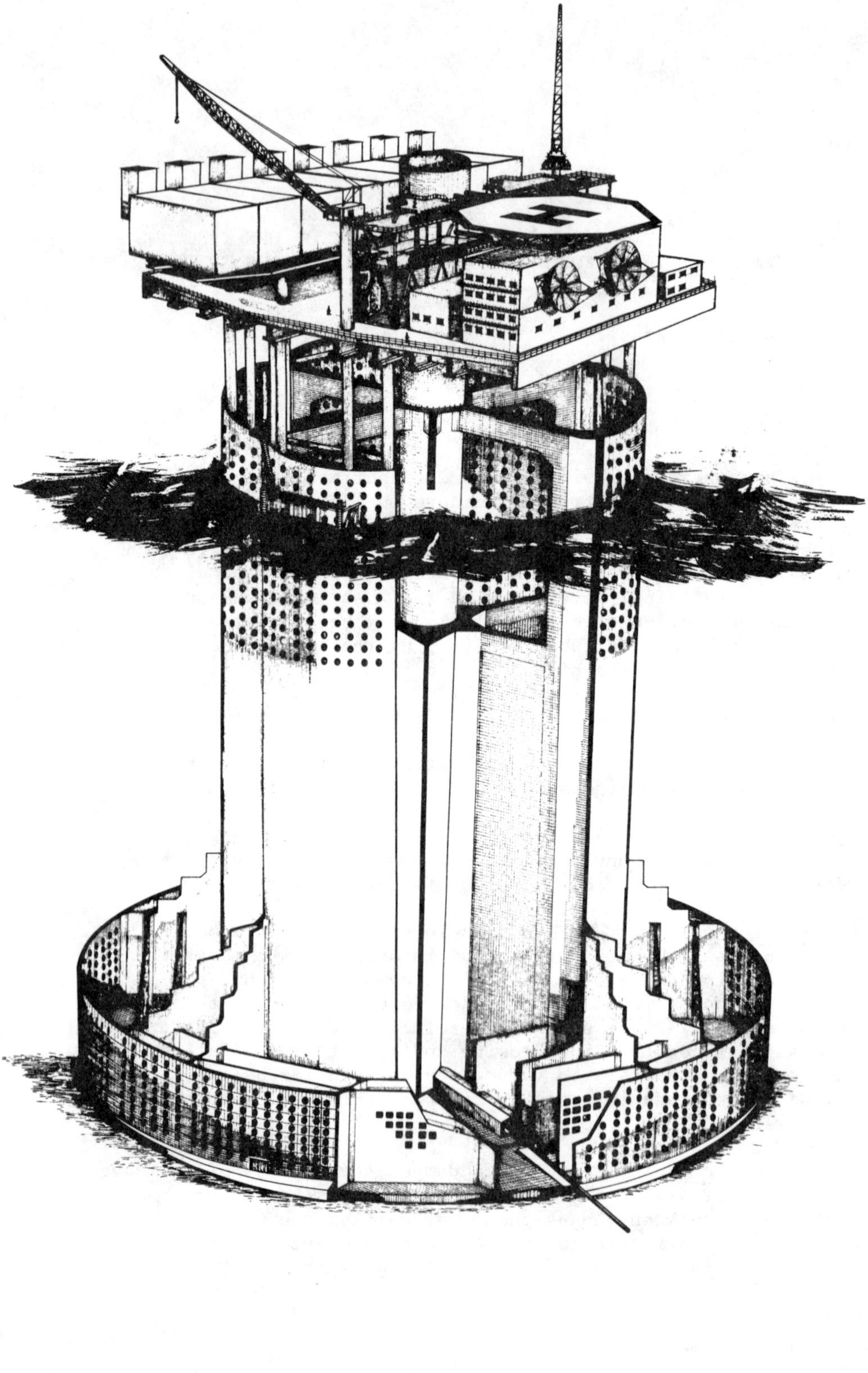

overturning moments. Furthermore it supports the deck columns.

A cylindrical and hollow central shaft, supporting the radial beams of the deck.

A substructure including one or several decks, which can have surface areas up to 10,000 m^2.

The oil exploration platforms first introduced some ten years ago in the North Sea have since permitted accurate observations of the waves at sea. They have evidenced the existence of waves of 20 metres, imposing on designers to take into account, platform dimensioning for a hundred year wave of 30 metres amplitude with a 18 seconds period. Such waves exert heavy horizontal loads on the structure which are generally cumulative with those created by the wind and which therefore make it necessary to install a protective device.

Hydraulic trials have led to the adoption of a uniformly perforated wall in the upper part of the structure also called Jarlan wall, the name of its investor, which partially breaks the waves by reflection phenomena as they approach the structure and procures an enclosure in which they can penetrate thus dissipating a good part of the wave energy by creating violent and multiple currents, turbulences and deflections inside of the resulting chamber.

The dimensioning and shape of the holes, their respective position and the thickness of the wall govern the proportion of the energy reflected outwards, the energy dissipated by turbulence and friction when crossing the wall, and the residual energy which penetrates into the chamber in the annual space. The dimensions of the chamber must take into account any possible resonance, the desired attenuation rates for the most probable waves in operating conditions and also for exceptional waves.

Under such conditions, the determination of this protection system requires taking into account numerous parameters in very complex calculations. The perforation coefficient adopted varies from 0.25 to 0.35 (void area with respect to the total surface). These calculations show that the perforated wall, in comparison with the solid wall, reduces noticeably the horizontal forces and the overturning moment of the platform, in the order of one half for the contemplated operating waves (wave height: 17 metres - Period: 12 seconds).

Such a protection against the sea wave loads makes it possible to construct these platforms in concrete, the walls of which are subjected to forces which cannot be compared with those imposed to the other types of platforms on piles or buoyant units.

ANTI-SCOUR SYSTEMS

If currents exist on the sea bottom, the installation of a large structure is likely to disturb their rate of flow, so generating local accelerations capable of putting sand in suspension and of creating scour. Furthermore, the disturbances imposed to the wave motions close to the external raft wall may create, at its base, turbulent currents which are added to the previous ones at ground level.

In order to prevent scouring, one may contemplate the construction, at the raft foundation periphery, of a low perforated wall which, in acting on the currents, has already given proof of being capable of reducing such a phenomena.

FOUNDATIONS

The wave and wind induced forces, per unit surface, exerted on very large structures generate considerable resultant forces and overturning moments which the foundations have to resist.

The following table shows the order of magnitude of wave induced forces exerted on various platforms and taken into account in the calculations:

SITE	WATER DEPTH (in ms)	EXTREME WAVE HEIGHT (in metres)	MAXIMUM HORIZONTAL FORCE (in tons)	MAXIMUM OVERTURNING MOMENT Tm = Ton×metre
EKOFISK 1	70	24.0	78,600	3.35×10^6
FRIGG CDP1	100	29.0	69,130	3.33×10^6
FRIGG MP2	97	29.0	69,150	3.13×10^6
NINIA CENTRAL	139	31.2	102,900	4.18×10^6

One of the primary conditions to be met by the raft foundation is that of non-lifting. This condition implies that the stabilizing moment increases simultaneously with the overturning moment.

As the platform is floating prior to being installed on the sea bed, its immersed weight is nil. It will be necessary to provide the adequate ballast for the required stabilizing moment, through ballasting operations using water and sand.

Forces and moments exerted on the platform foundations, combined with the poor quality of the soil generally encountered on the sea bed in the North Sea require a particular care to solve the soil mechanics problems.

The analysis of the foundation material is made under the limit conditions and allows for making sure that:

- The soil/structure contact is always maintained.
- The motions of the foundation, slipping and oscillation remain within the permissible limits.
- No liquefaction risk exists in the sand layers.

It should be noted that, in the case of foundations resting on clay layers, the surface layers show an insufficient slipping strength. The more resisting layers must therefore be reached using skirts installed under the raft which shall penetrate into the ground. The problem is to design skirts capable of transmitting the vertical and horizontal forces, although their thickness is limited to allow for the proper penetration.

RAFT FOUNDATION DESIGN

The raft foundation is designed as a reversed slab supported by the wall and diaphragms and loaded by the soil reactions which are likely to reach 40 to 80 tons/m^2.

Considering the uncertainties existing on the nature of the contacts which may be followed by an adequate instrumentation, the slab is designed so as to support much higher local pressures, sometimes reaching 200 tons m^2.

The solution retained for the junction of the vertical wall and the raft foundation consists in embedding the wall at the base, reinforced by structural systems designed to reduce the local distortions.

VERTICAL PLAIN WALLS

The hydrostatic pressures exerted on the vertical walls may vary from 50 ton/m^2 during the construction phase to as many tons/m^2 as the immersion depth on the final site, as for instance 145 tons/m^2 on the Ninian field.

The obligation to limit the draught during towing leads to search the minimum thicknesses of the walls so as to lighten the platform; only concrete showing high mechanical performance (400 kg/cm^2 minimal 28 days strength) and high tightness performances can withstand the exceptional stresses which are generated at various levels during the numerous phases of the life of the platforms.

Only an adequate design program will satisfy the stability conditions which depend on the geometry of the structure, the technological options as the boundary conditions, the construction tolerances and the elastic strains (deformations).

The high stress rates existing in the walls require the use of the multidimensional pre-stressing. The use of slip-forms on the major part of the platform generates numerous difficulties when placing ducts and cables, all the more so that as the wall rises at a rate ranging from 1 metre to 1.50 m/day.

VERTICAL PERFORATED WALLS

The dimensions and detailed shapes of the perforated wall or breakwater wall being essentially governed by hydraulic reasons, they must withstand wave induced forces which generate important bending, ovalisation and shearing moments, especially on the edges of the apertures.

Furthermore, the breakwater wall supports a part of the deck structure. They are calculated as curved partition beams, taking care to limit the weight inasmuch as possible so as to preserve the useful load on the platform. The perforated walls of the Ekofisk and CDP 1 Platforms have been constructed with precast high quality concrete elements, with a minimal 400 kg/cm^2 28 days compressive strength.

When the pre-cast elements have been placed side by side, a space is provided in between forming a vertical and horizontal duct network.

Upon installation of the prestressing cables in both directions, they will be concrete filled to form, with the pre-cast parts, a monolithic system.

For MCPO 1 platform, another technique has been used: it consisted in placing in the slipform, plastic forms which have the shape of the desired apertures and which are embedded in the concrete as the slip-form passes.

The rising speed of the wall is then over 1.50 metres/day.

For the Ninian platform the upper breakwater wall is prefabricated in 350 tons segmental elements each of which make up one quarter of the circular wall. These elements are installed using crane barges with a high lifting capacity. The horizontal connection is insured by a conjugated joint and the vertical joints are made with in situ concrete. Finally the overall integrity and rigidity of this construction is assured by prestressing.

TOWING

After its completion in the fjord, the platform is towed to its final installation site.

Surveys of the sea bed have been made previous to the towing all along the route to be followed and on the installation site.

A 2,000 HP slip equipped with ultrasonic sounders is used and measurements taken are corrected against tide tables. The route to be surveyed for the FRIGG-CDP1 platform was 370 nautical miles, long, i.e. 700 km approximately.

Prior to starting towing operations, the behaviour of the platform afloat is analysed under various wave and wind conditions. The optimum towing speed is defined on a model, so ensuring good stability conditions taking into account the wind actions.

The beginning of towing operations up to the open sea, and the installation operations on the site require favourable meteorological conditions.

Towing through the fjords is a difficult operation due to the narrow paths to be followed, which require an accurate knowledge of the position of the platform at all times. The boats have to operate quickly in the event of a relocation of the contemplated route; for this reason, the towing ropes are short and in narrow paths, two tug boats retain the platform at the rear so as to stabilise its lateral motions.

The period required to get out of the fjord was 72 hours for the CDP1 platform, so requiring a three days forecast of good weather.

Similarly, for the installation of the platform on site, a good weather window of 24 hours was necessary.

Concerning towing in open sea, the number and power of tug boats to be used is defined on model tests according to the draught of the platform, towing, wave and wind speed. The economical towing speed is about equal to two knots (3.7 km/h). In the case of FRIGG, with four tug boats totalling 45000 HP, the average speed was of two knots with maximum speeds up to 2.8 knots, which corresponds to the results obtained from model tests.

INSTALLATION

When approaching the selected site, the tug boats are delta arranged in order to control positioning of the platform. During the immersion phase, the platform is held within a 100 metres radius circle around the desired location.

The perforations of the breakwater wall have been obturated with steel plugs prior to starting towing operation; the plug tightness is obtained by flat, deformable seals, installed between the steel plug and the concrete.

The platform is immersed until it arrives in contact with the sea bed, by sea water ballasting. The descending speed is accurately controlled by submersible pumps installed on the outside of the main wall.

The platform is divided into compartments; the transfer of water from one compartment to another one allows the verticality of the system to be adjusted in the course of the immersion.

The following results have been obtained:

- EKOFISK - 19 m (distance from the desired centre)
  2.1° (angle with the desired orientation).
- CDP 1 - 14 m (distance from the desired centre)
  0.6° (angle with the desired orientation)
- MCP-01 - 7 m (distance from the desired centre)
  0.1° (angle with the desired orientation).

CONCLUSION

The development of the oil fields discovered in the North Sea has given rise to the construction of huge prestressed platforms of the gravity type, despite the extremely severe marine conditions encountered.

Such platforms show the following advantages:

- Safety of personnel and facilities;
- Towing to final site with deck and oil equipment already installed on deck which minimizes installation time and costs;
- Low maintenance costs of the substructure due to the nature of its base material once the structure is on site;
- Adjustable storage capacity;
- Protection against corrosion and chocks of the risers by placing them in the tunnels and central shaft;
- Large deck area;
- Possible access to the sea bed.

The importance of the programs, available means, number and variety of stresses have led to organise multi-valent teams capable to apply the most developed engineering methods and construction techniques.

The various aspects of the full success of the construction and installation of such gravity structures must be emphasized:

- Compliance with the construction schedule;
- Quality of the construction works;
- Conclusive experiment of the advantage shown by concrete as construction material for offshore structures;
- Possibility to adapt these gravity platforms to greater and greater water depths.

TOWING OF THE FRIGG MANIFOLD PLATFORM - 1975

OFFSHORE RISK AND ITS MANAGEMENT

E.M.Q. Røren, O. Usterud and C. Boe

Aker Engineering A/S and Royal Norwegian Council for Scientific and Technical Research, Oslo, Norway.

ABSTRACT

Offshore activities comprise numerous risks for people, environment and physical installations.

The increased size, complexity and costs of the installations, installed in deep water and hostile environment, has highlighted the need for implementation of modern risk management techniques.

Proper attention to risk identification, evaluation, measurement and program for control of risk may, if professionally carried out, limit risks to a level which is acceptable to society. Although statistical reliability data are missing to a large extent, for both components and systems, the risk analysis techniques have proven to be a useful communication tool during planning and organizing of design and fabrication, quality assurance review, operational procedures, personnel training and contingency planning.

Risk analysis is now increasingly being required by governmental approval agencies for review of offshore projects, both in the North Sea area and in US and Canadian waters.

The paper will describe status and possible future developments.

1. INTRODUCTION

Development of the offshore industry has been a great challenge to mankind from the early beginning of drilling for hydrocarbons in the marsh land along the Gulf of Mexico to the huge platform complexes presently being installed in the North Sea and other offshore areas around the world. Future developments may be even more challenging by development of floating production complexes, or complete sub-sea installation systems in deep and hostile waters far from shore, even in artic waters with drifting ice and limited service access.

The offshore industry has traditionally been a hazardous trade compared with most landbased industries. It is therefore obvious that improvements in management of risk for the great and complex installations with large crews and enormous investments should give a high pay-off, regardless of whether the value is measured in terms of human lives or money.

2. TYPICAL NORTH SEA INSTALLATIONS

2.1 Technical Presentation

The offshore industry has taken large steps beyond the boundaries of technology, regarding environmental forces, water depths, corrosion problems, size and complexity of platform design, construction and scheduling problems, as well as development costs.

The development projects have become multi-billion dollar investments with large cost and schedule overruns. Underestimating the problems of design, scheduling and fabrication of the ever increasing platform projects, together with the general inflation of costs for labour and materials, have added to these investments.

The developments in the North Sea area are, in many ways, typical for such large projects.

The illustrations, figs 1,2,3, are typical examples of the two different concepts of large platform developments in the North Sea, - the integrated single platform gravity structure versus the multi-platform jacket structure with modulized equipment.

The platform complexes are generally fully self-contained units, equipped with all necessary facilities for drilling and producing of hydro-carbons, reinjection of gas and water, transfer of hydro-carbons by pipeline or tanker, as well as all necessary utility systems, and living quarters, etc. Gravity structures are normally further equipped with large storage volumes for produced crude oil.

Tow-out and installation of the platforms, as well as most of the heavy offshore construction work, such as piling, module installation and pipelaying depends on good and reliable weather, and is thus limited to relatively short periods of predictable good weather during the summer season.

Delay in fabrication or installation schedules, or bad weather during the summer season may thus force the companies to postpone certain important offshore activities until the following

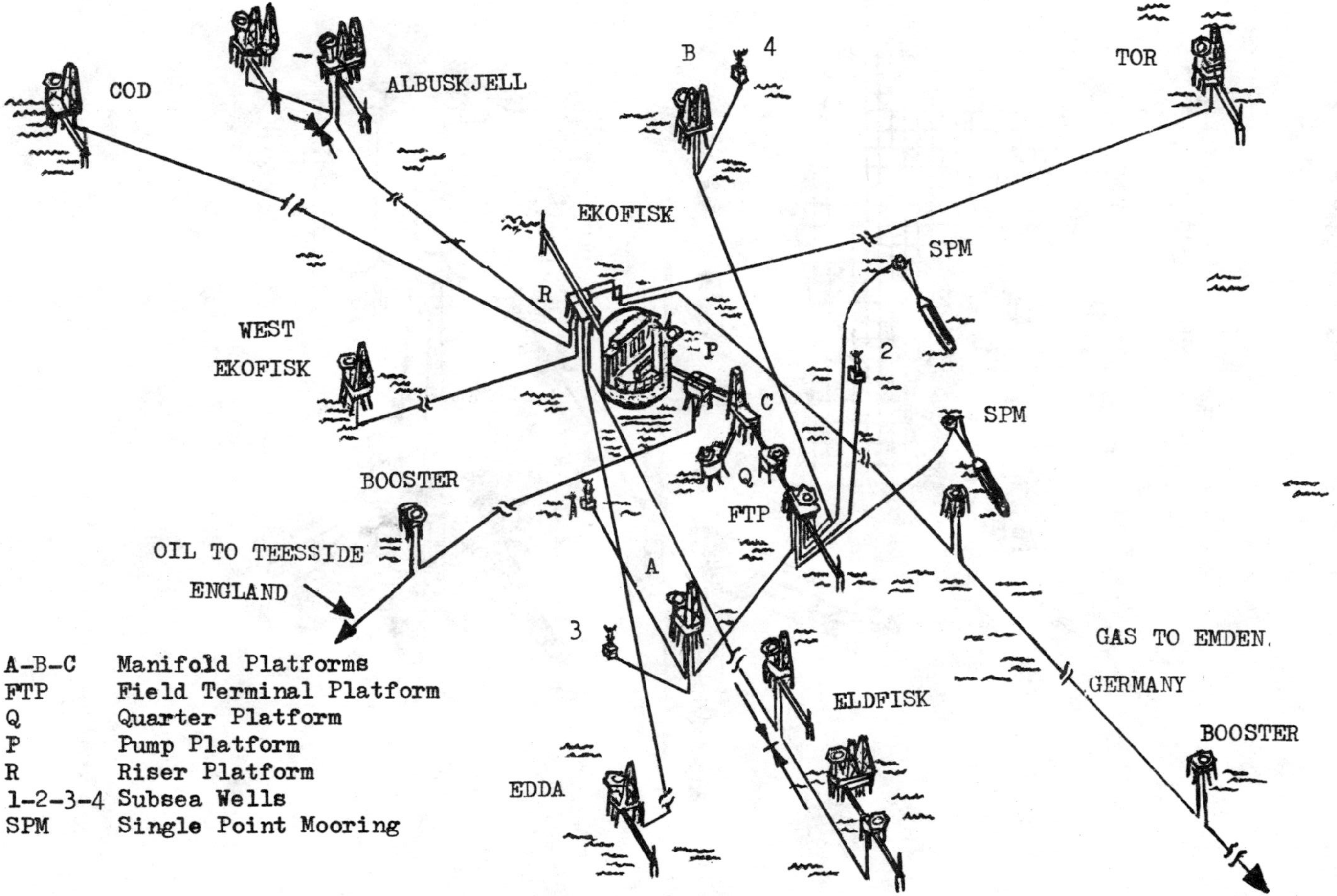

A-B-C	Manifold Platforms
FTP	Field Terminal Platform
Q	Quarter Platform
P	Pump Platform
R	Riser Platform
1-2-3-4	Subsea Wells
SPM	Single Point Mooring

Figure 1 EKOFISK Development

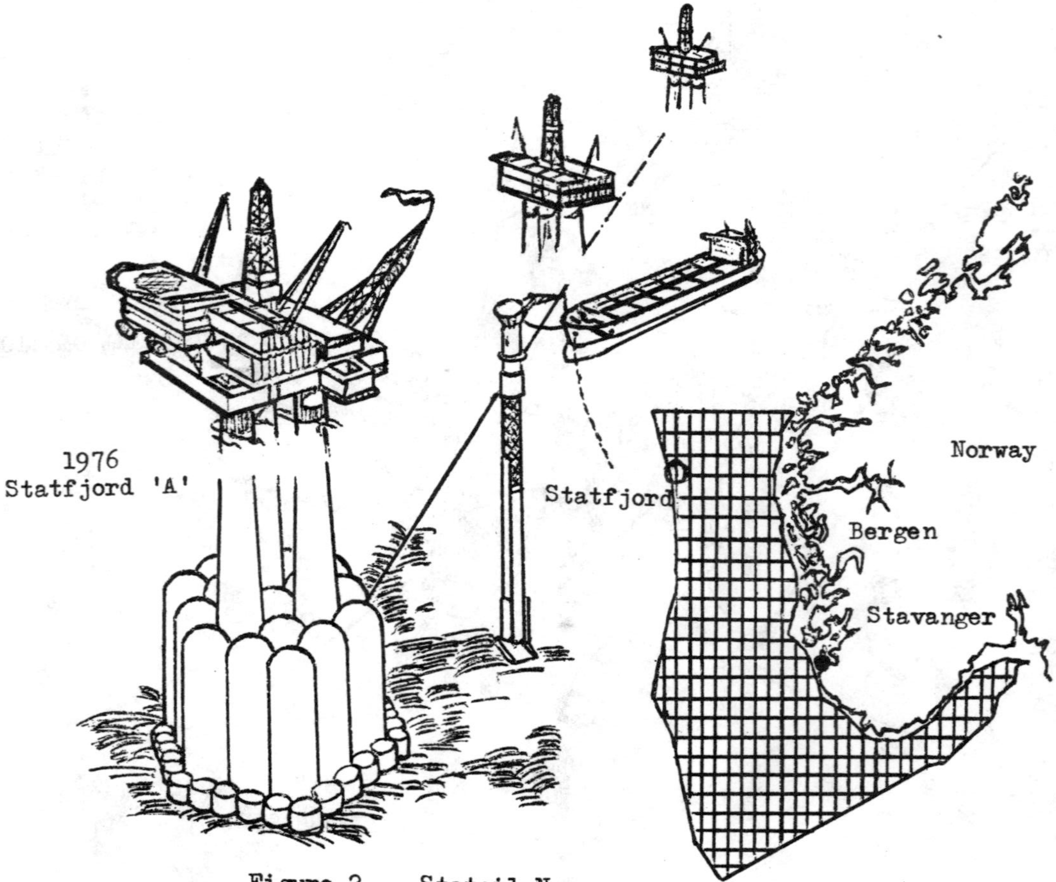

Figure 2 Statoil Norway

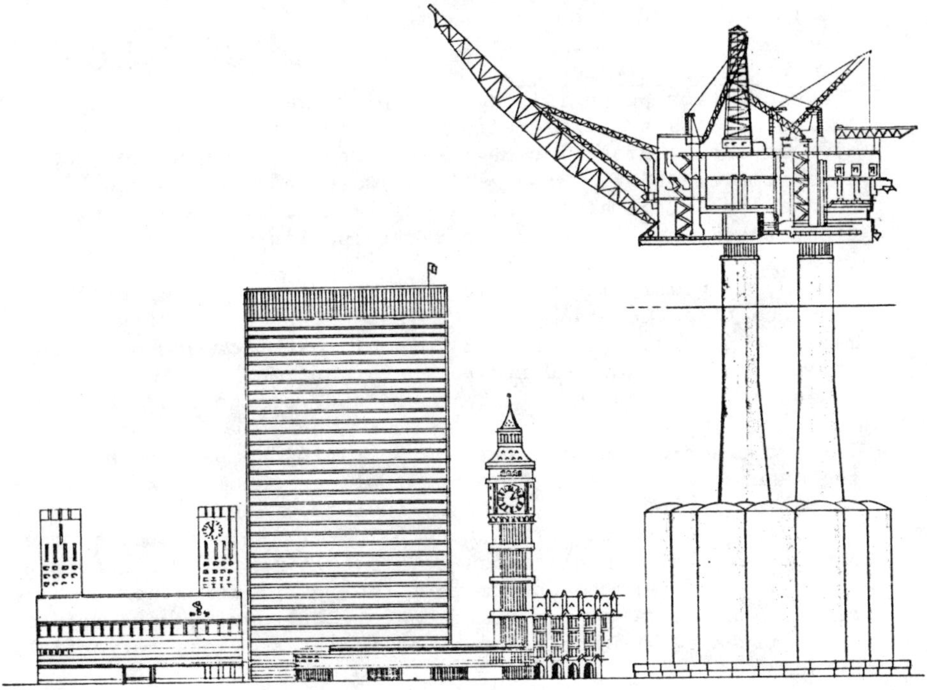

Figure 3

summer season. Subsequently extra costs are accumulated as a result of disrupted schedules for other parts of the project, and loss of production and cash flow.

2.2 Magnitude of Economy

The production potential from the platform complexes are enormous, - as an illustration - 200000 barrels of oil per day give an annual gross income for the operator, in Norway, of approximately US $700-800 mill, of which a major portion is channelled to the state as various forms of revenues (royalties, taxes, profit for state participation etc.).

A delay in platform completion and start-up, loss of production due to accidents or even a loss of an entire structure may thus not only be a severe loss of cash flow and profit for the oil company and its partners, but may also be a severe setback for the state income and foreign trade balance.

It is therefore not only for the operating company to evaluate the hazards and risks from a point of personnel risk, environmental pollution claims, company investment and income loss in case of accidents. The concern is certainly shared by the host governments on all these points. It is further required for any government to ensure that the installation has an acceptable safety standard and operational performance due to the influence from the revenues on the state gross national product.

Besides the cost for repair and cash flow losses from an accident, the operators have an unlimited responsibility for direct and consequential damages to third parties in event of environmental pollution.

Insurance is partly used to cover the cost for replacement or repair of structures and equipment, as well as clean-up costs and consequential damages to third parties.

With the increasing size of the platforms, with a single structure value exceeding US $1 billion, it is becoming more difficult to find acceptable rates and conditions for insurance. The major oil companies are also, to a great extent operating without insurance cover by underwriters, i.e. they carry self-insurance.

The total value of the repair cost carried by the operator and partners, and delayed cash flow, from the "BRAVO" blow-out in the North Sea this spring has been estimated to approx. US $200 mill, of which approx. 60-70% is a loss of revenue to the Norwegian Government.

3. NORMAL COVERAGE OF RISK

The offshore installations are protected against accidents and their consequences by following precautions:

- Safe design
- Operational procedures
- Automatic shut down systems
- Operator reliability
- Contingency plans

Although reports from the offshore industry, as from other comparable industries such as the on-shore process industry, and air and road traffic control, conclude that a major portion of all accidents are initiated by the human failure of operating personnel, most of the efforts to reduce the accident rate, and their consequences, have been dedicated to pure technical design improvements, in order to make the installation safer, and to reduce the consequences if - and when - an accident should occur. So far, the man-machine problems, and the training of personnel to react correctly and safely during both normal and abnormal critical situations, have not been devoted an equivalent attention. The pay-off might be greater in all respects: improved personnal safety reduced maintenance and repair costs, and increased annual production from a given installation.

However, let it be mentioned that a number of technical precautions has proven their efficiency through experience, and are generally accepted as requirements for safety and loss prevention by the operating companies.

These precautions have, furthermore, become mandatory requirements from the governmental approval agencies.

Examples of such safety precautions may be briefly listed as follows:

Environmental Loads Environmental load design crtieria for the North Sea are among the toughest in the world, with 100 years design wave of approx. 30 m and wind gusts at 120 knots. Fatigue loads are major design criteria due to frequent storms.

Materials and Quality Control Welding of very large thicknesses has emphasized the need for improved fabrication procedures, choice of materials and thorough quality control.

Platform Lay-out and Area Separation The complex platform installations with large number of wells, drilling equipment, separation - and gas compression equipment, reinjection trains for gas and water, as well as utilities and quarters requires special care during lay-out to obtain maximum separation and protection of the various activities.

Emergency Shut-down Systems Special emergency shut-down systems are installed for protection against blow-out from wells and isolation and bleed-off of each part of the process system.

Fire Proofing of Structural Steel Structural steel and gas-tight bulkheads are fire proofed against structural collapse from hydro-carbon fires.

Fire Extinguishing Systems Each area is generally protected by a multiple fire extinguishing system such as water spray/AFFF-foam/powder or halon/powder etc.,and two individual fire pump systems.

However,the most important design step in order to obtain a safe installation lies in the coordination of all safety aspects during the design in order to obtain a "safety-package" rather than a number of sub-optimized solutions where some key elements might be lost or did not get priority.

Due to the variety of parameters for a field development such as size of formation, gas-oil ratio, water depth and weather conditions, pipe line-versus offshore-loading, single structure versus multiple structure installation etc. there is no single solution for the optimal design.

A further complication is the difference in opinion among the oil companies regarding field development techniques, operational philosophy and experience.

Thus it seems likely that there will not be a uniform type of field development with standardized platforms and process plants, and the prime goal is to implement risk management and obtain the required safety standard regardless of the development techniques.

4. RISK MANAGEMENT

4.1 Risk Management - the Key to Risk Control

Originally risk management is the protection of economic assets, which explains why the term comes from the insurance business trades.

Offshore today, economic assets can be considered to comprise people and environment as well as the investment of installation and the income from the production. Therefore risk management means protection of any asset, which in turn can be handled through risk control based upon risk analysis.

The basic starting point for any risk management scheme offshore is that:

- Once a design solution or operational procedure has been decided and is implemented, the composite installation and its operation represents a level of risk to people, investment and environment dependent on the decisions made from conceptual design to commissioning.
- This level of risk is present, whether it is analysed or not, whether it is ignored or not, and it is an attribute like structural design or production capacity which can be appraised, changed and controlled.

This of course leads us at once to the inevitable question of how safe is safe enough, - or rather - to which extent do we have to, - and are we able to - control the risk involved?

In some cases seemingly straightforward answers are given. There are e.g. regulations given for the design load for a structure, yield and tensile strengths which should not be exceeded.

In other instances, regulations give distances of separation, maximum pressures etc. In these cases, a regulatory requirement covering technical details contributes to a risk level for that particular detail.

Put all the details together and one arrives at a resulting, overall risk level given by the regulations. The important thing to remember is that this resulting overall risk level is normally not covered by the existing regulations.

One may find that this resulting risk level is unsatisfactory, although the details are acceptable, because the regulations never conceived the installation as one common system, but as a mass of individual details.

One will also find that existing regulations ad procedures can not cover one of the main risk aspects, which for convenience is called the "human factor".

The main benefit from a risk management scheme is overall appreciation of risks associated to an installation. To apply risk management is to start a systematic process which has as its goal to implement actions which will cater for every conceivable risk and ensure satisfactorily risk control.

In every field development, decisions are taken from the very beginning of the conceptual design until the completion of construction and start-up which will influence the risk level of the installation.

The critical element is that the early decisions may set the boundary conditions for the later decisions, so that for instance decisions on platform configuration and design solutions will determine the framework of the risk spectrum for the rest of the operating life of the installation.

Risk management systemizes the coverage of the various decisions and options, and enables the management at an early stage of a development to foresee the scope of the necessary actions to be taken to ensure risk control and achieve an acceptable risk level.

Fig. 4 shows us that the basic types of actions which will ensure risk control comes within the three groups:

Personnal qualification - quality assurance - contingency planning.

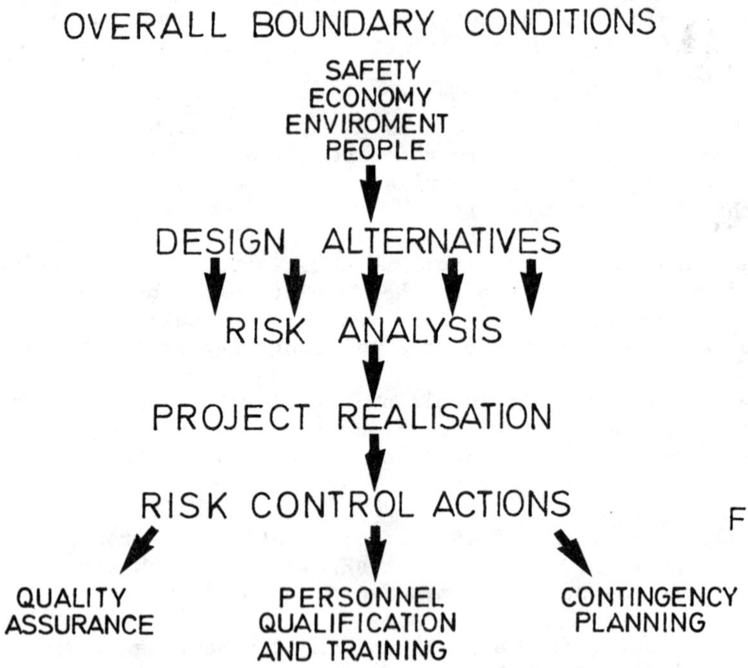

FIG.4

Fig. 4 also shows us that the extent and contents of these actions depend on the initial boundary conditions for the field development, and on a proper analysis of the risk associated with the selected alternative.

4.2 <u>Risk Analysis - the Core of Risk Management</u>
The heart of Fig. 4 is the reaction on risk analysis. Without proper and systematic analysis and assessment of the risks the entire concept of risk control and risk management becomes an illusion.

Risk analysis techniques are covered in several other papers and textbooks. Mostly they have their origin in reliability engineering and operation analysis, with the elements as follows:

- Identification of hazards or hazardous event in terms of their nature and consequence to personnel, environment and property,
- analysis of hazardous events in terms of causes - or causative patterns - and consequences,
- measurement of risk by quantification of event probabilities and corresponding consequences,
- assessment of risks and assembly of a risk control program containing the element of the three major types of action:

  Personnel qualification and training.
  Quality assurance.
  Contingency planning.

The analysis scheme described here is a simplified one. Difficult and time consuming tasks are hidden behind seemingly simple key words. The scheme is also an iterative one which is run through quite roughly at the first preliminary conceptual design stage, and more and more thoroughly as the design and operational procedures find their shape during the detail engineering, commissioning and start-up phases.

An example of how this iterative process may function is the following three-stage procedure:

<u>Phase 1 - Conceptual Stage Risk Analysis</u> The conceptual design risk analysis consists of a broad hazard identification on system level with corresponding analysis and assessment of the major risks to be found, as shown on fig. 5. Typical hazards may be as follows:

Blow-out.
Fire and explosions.
Hydrocarbon release.
Dropped objects.
Helicopter crash.
Ship collision.

Probability of hazards and consequences for personnel, facilities and environment, and operation of the facility are analysed and compared against the coverages of risk such as design precautions, operational procedures and contingency plans finalising into a conclusion of the risk level.

At this stage very little statistical data is required, statistical data which might be useful are very scarce anyway, and quantitative evaluation of assessment of hazards, consequences and preventive actions are judged by personal experience.

DEFENSE LINES
FOR
SAFETY

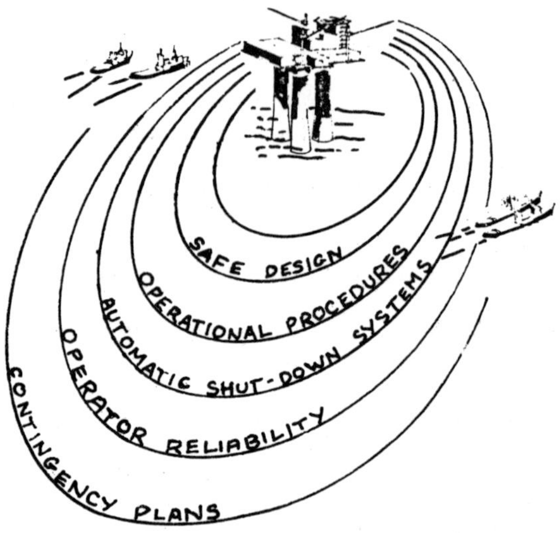

FIG. 5

This first step is performed when the design is still at the conceptual stage, and alternative solutions are assessed on an overall basis covering items such as preliminary main lay-out, process single line diagram, initial planning of the various construction phases as well as preliminary outline of operational procedures and their limits, especially with regards to simultaneous operations of any of the activities construction - drilling - production and maintenance.

Phase 2 - Detail design risk analysis The analysis, which is performed when the detail design is almost complete, is based upon detail review of systems, layout and structural design. At this stage the full hazard spectrum is identified and out-lined in necessary details, consequences are quantified and the analysis is directed towards finding causative patterns for the whole risk spectrum.

Phase 3 - Final risk analysis The third and final stage of the risk analysis is part of the final documentation issued for approval to obtain permission for start-up and operation.

The analysis is based upon "as built" drawings and specifica-tions, component specifications, operational procedures etc., and is a comprehensive and complete analysis which includes assessment of every component identified as critical, as well as the total installation as a system. At this stage detail failure effect analysis and reliability data for critical components and systems are used to obtain quantitative estimates of hard ware based risks. Gross estimates of operator reliability are used in contest with man-machine interfaces. Furthermore, detail review of operational pro-cedures and contingency plans are to be reviewed.

The final complete package issued for approval for start-up consisting of drawings, procedures, contingency plans and risk analysis, is to be the operators assurance towards the approval authorities that the required steps to obtain the desired safety for personnel, environment and installations have been obtained.

4.3 Risk Control - The Coverage of Risk Elements

Risk control means to take the actions required to reduce the identified hazards and their consequences to an acceptable level. Such actions may be improved design and quality control, stricter operational procedures, improved personnel qualification and training, or contingency planning.

As previously mentioned, the unprecise and often unfair statement "human failures during operation" counts for a majority of the accidents reported. Despite strict procedures and inspection systems, and personnel motivation and training, one can never exclude human failures if complicated and often unsuitable systems and procedures are accepted, since they represent an invitation to be by-passed for comfort of the operator, or in order to save time and money for the company.

Redesign of system may thus often be the only safe and acceptable way to avoid human failures.

The iterative process of risk analysis and risk control actions indicates the need to review design, operational procedures, personnel qualification and contingency planning as one complete package, whereby the safe design becomes the key element.

As a general statement may be said that safety built into the design is preferable compared with strict procedures or contingency plans which may not be used when required.

5. EXAMPLES OF RISK CONTROL ACTIONS

Experience has shown that dropped objects is one of the major hazards offshore, both with regards to the probability for accidents to take place , and risk for consequential personal injuries or loss of lives as well as damage to facilities.

Offshore structures are characterized by large heights, overcrowded installations and storage spaces, and nearly all transports are performed by cranes, both to- and from the structure, and within the structure.

Typical crane operations are:

- Unloading of food, drill pipes and general cargo from supply vessels, and backloading of garbage.
- Handling of drillpipes, casing, BOP and other drilling or wellhead equipment over the wellhead area.
- Internal transport of mechanical equipment to- and from areas during maintenance operations.

Dropped objects may kill personnel, damage equipment or structure or cause release of major hydro-carbon spill, fires, explosions or blow-out from ruptured vessels or piping.

The major reasons for dropped object hazards are the "human failures" such as:

- Ignorance of existing procedures in order to save time.
- Overloading of cranes.
- Incorrect use of slings.
- Lack of personal safety precautions such as waiting under swinging loads.
- Lack of general clean-up.

Other reasons may be unsafe design of cranes and other lifting devices, pipe routing or equipment located under or near drop areas, or inadequate operational procedures.

Examples of detail studies of dropped object hazards, and risk coverage through design - and operational procedures may be performed as briefly explained below:

1) Dropped object into the sea from a gravity structure
An analysis has been made on behalf of Condeep Group concerning the risk of damage to the upper domes of the oil storage caissons caused by dropped objects being lifted from a supply vessel to the platform.

Dropping of various items, which are regularly transported to and from the structure, were analysed, such as tubular goods (drill pipe, casing, drill collars), BOP-components and a complete mud pump. The analysis concluded that the drill collar was the worst case with regards to local damage to the concrete, with a maximum of 120 mm depth of damage, whilst the heavy mud pump was the worst case concerning the impact and moment acting on the spherical shell.

Possible means of reducing the consequences were reviewed, and the conclusion was that the domes of the CONDEEP structure may be effectively protected against failure from dropped objects by a correct choice of reinforcement, and a light weight concrete top cover.

2) Dropped objects into the wellhead area Dropped objects from the derrik or deck cranes into the wellhead area is regarded as one of the main reasons for blow-out or major hydro-carbon release.

The hazard is especially emphasized if simultaneous production and drilling/work-over is taking place within one wellhead area.

The hazards, which normally have their origin from "human failures" may not be solved by stricter procedures or redesign only.

Design review has indicated that hatches over each slot should be reinforced and extra protected by timber mats.

The hatches,however, are often designed as blow-out panels, and any reinforcement or timber mats may thus prevent the pressure relief, and thus increase the structural damage to bulkheads.

It is furthermore questionable whether the crew will keep to complicated procedures and always reinstal hatches and timber mats after use.

The solution may thus be to have the wellhead area designed with pressure relief in a bulkhead, and to avoid running flowlines under or near the hatches, as well as to impose strict procedures for shut-in of wells and depressurizing of flowlines next to the well in question during the entire period or parts thereof.

4. RISK MANAGEMENT IN THE FUTURE

Offshore activities comprise numerous risks for personnel, environment and property.

The increased size and complexity of the platforms installed in deeper and more hostile waters, the large number of personnel located on the platforms, the increased requirements for protection of the environment as well as the cost of the installations, their production potential, economic assets for the oil companies involved and the state revenues has highlighted the need for implementations of modern risk management techniques.

In spite of the fact that nearly all possible technical design knowledge has been implemented on the large platform structures in order to limit both the probability for hazardous events to take place, as well as to prevent minor accidents developing into major hazards for personnel, environment or installations, there is today no satisfactory answer to the consequences if major accidents like blow-out, major gas explosions or collision by large vessels should take place on a large single structure.

Risk analysis will thus become a mandatory requirement for approval of new installations on the continental shelf both in the United States, Canada and Norway.

U.S. Geological Survey has issued mandatory requirements for risk analysis of all new installations on the U.S. Outer Continental Shelf, and it is planned to implement similar requirements for existing plants.

The requirements which are limited to the drilling- and hydrocarbon processing portions of the installation are expected to be operational by 1978, and gives specific instructions for a "System Design Analysis" - procedure which is consisting of the following elements:

- Assessment of Design Features.
- Sizing, Setpoint and Timing Review.
- Failure Effect Analysis.

The Norwegian Petroleum Directorate (NPD) has also issued strict new rules with the requirement for the operators to prove to the extent possible through risk analysis that the projects do have an acceptable standard of safety.

Discussions regarding the new rules and their implementation on new projects have mainly focused on certain key paragraphs, as follows:

- Simultaneous drilling and production from one platform facility shall not take place unless special permission is given for each individual case.
- Living quarters shall be located in an area safely separated from the drilling and production areas, and shall, if necessary, be located on a separate platform.

The operators have so far been reluctant to accept these requirements, and also reluctant to present the conceptual design plans with risk analysis for approval prior to start of detail design and fabrication, as they are of the opinion that their "in-house" safety and loss prevention experience ensures acceptable safety standards.

Advanced risk management program as requested by the NPD for the offshore industry is still in its first steps.

The experience however, from risk management programs in other industries such as nuclear power plants, space- and aircraft industries has proven that it is possible, through advanced design review, risk analysis techniques and statistical reliability data banks to improve safety to meet the standards set by the society.

The first, and most important element for the operator is to get his design and operating people together and put down the conceptual design and operational limits. Thereafter through one of the standardized forms of risk analysis to review the hazards and precautions required to obtain an acceptable standard of safety.

Experience has shown that although statistical reliability data are missing to a large extent for both systems and components, the risk analysis technique has proven to be a useful tool for design review for both "conventional" and "prototype" installations, and for communication between designers, management and regulatory bodies.

Further improvements of risk management technique depend mainly on the following two factors:

- The method has to be fully accepted and implemented by by the parties involved as a suitable tool for design review and communication between the parties.
- The operators have to generate reliability data for components, systems and operational experience as well as accident rate, maintenance and repair problems in a

common databank, which should be available for use by all participants, similar to databanks for other industries such as "SYREL" operated by "Systems Reliability Services" in the U.K.

By utilizing the technical and operational experience from the oil companies, the design consultants, the contracting industry and the approval agencies, as well as common databanks, the risk management technique may be further developed to meet the increasing requirement for safety for both personnel and environment, as well as safeguarding investment and profit for the operating companies and revenues for the government.

7. CONCLUSION

The prime value of risk management based upon risk analysis is that it enables us to identify and evaluate the total risk for an installation, with regards to both technical and human factors, which has not been possible through traditional methods.

Risk management based upon risk analysis enables us to foresee the interrelations between the risks associated to details (components, structures, systems, procedures etc.) and the total risk from the details acting together as an operational unit. The method gives us the tool to identify the risks, and evaluate the necessary actions required to improve the risk level towards an acceptable limit.

Figure 6 Example of Conceptual Design Risk Analysis
DRILLING PLATFORM

AREA/CLASSIFICATION/ SOURCE OF EVENT	CONSEQUENCES FOR: PERSONS	FACILITIES &	OPERATIONS
Wellhead area/ Divl/high pressure hydrocarbon lines work-over and drilling operations.	-Normally 0.6 men subject to hazard within the area. -Normally 0.6 men subject to hazard when passing outside the area when in transit to/from bridge.	-Damage to equipment -Extensive fire or explosions may damage part of platform structure. -Fire or explosion may spread to other areas such as: -manifold area -drill floor/ shale shaker -mud pump/mud pit area -global platform fire -Fire or explosion may develop to blow-out on one or more wells, see: -Blow-out -Possibility of uncontrolled hydrocarbon spill	-Drilling ESD -Production ESD -Process platform ESD -Total shutdown of drilling and production operations until repairs are completed and approval is obtained to resume operations.

Figure 6 (Contd)

EVENT: FIRE AND EXPLOSION IN SEPARATE AREAS

COVERAGE OF RISK: DESIGN	OPERATIONAL PROCEDURES	CONCLUSIONS
-Wellhead area is open for natural draft ventilation. -Extra fire and gas detectors are fitted -Extra sprinkler, dual hose reels, hand extinguishers are fitted. -Bulkheads towards manifold and tank areas as well as area above is fire proofed to A-120. -ESD system will prevent feed of hydrocarbons into the area by wellhead wing valve and bottom hole valve. -Conductors are battered $0-\frac{1}{2}^{\circ}-1^{\circ}$ in order to increase distance between wells at bottom of conductor -Wellheads, flowlines and downstream equipment are protected against high/low pressure by automatic ESD-action from hi-lo-pilot up- and downstream of choke. -Wellheads, flowlines and down stream equipment are protected against excessive sand or formation erosion by wear type sand detectors with automatic alarm device. -Wellheads are protected from falling objects by removable I-beam frame.	-Guidelines for simultaneous operation of drilling, work-over and production: 1) All 20" casings will be set prior to completion of first wall and at 1200 ft. to provide diverter and DOP equipment asap. 2) All producing wells adjacent to the drilling well will be shut-in during rig-up or rig-down of drilling equipment, and installation of wellheads and flowlines. Wellhead and flowline pressure will be bled down to zero on these producing wells. 3) No hot-work will take place without total shutdown of drilling and production and bleed off of all hydrocarbon pressure to 0. 4) Movement of heavy equipment (dropping or swinging hazards) across wellheads or flowlines will not take place unless wellheads and flowlines are bled down to 0. 5) Any major problems with well control, drilling or producing will cause producing wells to be shut-in.	-Personnel have easy escape and re-entry for fire fighting in the area. -Fire fighting vessels may assist fire fighting through "open" side walls. -Personnel in the area may be subject to hazard from major gas releases. -Due to "open" side walls, there is small chance for major structural damage due to explosions. -Operational procedures for simultaneous drilling, work-over and production are preventing failure of one operation to cause hazard on other wells. -Open concept allows work to be performed from external areas.

6) Wire line work will utilize a BOP with sheer ram capability and be run through the subsurface safety valve.

7) Wire line lubricator and BOP will be tested prior to all operations.

8) BOP for drilling operations and wellhead ESD system will be tested at regular intervals and prior to start of operations.

PART VI POSITIONING AND INSTRUMENTATION

IMPORTANT CONSIDERATIONS OF OFFSHORE LOCATION CONTROL

H.P.J. Edge

Wimpey Laboratories, England.

1. INTRODUCTION

The aim of this paper is to discuss in general terms the very wide subject of Offshore Positioning and how it relates to the Offshore Construction Industry. I hope to draw attention to several important fundamental considerations which can all too easily be overlooked at the planning stage of an Offshore operation and which can lead to bad mistakes.

2. REQUIREMENTS OF LOCATION CONTROL IN OFFSHORE CONSTRUCTION

What are the various sorts of tasks carried out offshore and why do they need location control? This may sound a rather obvious question but it does no harm to consider the requirement from basic principles. Location control is required at all stages of development of offshore operations for a variety of reasons, some obvious, others less so. Some of these activities are listed below:-

- i) Exploration Geophysics
- ii) Exploratory Drilling
- iii) Site Investigation Drilling
- iv) Detailed Bathymetry and Seismic Profiling
- v) Submarine Investigations
- vi) Pipeline Route Surveying
- vii) Pipelaying
- viii) Underwater Placement of Modules
- ix) Construction of Port Facilities and Transshipment Facilities

In all these activities the operator needs to know where he is for a variety of reasons among which are the following:-

- i) The requirement to be able to return to the same spot on the seabed at a later stage in the operation
- ii) The legal requirement to demonstrate that work is being

carried out within the given concession limits and sometimes also that the work is within a specified International limit.

iii) Economic reasons of minimising wastage of materials, working time and material damage.

iv) Safety, usually in consideration of submarine operations.

v) Location control information is also often needed for use as steering information for the control of a task and in this connection there is a basic difference to be understood between location measurement and control of the operation. It very often happens in a Marine Environment that one knows where one is, what direction and how far one should move into the required position, but physical constraints (like manoeuvreability of the ship) make it impossible to realise that aim. I do not intend to digress onto that subject here but nonetheless I want to stress the distinction. Positioning is about providing the controlling information on position not about the mechanics of moving ships, platforms, pipes or whatever from one position to another.

3. THE FUNDAMENTAL CONSIDERATIONS

The fundamental considerations I mentioned when introducing the subject are I think best illustrated by the following questions:-

a) To what extent must we consider the Earth's Geodesy in fixing a frame of reference for positioning. In other words how do we represent the shape of the earth's surface and significant points on it onto a flat piece of paper i.e. a map?

b) What accuracy do we require, what accuracy is obtainable and what do we mean by accuracy in this context anyway?

c) How far offshore from the nearest land based survey triangulation network are we being asked to work and how does this affect accuracy?

d) What data rate is needed for the operation in question i.e. what time interval between position information are we asked for?

e) What is the operating environment of the positioning system going to be? How does reliability of the chosen positioning system affect the project? Do these considerations put constraints on our choice from other points of view?

3.1 Geodesy

Most people are aware that the earth is not a sphere it is more like a squashed sphere probably better described as an ellipsoid. Not so many people realise that the earth is not a perfect figure at all. Even if we smoothed out all the hills and mountains so that everything was at sea level there would still be undulations in the surface over large areas where the Geoid (the surface roughly equivalent to Mean Sea Level) was higher or lower than the reference ellipsoid by considerable amounts (sometimes hundreds of metres). The ellipsoid (sometimes called spheroid) is no more than a convenient mathematical model for the figure of the earth. Calculations of position for projection onto a map are more easily made using this model than one which more closely resembles the actual Geoid shape; for which projection calculations would be very complex. For historical reasons many continents and countries use different ellipsoid models as their "earth reference". Different radii of curvature and different flattening coefficients (which define the shape) have been used in localised attempts to arrive at a model which fits the country or continent in question fairly well. The attached diagram showing the relationship of Geodetic Surfaces (Fig. 1) is intended to illustrate this. Astronomers, Mathematicians and Surveyors have been trying since the time of Columbus to establish what the best ellipsoid dimensions are by measurements of very long geodetic survey networks on the ground. Some of these networks extend across continents.

Though not universally accepted, the so called "International Spheroid" gained sway as the best model for large areas of the world. In parallel with the acceptance of this model however, it was found necessary to make relative shifts of the origin, or centre, of the model so that the Ellipsoid surface would coincide more or less with the 'local' Geoid Surface. Local, in this context meaning local to a complete National Geodetic Survey Network. These shifts, the DX, DY, DZ shifts of the origin were initially relative to neighbouring area models so that where mapping in the two areas overlapped there was a way of relating the two meaningfully. With the advent of artificial satellites it has become possible to determine the Geoid Shape and dimensions much more accurately and to relate all "local" continental reference models to one truly International model. There have been a succession of models determined by satellite observations steadily improving in accuracy until the present one known as the World Geodetic System '72. The Datum of "Local" triangulation systems can now therefore all be related one to another through their relationship to the World Geodetic System. The datum of the local system being defined by the following constants:-

1. Constants defining the shape and size of the reference ellipsoid used, usually quoted as its Semi Major Axis in metres and the flattening coefficient.
2. The shift of the origin of the reference ellipsoid from the WGS72 ellipsoid centre. This is usually quoted in metres on three rectangular axes, X, Y, Z.

Why, the reader may ask, are we talking about this, what has it to do with offshore positioning? The reason is that we have

RELATIONSHIPS OF GEODETIC SURFACES

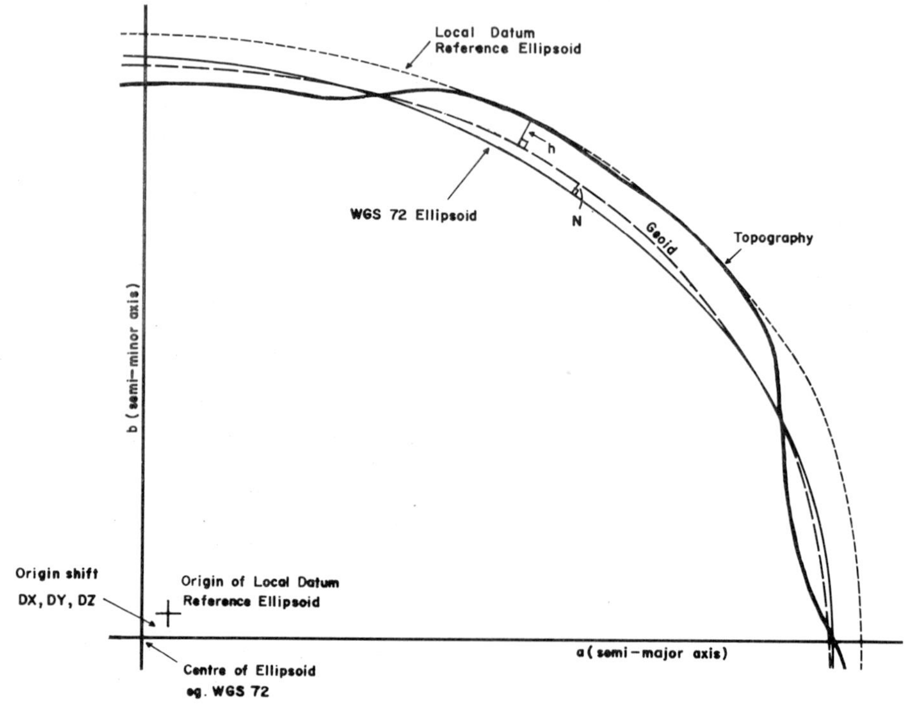

h = Elevation above Geoid

N = Geoid height

FIG.1

entered an era of global navigation where the relative accuracy of the navigation system is potentially greater than the shift of the local Datum which can itself appear as an error. The offshore engineer is drawn into this because his area of operations is often on the continental shelf at the extreme limit of local Geodetic control. Navigation systems now exist which carry control from continent to continent (viz SATNAV, LORAN, OMEGA and others) and we must be able to relate positions across these distances. We do have a means of doing this now by use of satellite navigation and it is important to consider the geodetic question at an early stage of any location control exercise because use of the wrong ellipsoil or origin could cause a mistake of about two hundred metres or so.

3.2 Projection of Reference Ellipsoid onto a Map

The next thing that must be considered is the projection and grid system to be used. Positions are often quoted in terms of Latitude and Longitude on the earth's surface. These are angular measurements of position from the plane of the equator to the points latitude and from the plane of the Greenwish meridian to the meridian of the point. This is a suitable measurement system for normal ship navigation use because the calculations of positions from star sites on the earth's surface can be made mode conveniently in angular measure. It is not however a very convenient system for the engineer to use because his systems are generally cartesian coordinate systems, that is measurements on the X, Y and Z axes from a chosen origin. A more common system used for positioning is the Transverse Mercator Grid System, a diagrammatic illustration of which is attached, Fig.2. This projection may best be thought of as a cylinder of paper which is wrapped around the surface of the earth so that its longitudinal axis is parallel to the plane of the equator. The inside surface of this imaginary cylinder touches the surface of the earth along a chosen meridian and the cylinder when rolled out flat, becomes ones map. Unlike the well known Mercators projection commonly used for navigational charts the Transverse Mercator projection has the advantage of retaining nearly the same scale at all latitudes so that for surveys covering a large extent of latitude one is not troubled by changing scales. On the other hand it suffers the disadvantage of covering only a limited extent of longitude. It is only practical to cover about 6° of the earth's surface in one 'zone of the projection before scale distortion away from the central meridian, becomes too significant. This projection has gained considerable international support from surveyors and cartographers to the extent that an international or universal Transverse Mercator Grid System (UTM) has been established which allows the surveyor to work in any part of the world on a unified grid system. This system divides the earth up into 6° segments of meridian which are numbered from Zone 1 covering 180°W - 174°West through to Zone 60 covering 174°E to 180°East. The grid distortion north of 80°N or south of 80°S becomes too great and in fact these are the northern and southern limits of

entered an era of global navigation where the relative accuracy of the navigation system is potentially greater than the shift of the local Datum which then itself appear as an error. The offshore engineer is drawn into this because his area of operations is often on the continental shelf at the extreme limit of local Geodetic control. Navigation systems now exist which carry control from continent to continent (viz SATNAV, LORAN, OMEGA and others) and we must be able to relate positions across these distances. We do have a means of doing this now by use of satellite navigation and it is important to consider the geodetic question at an early stage of any location control exercise because use of the wrong ellipsoid or origin could cause a mistake of about two hundred metres or so.

3.2 Projection of Reference Ellipsoid onto a Map

The next thing that must be considered is the projection and grid system to be used. Positions are often quoted in terms of Latitude and Longitude on the earth's surface. These are angular measurements of position from the plane of the equator to the points latitude and from the plane of the Greenwich meridian to the meridian of the point. This is a suitable measurement system for normal ship navigation use because the calculations of positions from star sites on the earth's surface can be made most conveniently in angular measure. It is not however a very convenient system for the engineer to use because his systems are generally cartesian coordinate systems, that is measurements on the X, Y and Z axes from a chosen origin. A more common system used for positioning is the Transverse Mercator Grid System, a diagrammatic illustration of which is attached (Fig. 2). This projection may best be thought of as a cylinder of paper which is wrapped around the surface of the earth so that its longitudinal axis is parallel to the plane of the equator. The inside surface of this imaginary cylinder touches the surface of the earth along a chosen meridian and the cylinder when rolled out flat, becomes ones map. Unlike the well known Mercators projection commonly used for navigational charts the Transverse Mercator projection has the advantage of retaining nearly the same scale at all latitudes so that for surveys covering a large extent of latitude one is not troubled by changing scales. On the other hand it suffers the disadvantage of covering only a limited extent of longitude. It is only practical to cover about 6° of the earth's surface in one 'zone of the projection before scale distortion away from the central meridian, becomes too significant. This projection has gained considerable international support from surveyors and cartographers to the extent that an international or universal Transverse Mercator Grid System (UTM) has been established which allows the surveyor to work in any part of the world on a unified grid system. This system divides the earth up into 6° segments of meridian which are numbered from Zone 1 covering 180°W-174°West through to Zone 60 covering 174°E to 180°East. The grid distortion north of 80°N or south or 80°S becomes too great and in fact these are the northern and southern limits of

DIAGRAMATIC REPRESENTATION OF THE UTM GRID SYSTEM

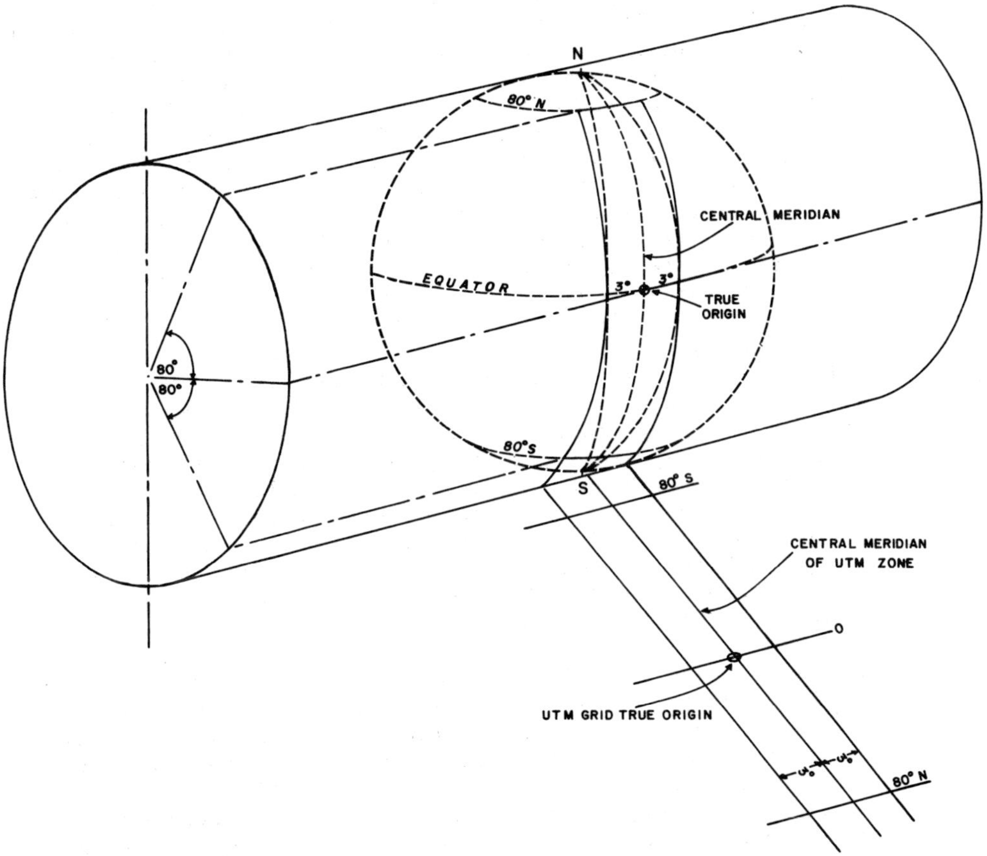

FIG. 2

the system. The true origin of all UTM grids is the point at which the central meridian of the zone in question intersects the equator. In order to make all coordinates positive however, a false origin is used for the North and South hemispheres such that the False Easting of the true origin is arbitrarily given the coordinate 500,000 East. In the Northern hemisphere no false Northing is applied and the "Northings" of a point are a measure of its distance from the equator in metres. In the Southern hemisphere the true origin is arbitrarily given the coordinate 10,000,000 metres north so that all points south of the equator have positive UTM northings. The reference ellipsoid used in the UTM system is usually, but not always, chosen to be the same as that used as the datum reference ellipsoid for the local geodetic network. So, for instance, in North America the Clarke 1880 Reference Ellipsoid would be used while in Europe the International Spheroid is used. So it can be seen that although the UTM is an Internationally accepted system it does not yet get us over the problem of reference ellipsoid changes between continents. It is important to bear this in mind in all Maritime Geodetic work where we are working in the divides between different countries.

3.3 Accuracy

The next fundamental consideration to have in mind is that of accuracy and this is an area in which there is more misunderstanding than any other in relation to offshore navigation. Someone once said that all things are relative and this was never more true than in the case of navigation equipment accuracy. Accuracy is a relative term and unless one is fully conversant with the system(s) against which another system is being compared for accuracy one really has no idea whether to believe the differences observed are a measure of accuracy or not.

We have already seen that when we quote a position on the earth's surface we are talking about its position relative to a chosen Origin and Reference Ellipsoid or Datum. The very first source of error therefore is the accuracy of the determination of the Reference Ellipsoid in relation to the Geoid surface. However, let us assume for a moment that we have determined some positions on the Reference Ellipsoid by Astro observations and let us call these Absolute Positions which correspond to the positions of the primary geodetic network which we can then use for measurements to other secondary positions. Then we can say that the difference between the coordinates of a point obtained by measurements to reference stations on this network and the true coordinates on the network is a measure of the absolute accuracy of the positioning system used to make the measurement. However, that is not the end of the story. There are several properties that are of interest to the surveyor which are sometimes loosely described as the "accuracy of the system". It is important to have the definitions of these established before discussing accuracy. These properties are:-

a) Absolute Accuracy
b) Relative Accuracy
c) Repeatability.

a) Absolute Accuracy is a measure of the ability of a system to obtain the true geodetic coordinates of a position on the Reference Ellipsoid. In reality however, absolute accuracy cannot be achieved, the best we can do is establish good Geodetic Networks as standards from which to work and do our best to eliminate systematic errors.

b) Relative Accuracy is a measure of the accuracy of a point in relation to other points used to position it from. Thus for example a point may have a very high Relative accuracy on an Engineers Site Grid but a very low Absolute accuracy if the relationship of that local grid system to the reference ellipsoid is not known with a high degree of (Absolute) accuracy.

c) Repeatability is a measure of the systems ability to return the observer at any time to the same spot on the earth's surface by using the same instrument readings. This is a useful property because there are often occasions when an observer is more concerned about being able to return to the same spot than with knowing what the absolute coordinates of that spot are. This is very often the case in offshore structure positioning.

Other properties and terms often used in connection with positioning systems which need defining are as follows. They also have a bearing on system accuracy. They are:

a) Resolution
b) Repetition Rate or Data Rate
c) Range.

a) Resolution is sometimes confused with accuracy particularly on some manufacturers brochures etc. It is no more than a statement of how many figures after the decimal point are displayed to the observer. A system must have sufficient resolution to take advantage of its potential accuracy. Excessive resolution on the other hand does nothing to improve accuracy and may sometimes mislead the observer into believing his system is achieving greater accuracy than it is.

b) Repetition Rate is the rate at which a system is capable of displaying the observer's position. It has great significance in the marine environment and a direct bearing on accuracy. The reason for this is that unlike the surveyor ashore, the Hydrographic Surveyor or Navigator is working on a constantly moving platform (a ship even with engines stopped is of course drifting with wind and current). Accuracy of position therefore deteriorates with time between fixes so the system with a high data rate offers a statistically better accuracy than one with a low data rate.

c) Range too is an important consideration. Apart from the obvious requirement that a system must have sufficient range to cover the working area it also has a direct bearing on accuracy. There are two significant effects which need explanation. Radio positioning systems are nearly all designed to enable the observer to measure distance from his position to two or more coordinated reference stations and they rely on an assumed velocity of propagation of radio waves. Changes in the propagation velocity which are affected by changing meteorological conditions and transmission paths cause errors in range measurements and thus in position. At short ranges this effect is minimised and can usually be calibrated out but unfortunately at long ranges where the effect is worst, calibration is most difficult. One must therefore be cautious in accepting accuracy figures quoted for long range systems because they are difficult to measure - or put another way they are difficult to disprove.
Another accuracy consideration of range is that of signal strength. When the signal to noise ratio of the system drops to near unity the resolution of the system suffers. The observer may then no longer be able to read the dials or displays of the system with certainty and accuracy is therefore lost.

4. POSITIONING EQUIPMENT

I hope by running through these fundamentals to have made the point that the whole question of location control for offshore positioning requires special consideration. Everything mentioned so far concerns theory only we have not even started to consider other more practical points like the working environment, the reliability of systems, requirements for permission to use radio frequencies, cost, etc., etc. May I make this a plea therefore for engineers working offshore to put positioning well up their list of priorities and ask them to consult with surveyors at an early stage in their project planning.

Let us look briefly at what equipment is available and how it is best deployed. I don't intend here to itemise every system available but to group systems into types and explain the differences and advantages group from group. (If the reader wants lists of specific equipment compared by performance he is recommended to consult either the I.H.B. special publication No. 39 or the F.I.G. Commission and Working Party Report on Position Systems).

I like to group systems as follows:-

1. Optical/Mechanical Systems
2. Radio Phase Comparison Systems
3. Radio Time Pulse Transmission Systems
4. Radar Time Pulse Transmission Systems
5. Acoustic Systems
6. Satellite Systems.

The first group needs little introduction to the engineer. I am referring here to theodolites, sextants, subtense staves etc. They must not be forgotten because they have one overriding advantage over all other systems. Having no electronic components they are inherently very reliable.

The second group has in common the technique used to measure range by comparing the phase of a received signal with respect to a standard comparison signal. This technique allows high accuracy of phase measurement and potentially high total range accuracy provided the number of whole wavelengths or 'lanes' between observer and shore stations are known. The lane count is unfortunately not always easily established and there is a fairly high risk of gross error if lanes, which can be as little as 40 metres across, are lost unnoticed by the observer. One very great advantage of this type however is that it has a virtually continuous data rate which allows high resolution, stability and statistically good accuracy. The Decca Hi-Fix and the Hastings Raydist systems are good examples of this type.

The third group, Radio Timed Pulse Transmission systems have the advantage of less lane ambiguity than the phase comparison type because identifiable pulse transmissions are spaced so that the observer only needs to know his position to a few kilometres. However, these systems suffer from having lower data rates than phase comparisons systems and statistically lower accuracy. This technique is used mostly for long ranges where peak power is only transmitted at the pulse repetition intervals of the system. Loran C and Pulse 8 are good examples of that type.

The fourth group covers systems operating around the Radar frequency bands and utilise active beacons set up on coordinated points as reference stations from which to obtain ranges to the observer. These systems give unambiguous range information being simply a means of timing a pulse from transmission to reception. Range is usually limited because working as they do at very high frequency levels they are limited to not much more than horizon line of sight range. Motorolla Mini Ranger, Decca Trisponder , and Syledis systems are good examples of these systems. The data rate of these systems is quite high and they are usually very portable.

The fifth group are Acoustic Systems employing underwater acoustic ranging techniques. These are of very limited range (only a kilometre or so in 150 metres of water) but they can provide high relative accuracy to a network of underwater seabed transponders and very high repeatable accuracy. Absolute accuracy is dependent entirely on what surface navigation system was used when deploying transponders. The Data Rate of these systems is quite low so they are best suited to relatively slow moving operations. They have one distinct advantage over radio positioning systems namely that no frequency clearance is required for their operation. The AMF ATNAV Acoustic Transponder Navigation system is a good example of one of these systems.

The last group, Satellite Navigation Systems as the name indicates rely on artificial earth satellites in orbit. The orbits of satellites (5 are currently in operation) are precisely known and a ship obtains a fix from any satellite above its horizon by measuring its range from the satellite at successive time intervals as it passes overhead.

The position can thus be established for the mean time of the pass provided the ship does not move during the pass or alternatively it knows how far it moved during the pass. The accuracy of a single SATNAV fix is only about ± 50 metres at best and the data rate is extremely low (less than 1 fix per 2 hours in Brazilian waters). Its advantage over other systems is that it is useable world-wide with no requirement to obtain frequency clearance. It is put to best use in obtaining Absolute Positions to extend Geodetic networks because when stationed for a lengthy period on a station to obtain what is called Multiple Pass Fix the statistical reliability of the position can be improved to about ± 7 metres. It is therefore useful for establishing the coordinates of a new transmission station for a Radio Positioning System or the Absolute coordinates of a newly positioned offshore platform.

All these groups have their advantages and disadvantages and must be considered and chosen according to the requirements. Perhaps it would be useful to mention a few of the sort of positioning tasks frequently encountered offshore and the suitability of various systems to meet the requirements. Generally speaking accuracy requirements increase as the project developes from preliminary exploration through site testing to field development. Initially, fairly coarse navigation may be used by Geophysical Exploration ships but when platforms are being located and pipelines laid between platforms, tolerances become quite fine.

Exploration Geophysics have to cover large areas and therefore need very long range positioning systems. As we have seen these tend to be less accurate systems but fortunately the Geophysical information is also fairly coarse and such systems as SATNAV and LORAN may be adequate for the purpose. When exploratory drilling is undertaken on the other hand there is a requirement for preliminary surveys to evaluate drilling hazards which needs more precise navigation, then there is often the need to fix the absolute position of the rig for legal purposes, to ensure that it is within the concession area for example. Multiple pass satellite navigation fixes are probably best for this latter purpose. The rig site - survey on the other hand requires survey lines to be run at quite close line spacing and to achieve this the survey ship must be fitted with a system giving position information at a high repetition rate and an accuracy significantly better than the line spacing required. Dependent upon range limitations, the Radar Time Pulse systems or Medium range Phase Comparison systems are probably best suited to this purpose. If the exploratory drilling is successful in proving oil of course, there immediately arises a keen interest in being able to re-locate the position of the hole. In other words the need for good repeatability in a positioning system becomes apparent.

At this stage in an oil field development, survey activity intensifies enormously. Detailed bathymetric and shallow geophysical surveys are needed. Soil sampling is needed to evaluate the load bearing nature of the seabed. Submarine investigations are often called for to inspect and film the seabed and remove obstructions such as lost drill casing, anchors, wires, etc., etc. It is at this stage that Acoustic Navigation systems are brought into use. With their inherently high repeatable accuracy the operator can feel sure that all ships operating on the site are in fact working on the same piece of seabed. They alone of course can be used for navigation of submarines around the seabed and when all this work is done the acoustic transponders can remain on the seabed until the offshore structure is put into place.

Surveying and positioning tasks do not end there however. Pipelines have to be laid to the platforms and loading facilities. These require route surveys, laybarge navigation and "as laid" surveys, all requiring a high degree of position accuracy for which the same criteria for choice of systems must be exercised.

5. SYSTEM RELIABILITY

A last word must be said on the reliability of positioning systems because the most accurate system in the world is certainly no help when it fails. The cost of buying and operating navigational systems is high and since they may only be needed for short periods in any operation it is often most economic to hire the services of a specialist company with the required system. The potential down time costs of the operation through failure of the positioning system are however very much higher than the cost of hiring equipment and location control services. It is wise therefore not to be too sparing in the provision of navigational systems. Experience in the North Sea and elsewhere have led operators to use two or even three back-up positioning systems on important projects often with good cause.

I would conclude by hoping that perhaps the length alone if not the arguments put forward in this paper may persuade the Offshore Civil Engineer that positioning offshore is a complex subject deserving careful consideration at an early stage in his project design.

A SATELLITE/ACOUSTIC INTEGRATED POSITIONING SYSTEM

C.D. Lynn and V. Hill

Decca Survey Ltd., United Kingdom

ABSTRACT

An integrated position-fixing system developed by Decca Survey Limited in cooperation with the Decca Navigator Company is described. The system uses a Decca Sat-Fix JMR-1 satellite receiver as primary position reference to set an Aqua-Fix/2 underwater transponder array on a geodetic datum worldwide entirely independently of any shore based radio positioning aids.

INTRODUCTION

Satellite doppler receivers have become a familiar part of the marine navigation scene in the last few years, as have underwater acoustic systems for precise offshore positioning. Individually, both of these types of aid, while extremely useful, have their drawbacks. The main ones are, for satellite receivers, the fact that the interval between fixes is of the order of hours rather than seconds and the fix accuracy is highly dependent on the precision with which the receiver's velocity is known. Acoustic systems are inevitably difficult to locate precisely on a geodetic datum as the 'transmitting stations' (the transponders) can hardly be surveyed in by conventional methods, and at extreme ranges from the nearest land cannot be accurately placed by normal radio-positioning methods.

The way round both of the satellite system's disadvantages just described, is to integrate it with another system; in the past this has usually been a radio navaid such as Decca Navigator Loran-C or Pulse/8 but for navigation purposes a doppler sonar and gyrocompass to give dead-reckoning fixes between satellite system provides a velocity input to the satellite fix computation. The DR systems are necessarily less accurate than those using radio fixes and suffer from a limitation in maximum water depth in which the doppler sonar will maintain bottom track.

The radio navaid integrated system is a better proposition for survey work but relies of course upon the operating area being within the cover of the chain used. Also the fix accuracy for these low-frequency systems is one or two orders of magnitude poorer than for a short range acoustic system.

Thus an almost insoluble problem has hitherto been to position oneself very accurately offshore without any reference to a shore-based aid. A point to make clear here is that an acoustic system alone will give a good repeatable fix relative to the acoustic net, but the position of the net itself may not be accurately determined on a geodetic datum. The integration of a satellite receiver with an underwater acoustic system such as Aqua-Fix/2 enables many of the disadvantages of the two components when used separately, to be overcome; the acoustic net is fixed by satellite, the survey vessel gets continuous fixes, and the acoustic system gives an accurate velocity vector for improving the satellite fix accuracy.

SYSTEM CONFIGURATION AND FUNCTIONS

The system integrates satellite data from a JMR-1 satellite doppler receiver with slant ranges to seabed transponders which are output from the Aqua-Fix/2 Control and Display Unit. Heading information is provided by a Sperry Mk 37 gyrocompass. The information is processed in a Hewlett Packard 2109 MX 'E' series computer with 32k of store. Output and operator entries are via a Texas Instruments 743 KSR keyboard terminal and a Perex Perifile 6041 cassette-unit is used as a temporary program store. An HP7210A plotter can also be connected.

The computer program, developed in-house, is crucial to the accuracy and ease of operation of the system. The majority of the program is in Fortran IV and may be readily modified to accept other inputs such as the radio aids mentioned earlier, speed-log etc.

The computer performs a variety of functions, mainly concerned with the calibration of the acoustic net. A description of these is probably the simplest way of describing the functioning of the system:

(a) It processes acoustic data to obtain the relative positions of the seabed transponders.
(b) It calculates drift and performs a rough orientation of the net with the aid of heading information. No satellite data is used during these two phases.
(c) It reads the satellite pass data into store, decodes the broadcast ephemeris and does the majority voting of the data.
(d) It calculates the fix, and by comparing it with the

acoustic fix, adjusts the position and orientation of the net to minimise residual errors.

(e) It outputs the results to the KSR printer and the plotter.

(f) It computes satellite alert tables if required.

CALIBRATION OF THE ACOUSTIC NET

Relative Geometry Phase

The transponders are deployed using dead-reckoning methods in an approximately square configuration (assuming 4 transponders for simplicity) around the area to be surveyed. The water depths at the drop points are measured or may be estimated if the water is deep; the depths are adjusted in the computation carried out on the slant ranges but the operator may overwrite the computed depths if he believes the measured depths to be more accurate.

Once the transponders are laid, the vessel steams around the net, keeping within acoustic range of all the transponders, collecting at least 140 sets of good slant ranges which the computer acquires automatically. It is good practice at this stage to return to the drop point of each transponder to ensure a balanced set of data. Operator entries include: a tide table to give correct antenna height (inaccuracies cause longitude positional errors), definition of fast and slow data rates for the Aqua-Fix/2 CDU (Control and Display Unit) which is computer controlled, and a velocity profile table to define the propagation velocity at various depths. The latter gives improved accuracy over a simple single-velocity model, as the propagation velocity varies with water temperature (and hence depth) and salinity.

Throughout this data collection phase, five consecutive sets of ranges are retained by the computer to predict the next range from each transponder and each new range is checked for validity against the prediction; it is rejected if it is outside a certain tolerance band. This software tracking filter, with the hardware pulse filtering in the CDU itself, ensures that erroneous data due to reflections etc. are rejected.

Having acquired the 140 data sets, the computer calculates a direct solution for the 3-dimensional positions of the transponders relative to each other, and then carries out a least-squares adjustment to obtain accurate x, y and z coordinates of the relative transponder positions. The preliminary direct solution provides a good estimate to ensure that the least-squares process converges to the correct solution.

Orientation Phase

The next phase of the calibration entails following 3 consecutive courses at different headings, using the gyrocompass,

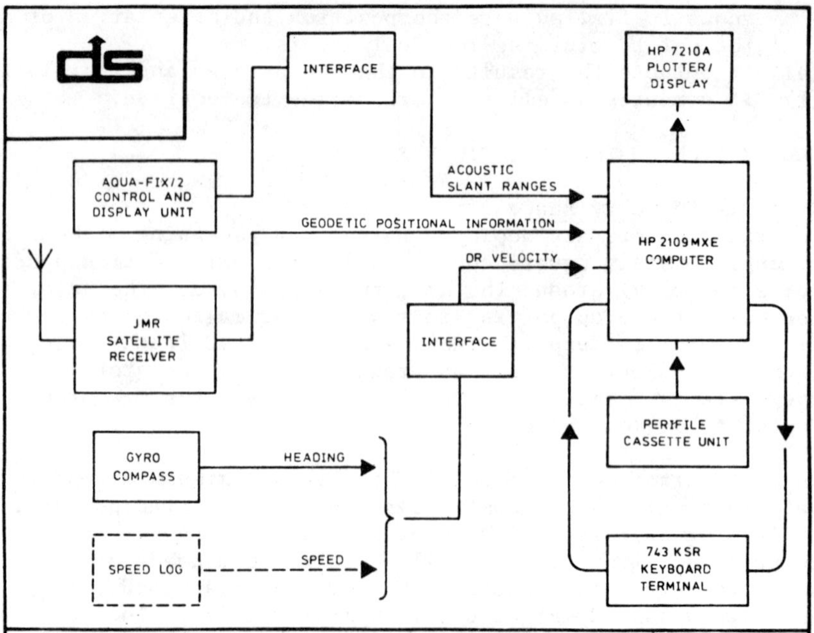

SAT-FIX/AQUA-FIX INTEGRATED SYSTEM

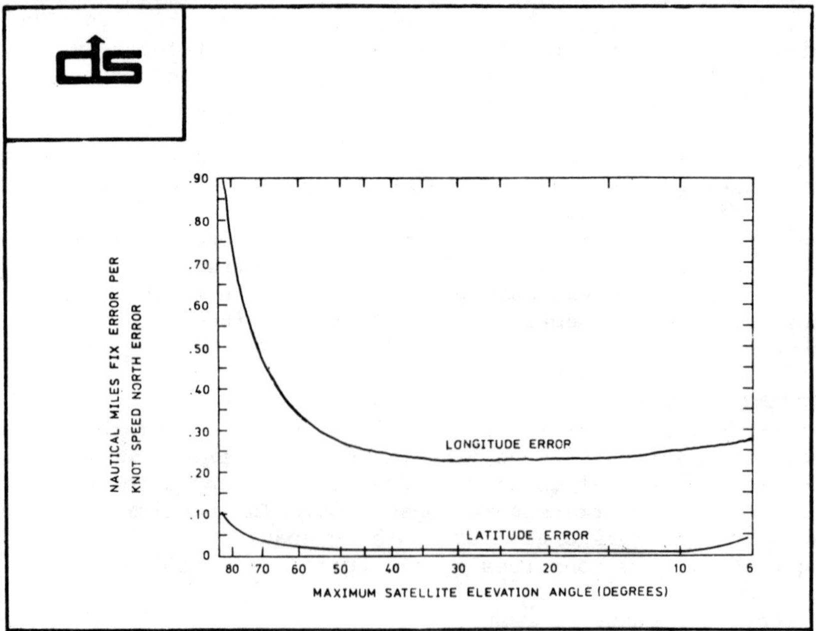

SENSITIVITY OF SATELLITE FIX TO SPEED NORTH ERROR

to determine the approximate orientation of the net. Magnitude and direction of drift, as well as the net orientation are output after this stage. The three legs should be at constant speed and preferably form three sides of a square, though this is not a rigid constraint. The computer applies a tracking filter to the slant ranges at this stage also, based on previous positions on the linear track.

Repositioning Phase

The final stage of the calibration operation is the taking of satellite data while within the acoustic coverage of the net in order to reposition the acoustic grid from its approximate origin (entered as approximate latitude and longitude of the first transponder) and orientation (derived as above) to an accurate geodetic referenced origin and orientation.

The computer accumulates satellite doppler counts at approximately 30 second intervals (actually alternate groups of 6 and 7 standard 4.6 second intervals). It concurrently takes in acoustic positions relative to the transponders. These acoustic positions, with the approximate orientation of the net derived during the previous phase, are used to obtain delta latitude and delta longitude between the 30 second intervals. A maximum of 37 sets of data are stored during a single 18 minute pass, though standard doppler count editing may reject some of these.

The set of acoustic positions is then shifted, retaining their relative latitude and longitude offsets until a minimum least-squares error between the calculated and observed range differences is obtained. The output at this stage is a satellite position and an acoustic position derived on a weighted mean basis to minimise errors due to mis-orientation of the net.

After the second and subsequent passes, the above process is repeated and the origin and orientation of the acoustic grid adjusted to obtain the best fit of the acoustic positions to the satellite positions. This process is circular in that an improvement in the net orientation gives an improved velocity vector for calculating the satellite fix, giving better fix accuracy and hence a further refinement to the net orientation.

OPERATION

After the calibration is complete, the operator has accurate latitude and longitude positions for all the transponders in the net, and the survey operation can commence. For complex operations where secondary surface or sub-surface vehicles are used in addition to the main survey vessel, the Aqua-Fix/2 system has the ability to provide simultaneous tracking of

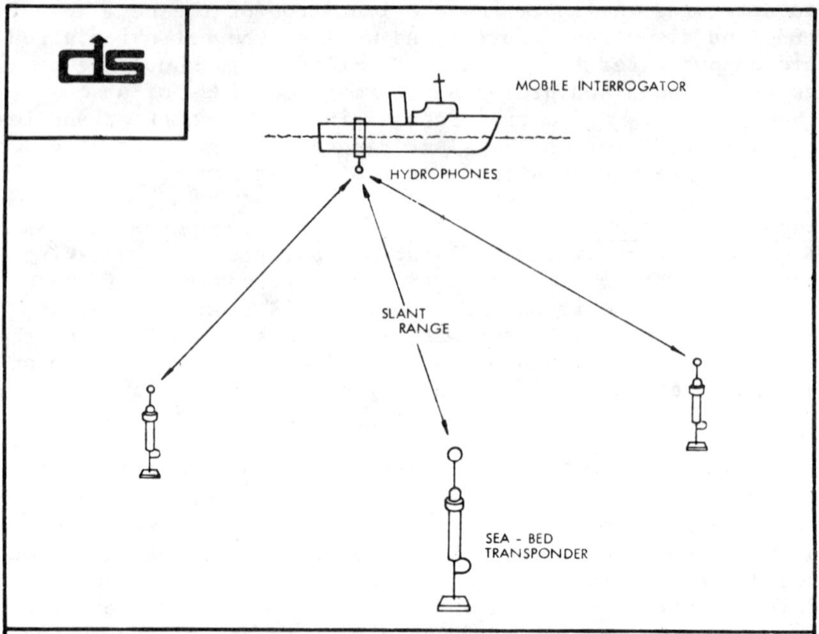

AQUA-FIX PRINCIPLE

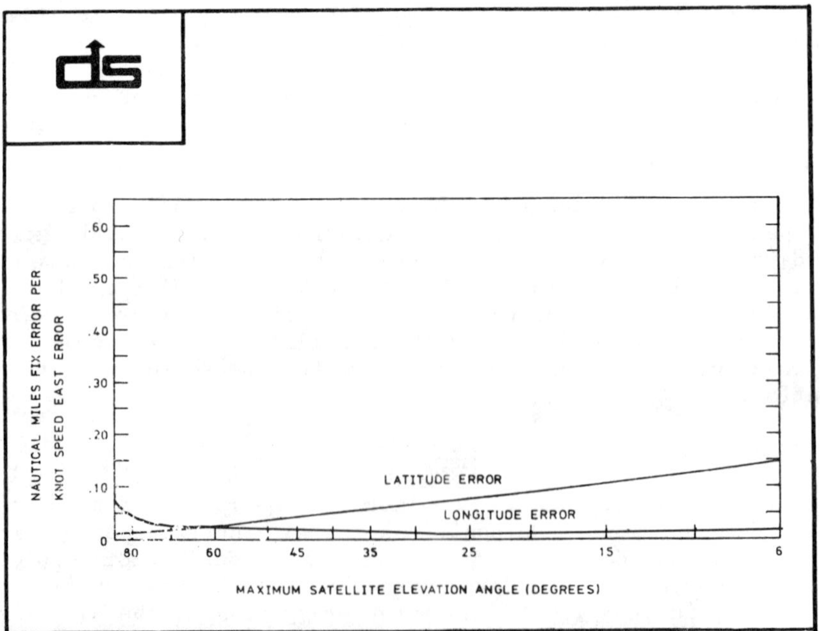

SENSITIVITY OF SATELLITE FIX TO SPEED EAST ERROR

two remote vehicles and the main vessel, using the relay mode of operation. In the case of a remote vehicle which is too acoustically noisy to reliably receive interrogations, it may be fitted with a 'pinger' and tracked using the hyperbolic mode of operation. This mode does not entail any restriction on the movement of the main vessel.

In order to make clear the operation of the integrated system, it may be helpful to describe a typical operation such as a site survey and subsequent positioning of a drill-barge in an area where there are either no shore-based radio aids available and the deployment of a temporary hyperbolic or 2-range chain is undesirable, or where the range to the nearest land is too great for accurate radio positioning.

The survey vessel, before sailing or on the way to the survey area, takes a series of satellite passes in order to generate a satellite alert table for the survey area. Suitable passes are chosen on the usual bases of elevation angle and non-interference and some idea of the amount of time needed to fix the net can be gained. The pass-selector in the satellite receiver can also be programmed to accept only the passes required.

Upon arrival in the operating area, it may be advisable to stop and take a satellite pass to check position. The vessel then moves to the first transponder drop point, takes another satellite pass and drops the first transponder. The other transponders are then deployed using dead-reckoning or, if desired, the towfish may be streamed and slant ranges used to assist the DR positioning.

Once the net is in operation, the ship steams around its perimeter at a low enough speed to ensure that propeller noise does not corrupt the data and the 140 sets of slant ranges are obtained. The time that this takes will depend of course upon the size of the net, but approximately 1-2 hours is a reasonable estimate. The orientation phase then requires that three constant headings be followed at constant speed, preferably at 90°-120° to each other, and lasting at least 5-10 minutes. In the final phase of the calibration it is preferable to steam slowly in a straight line for the duration of a satellite pass (approx. 18 minutes) and an east-west direction is preferable to north-south, as this minimises velocity errors. As the program uses the position at approximately the centre of the pass to re-position the net, the centres of the tracks steered during passes should be as widely separated as possible.

When sufficient passes have been completed to accurately position the net, the site survey itself may be carried out. On completion of the survey, the Aqua-Fix/2 CDU may be switched off or the survey vessel may depart and the underwater trans-

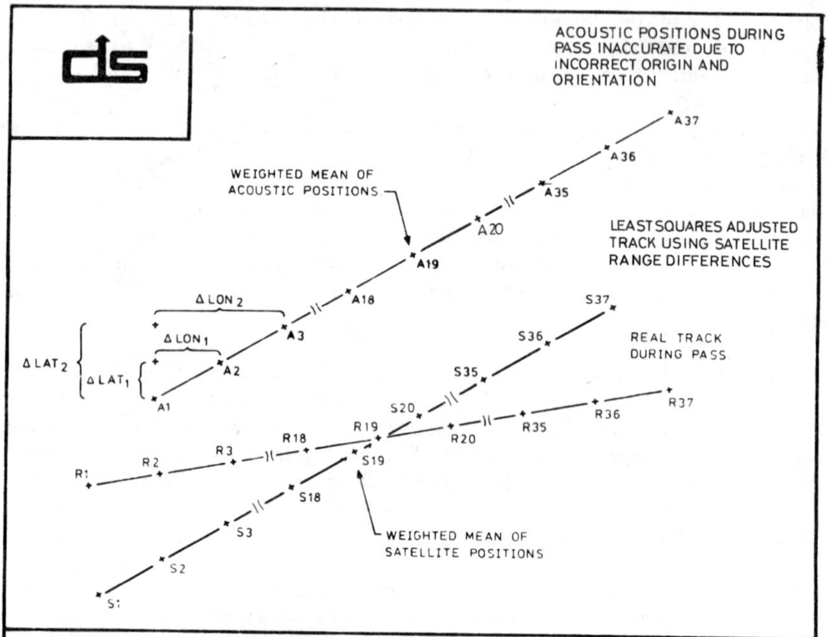

ADJUSTMENT OF ACOUSTIC TRACK USING SATELLITE DATA

ponders will revert to a listening mode, which they can maintain for up to 3 years.

When the drill barge is due to arrive the survey vessel returns, and using the previously derived positions for the transponders, uses the net to lay the marker buoys. It may also be thought advisable to verify the position of the net by re-calibrating. Upon arrival of the drill barge, the Aqua-Fix/2 CDU is transferred on board it from the survey vessel and its final position determined by the acoustic net. Once the drill barge is in position, the satellite receiver could also be transferred to it, and a confirmatory fix taken over several days, either as a single point fix, or using translocation.

CONCLUSIONS

The integrated system which has been described is at the time of writing (August 1977) undergoing trials at the Decca Trials Unit in Brixham and preliminary results indicate that the system meets, and in part exceeds, its design goals; a full report will be published shortly.

It is expected that the advantages of this type of integrated system will lead to its rapid acceptance by the offshore industry, and it will find applications in numerous situations; among these being rig positioning, site surveys, drill ship positioning (possibly dynamic) and well-head recovery. Applications in the near future may include sub-sea completion work, deep-sea mining and other activities off the continental shelf.

INSTRUMENTATION AND OBSERVATION OF OFFSHORE STRUCTURES

Tiago A.P. Lopes and Yosiaki Nagato

COPPE/UFRJ, Brazil.

ABSTRACT

This paper discusses the instrumentation and observation of offshore structures with the purpose of making a divulgation of this subject. Emphasis is given on the need for instrumentation of offshore structures either to monitor their installation or to observe their performance for the evaluation or the improvement of design methods.

Examples of types of instrumentation and general guidelines for the planning of experimental research are also given.

1. INTRODUCTION

Because of the complexity and importance of certain types of structures great care must be taken during their design as well as during the stages of their construction and operation, and experimental research on the behaviour of such structures is necessary so as to correctly evaluate their performance. The offshore structures for oil production are an example.

The need for instrumentation and observation of the behaviour of offshore structures is easily understood. However, the majority of people do not have an exact notion of what can or must be done in terms of experimental analysis of structures. Also, very few people in Brazil dedicate themselves to this field of work and the subject should be made better known in our technical circles.

Our comprehension of the problems associated with offshore structures being still in the initial phase, it is of vital importance that experimental research projects be executed to make it possible for us in the near future to correctly evaluate the methods used at present in the analysis and design of such structures and to obtain techniques that will not only result in greater safety, but probably will also contribute to the reduction of construction and operation costs.

2. NEED FOR INSTRUMENTATION

The experimental research allied with theoretical analysis have decisively contributed to the evolution of the techniques of design and execution of unconventional structures.

The construction of oceanic structures such as platforms for prospection and production of oil had its beginning when the exploration of deposits located on the continental shelf became feasible. The experience in the design of such structures being limited, great efforts have been made in instrumentation programs with the purpose of

- verifying the validity of the design assumptions;
- monitoring the safety of the structure during construction, transportation, installation and operation;
- collecting data to be used for the improvement of design procedures.

Experimental research can be undertaken either in laboratories or "in situ". In laboratory research, there is the possibility of testing the behaviour of materials under different loading conditions, of studying localized structural problems, of studying reduced models, etc. An experimental study made "in situ" provides the opportunity to obtain a series of informations which together with the data obtained in a laboratory permit a sound correlation of theoretical and practical results.

The need for instrumentation of fixed platforms for oil exploration is justified by the fact that it is not possible to make a prediction of their behaviour with the same reliability as in the case of conventional structures. Factors such as topology, installation method, loading conditions and operational requirements are such that it is extremely difficult to estimate the behaviour of these structures on the basis of data provided by conventional structures. As an example we may mention the platforms installed in the North Sea some years ago. The period from the time they were installed till the present day being short in relation to their lifetime, the quantity of available data is still quite limited.

In order to better emphasize the necessity of experimental analysis of offshore structures, we should like to mention some reports, papers and rules that evidence this necessity.

a) In a report [1] made by G. Somerville and H.P.J. Taylor of the Cement and Concrete Association under the title of "Concrete in the Oceans", the instrumentation of the first concrete platforms is strongly recommended, with emphasis on the importance of data collected from structures "in situ" at the present stage of outright expansion of the techniques of design and construction of this kind of structure. The report was finished in 1974 and emphasizes the necessity of consulting specialists in the design and testing of concrete structures to assure proper planning of instrumentation for performance control of such structures.

b) In the Recommendations for the Design of Concrete Sea Structures - FIP/74 [2], items R.3.4.3 and R.3.4.4 state:
- A design may be deemed satisfactory on the basis of results from an appropriate model test coupled with the use of model analysis to predict the behaviour of the actual structure, provided that the work is carried out by engineers with relevant experience and using suitable equipment.
- A design may be deemed satisfactory if the analytical or empirical basis of the design has been justified by development testing of prototype units and structures relevant to the particular design under consideration.

c) In the Rules for the Design, Construction and Inspection of Fixed Offshore Structures - DET NORSKE VERITAS/74 [3], in the part of general rules, the item "Instrumentation" is included and reads as follows:
- By instrumentation is meant the use of special devices for observation and monitoring of the performance of a structure during construction, transportation, installation or operation.
- Instrumentation may be required when visual inspection or simple measurements are not considered practicable or sufficiently reliable, and available design methods and previous experience are not satisfactory for a reliable prediction of the performance of the structure and its foundation.
- The required instrumentation will in general be limited to monitor parameters which may influence the procedures for construction, transportation, installation or operation of the structure.

d) In a paper [4] presented at the international conference on Behaviour of Offshore Structures - BOSS'76 the general characteristics and the instrumentation of the gravity platforms of the North Sea are described. Practically all the platforms are instrumented, being equipped with an average of 50 strain gauges, 20 soil pressure gauges, 10 pore pressure gauges and 10 accelerometers, besides other measuring instruments (Figure 1).

Another paper [5] presented at that same conference describes the characteristics of a hybrid platform installed in the North Sea exclusively for research purposes, and

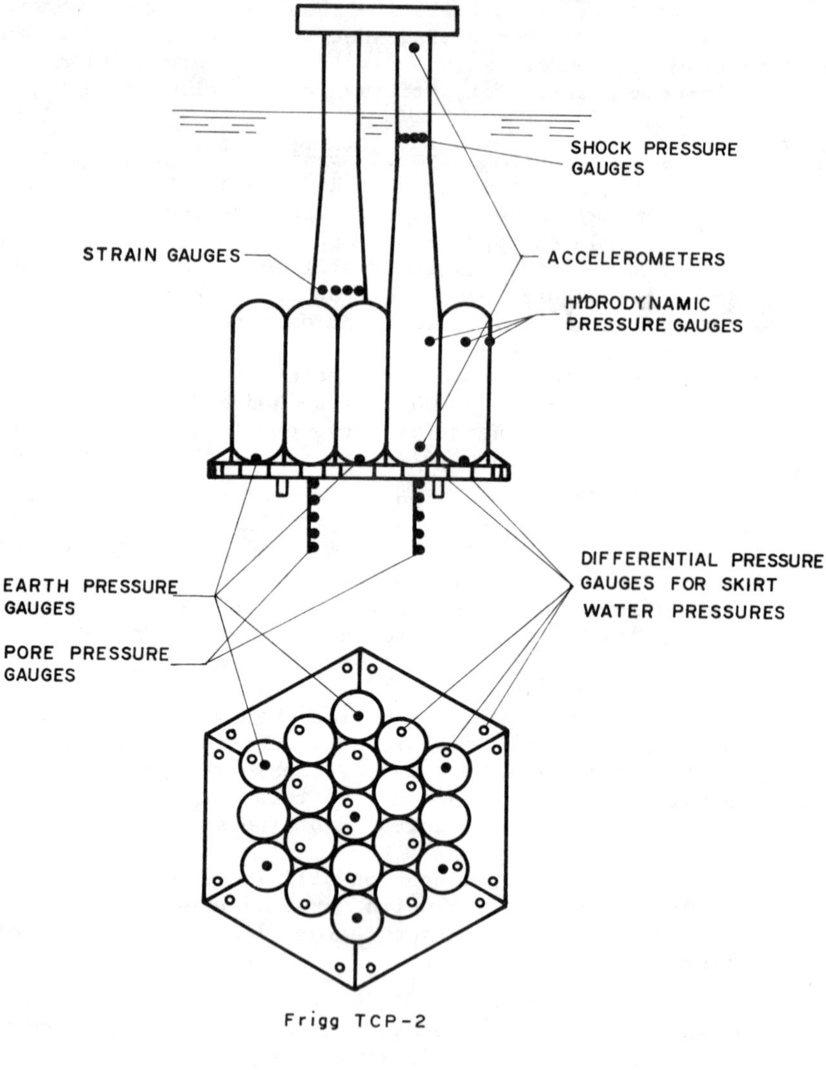

Fig. 1 Instrumentation of a gravity platform in the North Sea[4]

gives some details on the instrumentation of the platform and some comments on the first results obtained since the beginning of 1976 (Table 1).

A third paper [6] presented at that conference discusses the need for instrumentation of gravity-type offshore structures. A general discussion on the usefulness of instrumentation programs for installation and for performance observation of the structures is also presented.

3. TYPES OF INSTRUMENTATION

The instrumentation to be installed in an offshore structure depends on several factors as the type of structure, the purpose of the instrumentation, and others. Generally speaking, we can divide the instrumentation into two main groups:

- Instrumentation for monitoring the transportation and installation.
- Instrumentation for performance observation.

3.1 Instrumentation for monitoring the transportation and installation

This type of instrumentation is used to accompany the transportation and installation of a platform, for the purpose of verifying if its structure or the soil are damaged during the process. For the sake of example, we may list the monitoring of the following data:

- movements of the platform - using accelerometers;
- behaviour of the sea - using buoys anchored near the platform;
- distance from the base of the platform to the bottom of the sea - using echo-sounders;
- draught of the platform - by measuring hydrostatic pressure at the base;
- level of ballast in the tanks;
- strains at the bottom of the cells - using electrical strain gauges;
- inclination - using inclinometers;
- contact pressure at the bottom - using pressure transducers;
- others.

3.2 Instrumentation for performance observation

This kind of instrumentation permits collection of information both on the structure and on the environment. This information can be used to verify the performance of the structure and for the improvement of design methods. This kind of instrumentation can be divided into two groups:

Group 1 - measurement of oceanographic and meteorological data:

- direction and velocity of the wind;
- air humidity, pressure and temperature;
- height and direction of waves;

Table 1 Instrumentation of the Research Platform NORDSEE [5]

MEASUREMENT		INSTRUMENTATION	RECORDING FACILITIES
Instrumentation for the measurement of the environmental conditions	- air pressure	barometer	1)
	- air temperature	thermometer	1)
	- humidity	hygrometer	1)
	- wind velocity	cup-type anemometer with measuring generator - Type 4011 of Fa. Friedrichs	1) 3) 4)
	- wind direction	wind indicator with ring potentiometer Type 4111 of Fa. Friedrichs	1) 2)
	- visual range	light scatterer and recorder (AEG)	2)
	- tide level	pressure measurements of the column of water using the system of streaming air bubbles. Type Omega of Fa. Seba	2)
	- water/steel temperature at -4 and -19m chart 0 (steel temperature taken at + 15.5m)	resistance thermometer PT 100 with digital indication, Type NUR of Fa. CoreciNumecor	1) 4)
	- wave height	a) sonic method - oscillator Type C of Fa. Fahrenholz b) pressure sensor: principle of vibrating wire, Model WS - 704 of Fa. Ocean Applied Research	2) 3) 4)
	- current velocity	electro-magnetic 2 component sensors Fa. Colnbrook	3) 4)
	- pressure	pressure sensor BAW with 240 strain gauges, full bridged, measuring range 50 mW	3) 4)

Table 1 (Contd.) Instrumentation of the Research Platform NORDSEE [5]

	MEASUREMENT	INSTRUMENTATION	RECORDING FACILITIES
Instrumentation for the measurement of the structural behaviour	- strains in the members of the tubular structure	spot welded, uniaxial strain gauge bridges Type HBM 360.01-2001 of Fa. Hottinger	3) 4)
	- strains at chosen points in the joints	adhesive strain gauge rosettes Type 1520 Ry 11 - Fa. Hottinger	3) 4)
	- strains in the reinforced concrete foundation body	strain gauges, principle of vibrating wire, Type MDS 53a of Fa. Maihak, measuring length 250 mm	1) 2)
	- acceleration of deck and joints	servo-accelerometer - horizontal direction: Systron Donner system 4310A-1X33 measuring range ± 1 g - vertical direction: System Endevco QA-116-17, measuring range ± 10 g	3) 4)

Recording facilities: 1) notation; 2) integrated recorder or printer; 3) separate analog recorder; 4) PGM - magnetic tape recorder

- direction and velocity of currents;
- others.

Group 2 - measurement of platform data:

The measurement of data related to the platform is different, depending on whether the platform is made of steel or concrete. In this work, however, no details will be given on this matter.

Typical measurements are:

- linear accelerations and displacements at the base, half height and deck of the platform;
- angular accelerations and displacements at the base and deck level;
- pore pressure in the soil;
- strains in regions considered important;
- others.

Figure 1 illustrates the instrumentation of a gravity platform located in the North Sea.

4. PLANNING OF EXPERIMENTAL RESEARCH

First of all, it should be emphasized that the experimental study is not to be considered as an emergency solution to be remembered when problems arise during construction or utilization of a structure. The scope of the experimental study as research for development is much wider and comprises research in laboratory as well as on structures "in situ".

Experimental research is team work, involving specialized personnel and usually expensive equipment. It must be carefully planned and executed in order to render results that can be effectively used.

Depending on whether it involves research in a laboratory or instrumentation of structures "in situ", the planning of an experimental research program is somewhat different. Generally speaking, however, the planning demands joint efforts of designers, analysts, specialists in instrumentation and in interpretation of experimental results, builders and owners of the structures.

When planning instrumentation of structures "in situ", it should be clearly determined what shall be measured, where, how and when the measurements are to take place and how the results shall be used. It is necessary that such planning be made well in advance and be integrated in the project as an important component.

In principle, the larger the number of measuring instruments

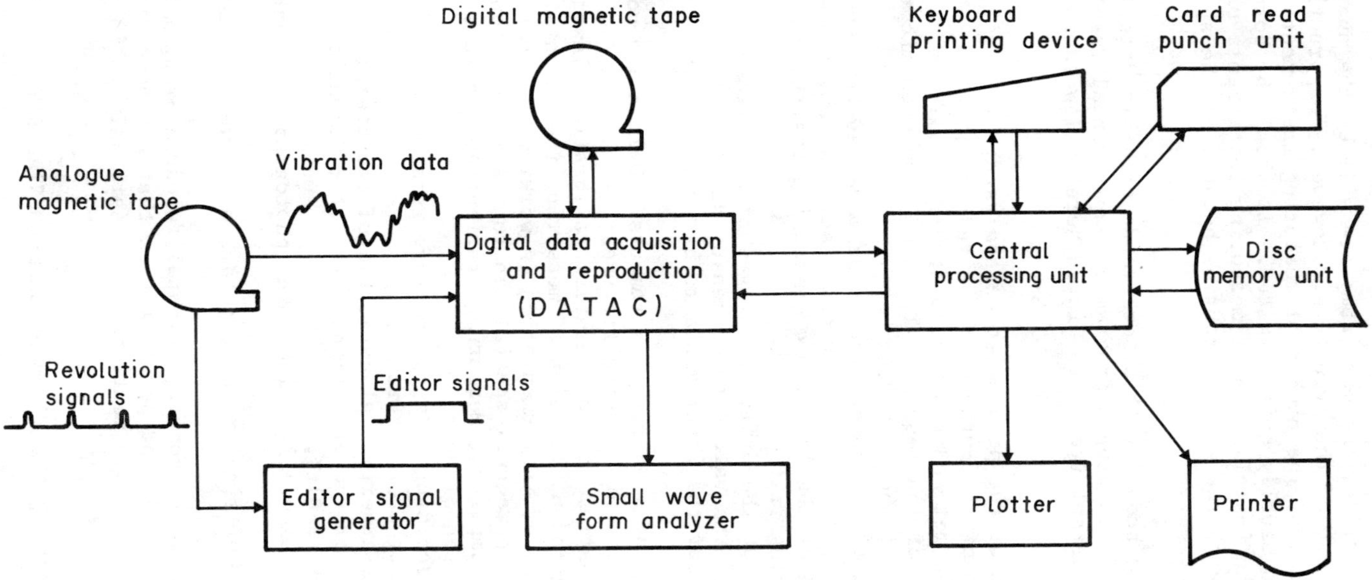

Fig. 2 Data processing system using a small digital computer(7)

properly installed, the more complete will be the analysis of the performance of the structure. However, for practical and economic reasons, the number of measuring instruments to be installed is usually quite limited. It should be borne in mind that the installation of measuring instruments on a site interferes with the working routine because of the special care it requires.

It is not the scope of this work to describe in detail all the phases of planning a program of experimental research. The more important items are listed hereinbelow as an illustration:

- Definition of the problem to be analysed;
- Definition of the purpose of the research;
- Terms to be met and funds needed;
- Limitation of the research. Technical and economic reasons;
- Definition of the work team;
- Definition of the quantities to be measured (in the structure, in the foundations, external actions, environment);
- Definition of the location of measuring instruments and of the measuring station;
- Definition of the different stages of measurements;
- Estimate of the values to be measured;
- Definition of the instruments to be utilized; static and dynamic
- Acquisition of measuring instruments. Design and construction of equipment for specific uses;
- Study of instrumentation details: calibration, installation, protection and utilization of measuring instruments;
- Interference of instrumentation in working routine;
- Training of personnel specialized in instrumentation;
- Measuring campaigns; Measuring station; Automatic stations; Identification of data;
- Adequate recording of supplementary information;
- Quality control of materials;
- Analysis of experimental data; Use of computers;
- Utilization of the experimental results; Comparison with theoretical results;
- Research documentation, including photographic documentation;
- Preparation of final report, detailed and well illustrated.

It is obvious that each experimental research program has its own characteristics and peculiarities that influence its planning and the above mentioned items are only valid as general guidelines [9].

5. HANDLING AND UTILIZATION OF EXPERIMENTAL DATA

In order to obtain the more significant data in a short time and in a form easy to be interpreted, the system of data

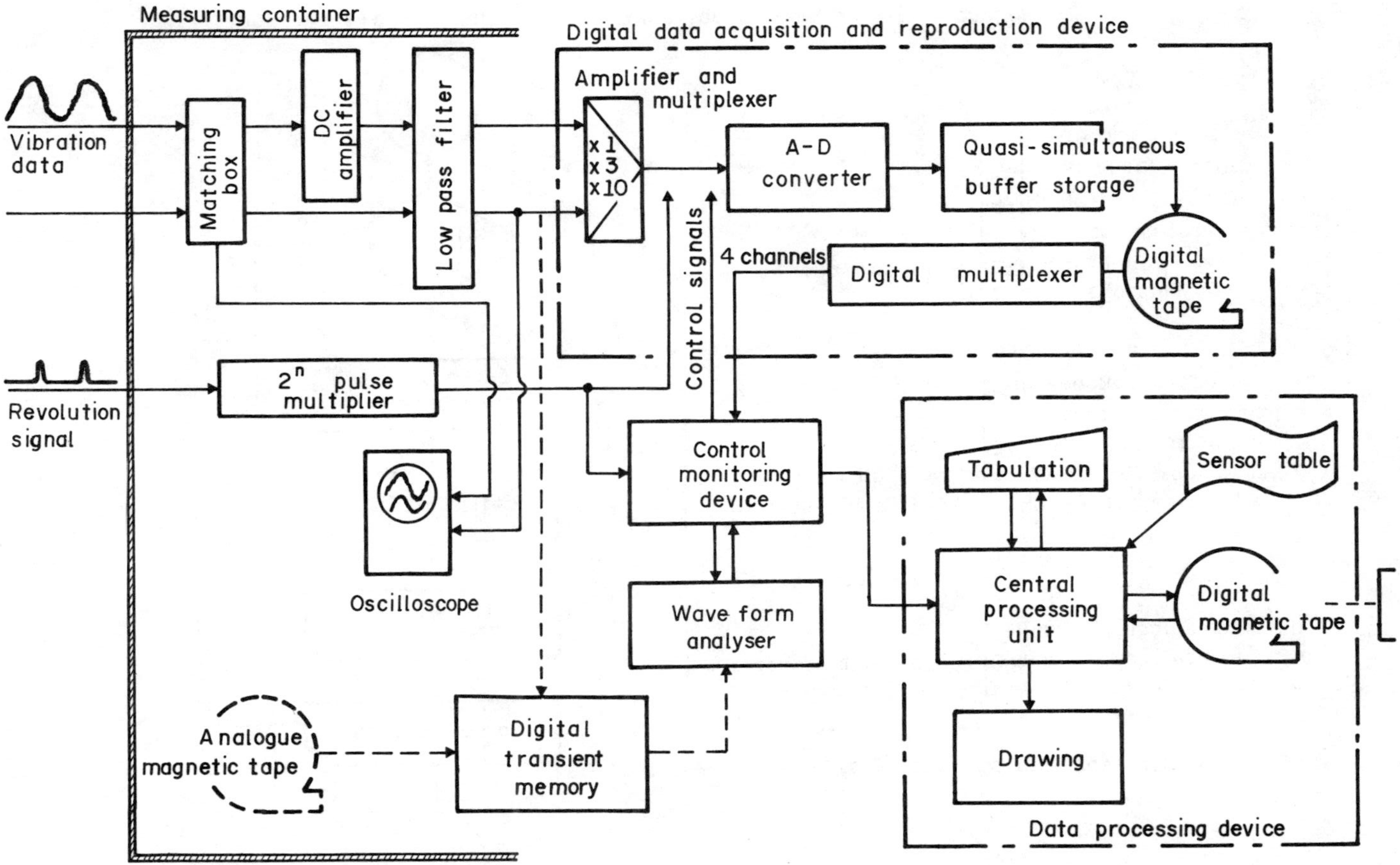

Fig. 3 Block diagram of on-board data acquisition and processing system [7]

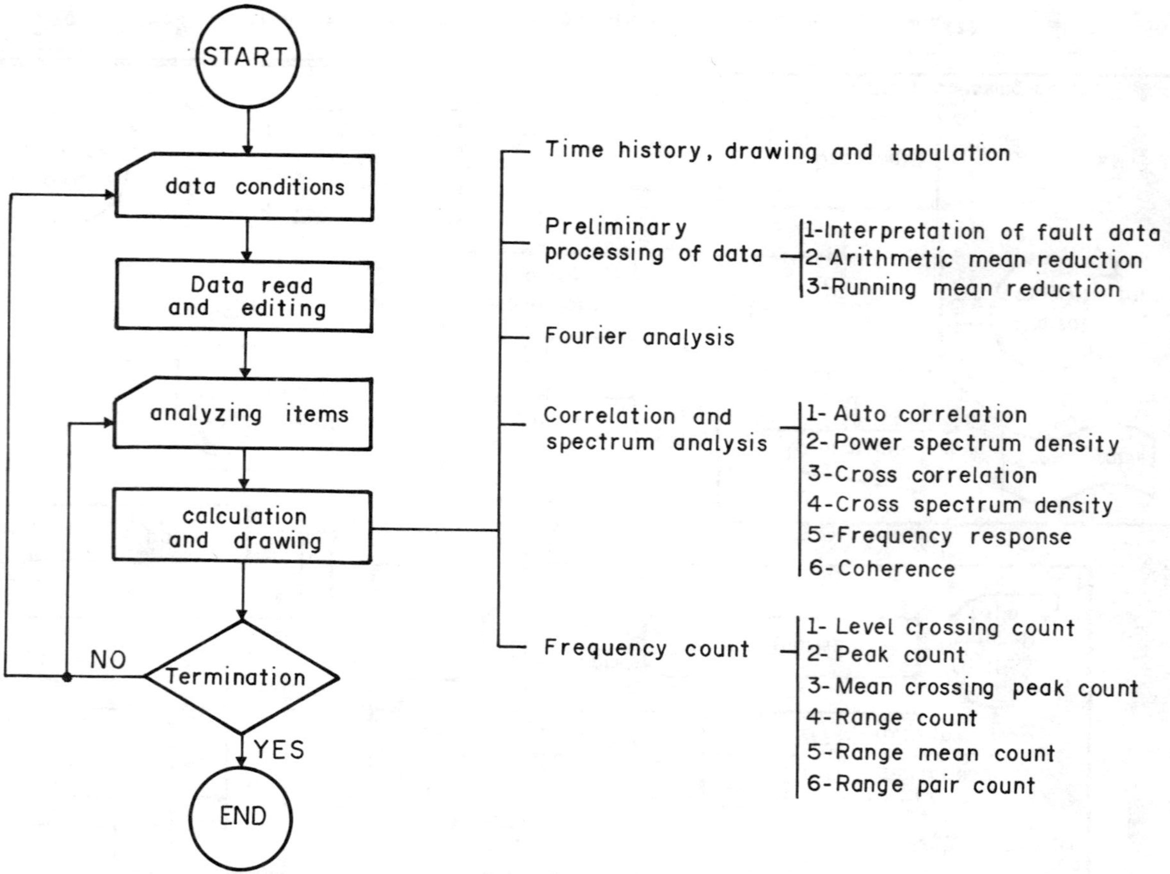

Fig. 4 Computer programs for random vibration analysis(7)

acquisition for this type of installation should be digital.

Usually the number of channels in an instrumentation system for offshore platforms is above 50 and requires an efficient system of data processing.

Figures 2 and 3 show two systems for acquisition and handling of data developed in [7]. Figure 4 shows a diagram for analysis of vibrations [7],[8].

6. CONCLUDING REMARKS

The instrumentation and observation of offshore structures is necessary both for safety control during installation and for collecting data for the evaluation or for the improvement of design methods.

In order to obtain significant and reliable data the instrumentation must be carefully planned and executed. Specialized personnel and efficient data acquisition and data reduction systems are essential.

The subject of instrumentation and observation of offshore structures is a very important one, and exchange of experience and of data related to this subject should be encouraged.

REFERENCES

1. Somerville, G. and H.P.J. Taylor, "Concrete in the Oceans", Cement and Concrete Association, 1974.

2. FIP - "Recommendations for the Design of Concrete Sea Structures", 1974.

3. DET NORSKE VERITAS - "Rules for the Design, Construction and Inspection of Fixed Offshore Structures", 1974.

4. Foss, I. "Instrumentation for Operation Surveillance of Gravity Structures", BOSS'76, Vol. I.

5. Longrée, W.-D. "Aspects of the Instrumentation and Measurement Performance of the Research Platform NORDSEE", BOSS'76, Vol. I.

6. DiBiagio, E., Myrvoll, F. and Hansen, S.B. "Instrumentation of Gravity Platforms for Performance Observations", BOSS'76, Vol. I.

7. Fujii, K. "Advanced Method of Ship Vibration Measurement", COPPE/UFRJ, 1977.

8. Lopes, T.A.P. and R.R. Tacques, "Análise Experimental das Vibracoes em Navios e Plataformas Oceânicas: Técnicas e Aplicacões", Simposio sobre Tendências Atuais no Projeto e Execucão de Estruturas Marítimas, COPPE/UFRJ, 1977

9. Nagato, Y. "Análise Experimental de Estruturas e Compontes de Obras Marítimas", Simpósio sobre Tendências Atuais no Projeto e Execucão de Estruturas Marítimas, COPPE/UFRJ, 1977.